Biometric Authentication in Online Learning Environments

A.V. Senthil Kumar
Hindusthan College of Arts and Science, India

A volume in the Advances in Educational Technologies and Instructional Design (AETID) Book Series

IGI Global
DISSEMINATOR OF KNOWLEDGE

Published in the United States of America by
IGI Global
Information Science Reference (an imprint of IGI Global)
701 E. Chocolate Avenue
Hershey PA, USA 17033
Tel: 717-533-8845
Fax: 717-533-8661
E-mail: cust@igi-global.com
Web site: http://www.igi-global.com

Library of Congress Cataloging-in-Publication Data

Names: Kumar, A. V. Senthil, 1966- editor.
Title: Biometric authentication in online learning environments / A.V. Senthil Kumar, editor.
Description: Hershey, PA : Information Science Reference, [2019] | Includes bibliographical references.
Identifiers: LCCN 2018036324| ISBN 9781522577249 (h/c) | ISBN 9781522577256 (eISBN)
Subjects: LCSH: Web-based instruction--Design. | Biometric identification.
Classification: LCC LB1044.87 .B56 2019 | DDC 371.33/44678--dc23 LC record available at https://lccn.loc.gov/2018036324

This book is published in the IGI Global book series Advances in Educational Technologies and Instructional Design (AE-TID) (ISSN: 2326-8905; eISSN: 2326-8913)

British Cataloguing in Publication Data
A Cataloguing in Publication record for this book is available from the British Library.

For electronic access to this publication, please contact: eresources@igi-global.com.

Advances in Educational Technologies and Instructional Design (AETID) Book Series

Lawrence A. Tomei
Robert Morris University, USA

ISSN:2326-8905
EISSN:2326-8913

Mission

Education has undergone, and continues to undergo, immense changes in the way it is enacted and distributed to both child and adult learners. In modern education, the traditional classroom learning experience has evolved to include technological resources and to provide online classroom opportunities to students of all ages regardless of their geographical locations. From distance education, Massive-Open-Online-Courses (MOOCs), and electronic tablets in the classroom, technology is now an integral part of learning and is also affecting the way educators communicate information to students.

The **Advances in Educational Technologies & Instructional Design (AETID) Book Series** explores new research and theories for facilitating learning and improving educational performance utilizing technological processes and resources. The series examines technologies that can be integrated into K-12 classrooms to improve skills and learning abilities in all subjects including STEM education and language learning. Additionally, it studies the emergence of fully online classrooms for young and adult learners alike, and the communication and accountability challenges that can arise. Trending topics that are covered include adaptive learning, game-based learning, virtual school environments, and social media effects. School administrators, educators, academicians, researchers, and students will find this series to be an excellent resource for the effective design and implementation of learning technologies in their classes.

Coverage

- Hybrid Learning
- Social Media Effects on Education
- Educational Telecommunications
- Virtual School Environments
- Bring-Your-Own-Device
- Adaptive Learning
- Game-Based Learning
- Instructional Design Models
- Collaboration Tools
- Classroom Response Systems

IGI Global is currently accepting manuscripts for publication within this series. To submit a proposal for a volume in this series, please contact our Acquisition Editors at Acquisitions@igi-global.com or visit: http://www.igi-global.com/publish/.

The Advances in Educational Technologies and Instructional Design (AETID) Book Series (ISSN 2326-8905) is published by IGI Global, 701 E. Chocolate Avenue, Hershey, PA 17033-1240, USA, www.igi-global.com. This series is composed of titles available for purchase individually; each title is edited to be contextually exclusive from any other title within the series. For pricing and ordering information please visit http://www.igi-global.com/book-series/advances-educational-technologies-instructional-design/73678. Postmaster: Send all address changes to above address.

Titles in this Series

For a list of additional titles in this series, please visit: www.igi-global.com/book-series

Handbook of Research on Assessment Literacy and Teacher-Made Testing in the Language Classroom
Eddy White (University of Arizona, USA) and Thomas Delaney (University of Oregon, USA)
Information Science Reference • copyright 2019 • 410pp • H/C (ISBN: 9781522569862) • US $255.00 (our price)

Interdisciplinary and International Perspectives on 3D Printing in Education
Ieda M. Santos (Emirates College for Advanced Education, UAE) Nagla Ali (Emirates College for Advanced Education, UAE) and Shaljan Areepattamannil (Emirates College for Advanced Education, UAE)
Information Science Reference • copyright 2019 • 346pp • H/C (ISBN: 9781522570189) • US $195.00 (our price)

Bayesian Networks for Managing Learner Models in Adaptive Hypermedia Systems Emerging Research and Opportunities
Mouenis Anouar Tadlaoui (Abdelmalek Essaadi University, Morocco) Mohamed Khaldi (Abdelmalek Essaadi University, Morocco) and Rommel Novaes Carvalho (University of Brasília, Brazil)
Information Science Reference • copyright 2019 • 199pp • H/C (ISBN: 9781522574132) • US $165.00 (our price)

Handbook of Research on TPACK in the Digital Age
Margaret L. Niess (Oregon State University, USA) Henry Gillow-Wiles (Southern Oregon University, USA) and Charoula Angeli (University of Cyprus, Cyprus)
Information Science Reference • copyright 2019 • 490pp • H/C (ISBN: 9781522570011) • US $255.00 (our price)

Educational and Social Dimensions of Digital Transformation in Organizations
Paula Peres (Polytechnic of Porto, Portugal) Fernando Moreira (Universidade Portucalense, Portugal & Universidade de Aveiro, Portugal) and Anabela Mesquita (Polytechnic of Porto, Portugal)
Information Science Reference • copyright 2019 • 372pp • H/C (ISBN: 9781522562610) • US $185.00 (our price)

Handbook of Research on Curriculum Reform Initiatives in English Education
Christopher Denman (Sultan Qaboos University, Oman) and Rahma Al-Mahrooqi (Sultan Qaboos University, Oman)
Information Science Reference • copyright 2019 • 372pp • H/C (ISBN: 9781522558460) • US $225.00 (our price)

Global Perspectives on Work-Based Learning Initiatives
Jon Talbot (University of Chester, UK)
Information Science Reference • copyright 2019 • 344pp • H/C (ISBN: 9781522569770) • US $185.00 (our price)

IGI Global
DISSEMINATOR OF KNOWLEDGE

701 East Chocolate Avenue, Hershey, PA 17033, USA
Tel: 717-533-8845 x100 • Fax: 717-533-8661
E-Mail: cust@igi-global.com • www.igi-global.com

Editorial Advisory Board

Table of Contents

Detailed Table of Contents

The biometric authentication in the online learning environment (OLE) is still exploratory and, despite an increase in keystroke dynamics biometrics research, many challenges remain in designing this authentication system due to the fact that it is economical and easily integrated with existing computer security system in OLE. Existing research in keystroke dynamics tends to focus on finding how keystroke dynamics of users can support non-intrusive authentications of users in OLEs. However, there is little evidence that researchers have approached the issue of unauthenticated users to take the role of authenticated users and perform tasks in the OLE with the intent of building models based on the keystroke dynamics of users. In a nutshell, the aim of this chapter is to provide an overview of the existing applications of keystroke dynamics as biometric authentication in the OLE, keystroke dynamics framework being designed for the OLE, advantages and disadvantages of a keystroke dynamics biometrics approach, as well as offering suggestions and possible future research directions.

The number of users using the internet has drastically increased. Due to the large number of online users, demand has increased in various fields like social networks, knowledge sharing, commerce, etc. to protect the user's private data as well as control access. Unfortunately, the need for security and authentication for individual data also increased. In an attempt to confront the new risks unveiled by the networking revolution over the recent years, we need an efficient means for automatically recognizing the identity of individuals. Biometric authentication provides an improved level of security and paves the way to the future. Further, biometric authentication systems are classified as physiological biometric and behavioral biometric technologies. Further, the author provides ideas on research challenges and the future of authentication systems.

The objective of this chapter is to verify the identity of the claimed learner by extracting the prosodic features of the speech signal. TIMIT Acoustic-Phonetic Continuous Speech Corpus is used for learner verification using prosodic and articulation features such as energy, pitch, and formants. The prosodic feature includes pitch (F0), and articulation feature includes formants (F1-F7). From this database, for this project in the training phase, 200 learners were used and in the testing phase 160 learners were used. The pitch and formants were extracted using linear predictive analysis. The first seven formants were used for verification purpose. The feature set consists of eight features. The features are fed into the Guassian mixture model. In the Gaussian mixture model, parameters are estimated from the training and testing data using the iterative expectation-maximization. Log likelihood score is computed using these parameters, and then these scores are normalized to make decisions. The decision is made based on the threshold.

Foreword

In the past twenty years the use of Internet technologies has been substantially growing for delivery of educational content. E-learning environments have been incorporated in many universities for the delivery of e-learning courses. However, opponents of e-learning claim that a central disadvantage of such teaching medium is the growing unethical conduct in such environments. With the rapid proliferation of online learning, students are increasingly demanding easy and flexible access to learning content at a time and location of their choosing. In these environments, remote users connecting via the public internet or other unsecure networks must be authenticated prior to being granted access to sensitive content such as tests or personal/private records.

Today, the overwhelming majority of online learning systems rely on weak authentication mechanisms to verify the identity of remote users only at the start of each session. One-time authentication using password, personal identification number (PIN), or even hardware tokens is clearly inadequate in that it cannot defend against insider attacks including remote user impersonation or illegal sharing or disclosure of these authentication secrets. As such, these methods are entirely unsuitable for circumstances where the outcome of an online assessment or a course of study is the granting of a formal degree, professional certification, or qualification or requalification for a particular skill or function.

Biometrics allows a person to be identified and authenticated based on a set of recognizable and verifiable data, which are unique and specific to them. Biometric authentication is the process of comparing data for the person's characteristics to that person's biometric "template" in order to determine resemblance. The reference model is first store in a database or a secure portable element like a smart card. The data stored is then compared to the person's biometric data to be authenticated.

Biometric authentication is a security process that relies on the unique biological characteristics of an individual to verify that he is who is says he is. Biometric authentication systems compare a biometric data capture to stored, confirmed authentic data in a database. If both samples of biometric data match, authentication is confirmed. Typically, biometric authentication is used to manage access to physical and digital resources such as buildings, rooms and computing devices. Biometric authentication is incredibly popular in mobile devices nowadays. Various biometric techniques used are facial recognition, fingerprints, voice recognition, retina scans, palm prints and more. Also biometric authentication in online learning environments plays a vital role nowadays.

The objective of this book is to provide the latest and advanced biometric authentication technologies used in the field of online learning environments. This book will be very useful to individuals, researchers, inventors, companies, scientists, libraries and development practitioners in the field of biometric authentication in online learning environments.

Alok Prakash Mittal
All India Council for Technical Education, India

Alok Prakash Mittal *started his teaching career at REC, Kurukshetra as a Lecturer in 1985, joined REC Hamirpur (Himachal Pradesh) as Assistant Professor in 1987. Dr. A.P. Mittal returned to REC Kurukshetra as Assistant Professor in 1989, joined Chotu Ram State College of Engineering (CRSCE) (present DCRUST), Murthal as Professor of Electrical Engineering in 1997. In 2001, Prof, A.P. Mittal joined NSIT, Delhi as Professor and Head of Instrumentation and Control Engineering Division. Prof. A.P. Mittal worked as Dean, Student's Welfare, Dean, Industrial Research and Development, Dean, Post Graduate Studies and Research, Dean, Faculty Affairs and Dean Alumni Affairs at various durations from 2003 to 2016. Prof. A.P. Mittal also worked as Member of Accreditation teams for accreditation of different technical institutes. Prof. A.P. Mittal was Chairman of Admission Committee at NSIT during 2005, 2006, 2010 & 2011. Prof. A.P. Mittal worked as General Chair for IEEE Conference on Power Electronics-2010. Prof. Alok Prakash Mittal, Professor of Instrumentation and Control Engineering(ICE) at Netaji Subhas Institute of Technology (NSIT), New Delhi, joined AICTE as Member Secretary on 22nd April 2016. Prof. Alok Prakash Mittal graduated from Madan Mohan Malviya Engineering College, Gorakhpur, Uttar Pradesh in Electrical Engineering in 1978, post graduation from Electrical Engineering Department, University of Roorkee in 1980 and Ph.D from IIT, Delhi in 1991. He has guided 17 students for Ph.D and published more than 128 research papers in Journals and Conferences.*

Preface

The growth of Internet has largely evolved teaching and learning from a conventional classroom into an invaluable educational resource accessible remotely from disperse geographical locations, beyond physical boundaries. The online learning environments are likely to be accessible, available, updatable, resource efficient, usable and economical (Ruiz, Mintzer, & Leipzig, 2006). The dramatic growth in distance and e-learning is evident in the number of institutions that offer e-learning. The growth in e-learning is the new dynamic learning models and is leading the academic market to a significant paradigm and cultural change (Gunasekaran, McNeil, & Shaul, 2002).

Courses and entire degree programs are delivered via the Web anywhere at anytime. In addition, e-learning courses are offered by private, public as well as corporate universities. As a result, new resources such as e-books, books on CD-ROMs and e-exams haven adopted to e-learning courses. As a result numerous academic institutions are planning to increase the number of e-learning courses to meet the growth in this demand. However, security issues related to e-learning systems have been raised by several scholars (Raman & Levy, 2006).

Online examination is a fundamental and integral part of the learning environment. In the online learning and examination, students interact and submit their work remotely and therefore, building confidence and trust is of vital importance (Karvoren, 1999). The online teaching and learning approaches has largely changed student assessment methods. In spite of the anticipated benefits on online learning, security remains one of the major issues and a threat to the success of online learning (Alwi & Fan, 2010).

Yu and Tsao (2003) discussed security challenges of e-learning environments. However, their exploration focused on shielding the technology infrastructure against unauthorized users. Current security practices in e-learning systems relay principally on the utilization of passwords authentication mechanisms. Huang, Yen, Lin, and Huang (2004) discussed aspects of security in e-learning systems and suggested attention in two layers when securing e-learning systems. The first layer addresses security of the technology infrastructure used to facilitate e-learning (i.e., hardware, networks, etc.) and the second layer addresses the various applications employed in enabling e-learning (i.e., learning management systems, rich media communications tools, etc.). Hugl (2005) noted numerous security related technologies that are not currently employed in e-learning. One such solution can include biometrics technologies that may potentially become an integral part of e-learning systems. Biometrics is a recognition system that relies on individual humane identities such as DNA, Voice, Retinal and Iris, Fingerprints, Facial images, and Hand prints. Essentially, biometrics technologies operate by scanning a physical characteristic and matching it with the stored data.

Reliable student authentication is extremely relevant to online learning environments because of high stakes being involved. Authentication attempts to verify that the users are who they claim to be. Biometrics is defined as the identification of an individual based on physiological and behavioral characteristics. Commonly used physiological characteristics include face, hand (fingerprint, hand geometry, palm print), eye (iris and retina), ear, skin, odor, dental and DNA. Commonly used behavioral characteristics include voice, gait, keystroke, signature, mouse movement, and pulse. Two or more of the aforementioned biometrics can be combined (Multimodal) in a system to improve the recognition, accuracy. In addition, some soft biometric traits like gender, age, height, weight, ethnicity, and eye color can also be used to assist in identification.

This book, focusing biometric authentication in online learning environments, has tried to provide the techniques and best practices in the field of security in online learning environments. The main objective of this book is to provide security to improve e-learning and on-line examination by utilizing biometric based authentication techniques.

The audience of this book will widely vary from business people, housewives, professors, tutors, individuals, researchers, scientists, academicians, students and libraries. This book will generate tremendous solutions to provide effective ways to improve learning, and to explore and examine the use of agents in online learning systems to achieve more personalization and propose a framework for more general system architecture of online learning system.

ORGANIZATION OF CHAPTERS

This book has been divided into three sections, "Concepts and Approaches," "Methodologies and Applications," and "Innovations and Technologies." Altogether there are eleven manuscripts covering wide range of concepts, methods, and applications.

Chapter 1 presents a framework for defining a quick and secure authentication approach that can be followed in e-learning process which doesn't require any specialized hardware, except, the keyboard for the applications in the highly-connected world.

Chapter 2 describes the importance of keystroke dynamics in e-learning. Keystroke dynamics along with other biometrics security features will be a self sufficient, adaptive and impenetrable security system. The chapter suggests that keystroke dynamics could be paired with mouse dynamics to increase user authentication data and improve the identification authentication process.

Chapter 3 provides detailed insights of how do the main cheating risks for educational assessments differ between online and on-campus tests, as well as between different types of assessments, for which of these cheating risks will biometric authentication provide significant mitigation and for those risks where biometric authentication alone is not sufficient mitigation, how can biometric authentication be combined with other means to make the exam more secure.

Chapter 4 discusses a proposed solution to enhance the authentication of LCMS various techniques available authentication like voice based authentication, face recognition and signature based methods. It also discusses the image retrieval mechanism required signature and face based authentication.

Chapter 5 provide an overview of the existing applications of keystroke dynamics as biometric authentication in the Online Learning Environment, Keystroke dynamics framework being designed for the OLE, advantages and disadvantages of a keystroke dynamics biometrics approach, as well as offering suggestions and possible future research directions.

Chapter 6 suggests that due to the different authentication techniques available in the society, it facilitates and reduces the identity threat in various fields like government, public sectors, health sectors, financial transactions, forensics etc. Also strengths, weakness and challenges of various biometric authentication systems have been discussed comprehensively. Additionally, the unimodal and multimodal biometric systems along with advantages and drawbacks have been discussed elaborately.

Chapter 7 presents a work to verify the identity of the learner in online learning environment by extracting the prosodic features of the learner speech signal which shows that the different prosodic and articulation features can produce good results in learner verification.

Chapter 8 presents the various biometric technologies used for authentication, the methods employed by the students for cheating during the online exams, keystroke dynamics as a solution for strong authentication model during the online exams and their applications on different domains.

Chapter 9 proposes a wrapper-based classification technique using PSO-ELM-ANP method for keystrokes obtained from laptop. The same proposed wrapper-based classification technique using PSO-ELM-ANP method can be tested for touch screen gadgets and mobile environments.

Chapter ten proposes a very creative and considers both the estimation of probabilities and correlation of layouts invariants to interpretation/revolution to perform characterization, an unexplored path till now as anyone is concerned.

Chapter eleven provides an overview of security in online learning environments which are by their nature subject to many of the attacks that standard online systems are vulnerable to also.

CONCLUSION

E-Learning offers a lot of opportunities for the expansion of education and learning beyond the classroom, particularly, in developing countries. The major challenge in e-learning today is the issue of security. Lack of attention to security issues in e-learning is a problem because important issues of student and staff privacy are at stake and also online learning credibility is at stake. Security elements, such as availability, integrity and confidentiality should be given more attention in order to prevent security breaches before it is too late. As more and more people are using online learning environments, more attention and efforts are needed from online learning providers and practitioners to prevent possible security breaches in online learning before it is too late.

A. V. Senthil Kumar
Hindusthan College of Arts and Science, India

REFERENCES

Alwi, N. H. M., & Fan, I. S. (2010). Threat analysis for e-learning. *International Journal of Technology Enhanced Learning*, *2*(4), 358–371. doi:10.1504/IJTEL.2010.035738

Gunasekarn, A., McNeil, R. D., & Shaul, D. (2002). E-learning Research and applications. *Industrial and Commercial Training*, *34*(2), 44–54. doi:10.1108/00197850210417528

Huang, W., Yen, D.C., Lin, Z.X., & Huang, J.H. (2004). How to compete in a global education market effectively: A conceptual framework for designing a next generation eEducation system. *Journal of Global Information Management, 12*(20), 84-107.

Hugl, U. (2005). Tech-developments and possible influences on learning processes and functioning in the future. *Journal of American Academy of Business,* 6(20), 250-256.

Korvoren, K. (1999). Creating Trust. *Proceedings of the Fourth Nordic Workshop on Secure IT System.*

Raman, M., & Levy, Y. (2006). Securing e-learning system: A case of insider cyber attacks and novice IT management in a small university. *Journal of Cases on Information Technology*, *8*(4), 24–34. doi:10.4018/jcit.2006100103

Ruiz, J. G., Mintzer, M. J., & Leipzip, R. M. (2006). The impact of e-learning in medical education. *Academic Medicine*, *81*(3), 207–212. doi:10.1097/00001888-200603000-00002 PMID:16501260

Yu, C., & Tsao, C. C. (2003). Web Teaching: Design, security and legal issues. *Delta Pi Epsilon Journal*, *45*(3), 191–203.

Acknowledgment

I would like to thank IGI Global for offering me the opportunity to edit this book on Biometric Authentication in Online Learning Environments. I am grateful to Editorial Advisory Board Members and all the valuable contributing authors of this book for the support, assistance, suggestions and contribution. My sincere thanks extends to all reviewers for their extended support during the tiring review process.

I will not have completed this work in time unless continuous support came from my wife, daughter and family members.

Finally, my regards go to the colleagues and friends for their supports and encouragements, without which, I would not have been able to complete this research publication.

A. V. Senthil Kumar
Hindusthan College of Arts and Science, India

Section 1
Concepts and Approaches

Chapter 1
Keystroke Dynamics in E-Learning and Online Exams

Jagadamaba G.
Siddaganaga Institute of Technology, India

Sathish Babu B.
R. V. College of Engineering, India

ABSTRACT

In the internet era, the online examination has become an integral component of online processing and online learning. Student assessment in the online education system is submitted remotely without any face-to-face recognition and interaction. However, student authentication is the significant challenge in online education and examination. This chapter aims to examine various authentication systems, potential threats, and solutions to student authentication in the online examinations and learning. In this chapter, a keystroke-based authentication system is discussed for online examinations. Keystroke-based authentication does not require any additional investments as compared to the other existing authentication approaches such as face recognition, iris recognition, fingerprint, and so on.

INTRODUCTION

E-learning systems are becoming one of the largest growing segments of Web-based systems (Shauna Beaudin, Yair Levy, James Parrish and Theon Danet, 2016; Alwi & Fan, 2010). E-learning platform uses a wide range of learning activities such as to meet learning outcomes (Levy, 2008). However, the prevalent uses within academic institutions, e-learning systems are a strategic way for institutions/organizations from various institutions/industries to deliver training to students/employees in order to improve their course content or skills or obtain certifications. In the current scenario, users interact with the e-learning systems through a variety of learning activities. But according to Bailie (Bailie & Jortberg, 2009) e-learning activities can be categorized as formative or summative e-assessments. However, the formative e-assessments are used to identify the gap between current understanding and the desired goal by providing feedback, dialogue, and non-assessed activities. Hence, e-learning systems must ensure that users completing e-learning activities are genuine and secured (Peres, Lima, & Lima, 2014). The

DOI: 10.4018/978-1-5225-7724-9.ch001

Joint Information Systems Committee (JISC) (2006) has defined the e-assessments as, "the end-to-end electronic learning and assessment processes where [Information & Communications Technology] ICT is used for the presentation of learning or assessment activity. A risk to e-learning systems is when users deliberately reveal their authentication details to allow another user to impersonate them (Apampa, Wills, & Argles, 2010). Impersonation is considered the intentional collaboration between users with the intent to commit a fraudulent behavior by the misrepresentation of identity potentially undermining the value of the system (Apampa et al., 2010; Gathuri, Luvanda, & Kamundi, 2014).

To understand the challenges associated with an e-learning environment, we should know the definition of what the term e-learning means. Clark and Mayer help define e-learning as instruction delivered by any technological mode intended to promote learning (2011). Teaching and learning in an e-learning environment happen differently than in the traditional classroom and can present new challenges to instructors and learners participating in the online learning environment. Technology-assisted learning tools are quickly changing the face of education, transitioning of the classroom learning environment to an online-only or blended online learning experience (http://biometrics.pbworks.com)

The offline exam is chosen as an evaluation method for both online and off-line courses, but online course examinations have also expanded rapidly (http://polinance.blogspot.in) and are used to evaluate the student's knowledge of using modern computer technology without any effects on the traditional university course exam that uses pens, papers and invigilators. Online exam can improve the standards of the student's examination whereas the traditional examination. Thus, the online examinations are considered as an essential source for university exam, and the development of network technology polices has given the possibility to conduct the exams online that can benefit a large number of student community. In the new technology, the computer has been generally useful to the fields of education and gives the professors/instructors the advantage of an adequate assessment using the new era e-learning and e-exam procedure. The study conducted on online exam and traditional exam indicates that online learning has better results than traditional exams (R. Joyce, G. Gupta, 1990; Eros Desouza, Matthew Fleming, 2003). On the other side of online learning and examination conduction includes many challenges. A study suggested that about 73.6% of the students in the sample held the perception that it is easier to cheat in an online versus traditional course or e-examinations (King, Guyette, & Piotrowski, 2009). Among them, authentication of the user who involved in the process is a major one. However, many solutions were also proposed to avoid the online cheating and are as follows:

1. Comparing the IP addresses to see if two students are in the vicinity of each other.
2. We can make use of plagiarism detection tool Turnitin and search engine to check some questions for possible dishonesty (Yong Sheng, Vir. V. Phoha, Steven, and M, Rovnyak, 2005).
3. Watchful eye of a webcam or with software profiling your keystrokes or your syntax to see whether it really is you answering the questions (https://elearningindustry.com)
4. Commercial products for proctoring E-exams: The first one is named Secureexam, a remote proctor made by Software Secure; the second one is named Webassessor, made by Kryterion; and the third one is named ProctorU, made by Axicom (Yong Sheng, Vir. V. Phoha, Steven, and M, Rovnyak, 2005).
5. Effective authentication methods (Fabian Monrose, Aviel D. Rubin, 2000; Romain Giot, Bernadette Dorizziy, Christophe Rosenberger, 2011; Ramu, Dr.T. Arivoli, 2013)
6. Usage of microphones that visually record the student's testing environment, movements, and background noise (Patricia A. and Robert I, 2009)

7. Onsite, remote proctoring to monitor the student taking the exam at the student's location (Patricia A. and Robert I, 2009)
8. "out-of-wallet" data mining of personal data that randomly requires students to answer personal questions before or during the examination time (Patricia A. and Robert I, 2009)
9. Usage of lockdown browsers that prohibit students from Internet browsing or instant messaging service while taking an exam (within a learning management system) (Patricia A. and Robert I, 2009)
10. Secure password and identification systems (Patricia A. and Robert I, 2009)

BACKGROUND

Reliable student/user authentication is extremely relevant to online examinations because of high stakes being involved. Authentication attempts to verify that the users are whom they claim to be. In an online examination scenario, authentication aims to verify the identity of online students and plays a crucial role in security (R. Joyce, G. Gupta, 1990). Over the past few years, varied approaches to user authentication in the online learning environment have been tested and are now being incorporated into online learning ventures. In many cases, such alternatives have served as substitutions to the conventional means of on-site, face-to-face examination sessions. The mainstream authentication techniques are based on the user's knowledge, objects possession, and biometric features. Even with these available alternatives, the pressure continues to find a trustworthy yet cost-effective use of proctored evaluations administered through authorized testing centers (Jeffrey L. Bailie and Michael A. Jortberg, 2009). However, the use of advanced technology intended to validate an individual's biometrics including fingerprint readers, signature, facial, or voice recognition programs. The methods of various authentications in the online examination are:

1. Knowledge-based authentication (KBA): username and password.
2. Object-based authentication (QBA): RFID, chip cards, etc.
3. Profile-based authentication: username and password, profile.
4. Biometric-based authentication: iris, fingerprint, palm, voice, keystroke, etc.

Knowledge based authentication (KBA) identifies the end users by questioning the users to answer specific security questions in order to provide accurate authorization for online or digital activities. The main idea behind the KBA is that by selecting questions that only the target individual would know the answers to, systems can verify whether a user is the legitimate owner of a password-protected area or not. Although KBA can be an effective way to manage authorization for individual users, there are also critical concerns about privacy that have been raised around the idea of using this kind of personal information for online or network security (https://www.techopedia.com/definition). KBA questions can be either static or dynamic. But, both static and dynamic schemes rely on the assumption that if someone knows the correct answers to the secret questions, their identity can be confirmed. KBA is often used as a component in multifactor authentication (MFA). Object based authentication (QBA) is the authentication system which verifies the identity of the end user with the object the user is having. The objects can be RFID cards, chip cards, etc. But the QBA is combined with the KBA or profile based authentication. Profile based authentication (PBA) is based the role the user is playing. It is a risk based authentication which takes into account the profile of the end user.

Biometric-based authentication (BBA) was one of the methods which can be used for identifying the user based on the user biological evidences. The biometrics authentication is performed by the verification of an individual's physical or behavioral characteristics. Generally, a biometric system is designed to solve a matching problem through the live measurements of human body features which are more appropriate to the online learner verification. It operates in two stages. First, a user must register him/herself with a biometric in a system where biometric templates are stored. Second, the person must provide the same biometric for new measurements. The output of the new measurements will be processed with the same algorithms as those used at registration and then compared to the stored template. If the similarity is greater than a system-defined threshold, the verification is successful otherwise it will be considered unsuccessful. But, altogether biometric frees individuals to memorize passwords and carry cards, as the person is the key for identification. Many biometric authentication features have been evolved from recent research (C. Rogers, 2006, https://www.bbc.com/news, C Adams and A. A. Armstrong, 1998; Sarrayrih, M.A. and Ilyas, M., 2013; Joosten-Ten Brinke D., Van Bruggen et al., 2007) and implemented in online learning systems including fingerprint, video authentication, face recognition, audio recognition or combination of these features in the form of multi-modal biometrics (R. Joyce, G. Gupta, 1990). Table 1 lists few biometrics and their features for authentication with error correction methodology to support the knowledge about the biometric techniques used in the security systems (R. Joyce, G. Gupta, 1990; Yong Sheng, Vir. V. Phoha, Steven, M, Rovnyak, 2005).

The biometric identification based systems also include hand geometry, thermal patterns in the face, blood vessel patterns in the retina and hand, finger and voice prints, and handwritten signatures (Fabian Monrose, Aviel D. Rubin, 2000; Ramu, Dr.T. Arivoli, 2013). However, some of the techniques being deployed in commercial products such as fingerprint are so easy to fool, while others like iris pattern recognition are too expensive and invasive (Byeong Ho and Hyejin Kimand, 2015). Among the identified biometric types in Table 1, keystroke dynamics was identified as the best one as it relies on pattern recognition to provide a passive, inexpensive and more foolproof method than traditional passwords for verifying an individual's identity.

Table 1. Biometrics and their features for authentication with error correction

Reference	Biometric Types	Feature Extraction	Error Correction
(http://www.blackboard.niu.edu/blackboard)	Keystroke	Duration of latency for typing patterns	Discretization
(G. Jagadamba, B. Sathish Babu, 2015)	Voice	Text dependent or text independent speaker utterance units	Discretization
(C. Rogers, 2006)	Signature	Dynamic signature features	Averaging
(Byeong Ho and Hyejin Kimand, 2015; https://www.bbc.com/news)	Face	Facial features, position, sizes angles, etc.	RS code
(C Adams and A. A. Armstrong, 1998; Byeong Ho and Hyejin Kimand 2015)	Iris	Iris image representation with wavelet	RS code
(Sarrayrih, M.A. and Ilyas, M., 2013).	Finger print	Features such as ridge ending and ridge bifurcation etc.	Quantization
(Joosten-Ten Brinke D., et al., 2007)	Palm print	Features such as principle lines, wrinkles, delta points area of palm etc.	RS code

(Fabian Monrose, Aviel D. Rubin, 2000) strongly argue that the use of keystroke rhythm is a natural choice for computer security. This argument stems from observations (Romain Giot, Bernadette Dorizziy, Christophe Rosenberger, 2011) that similar neurophysiological factors that make written signatures unique are also exhibited in a user's typing pattern. When a person types, the latencies between successive keystrokes, keystroke durations, finger placement and applied pressure on the keys can be used to construct a unique signature (i.e., profile) for that individual. For well-known, regularly typed strings, such signatures can be entirely consistent. Furthermore, recognition based on typing rhythm is not intrusive, making it entirely applicable to computer access security and e-learning authentication as users will be typing at the keyboard. The detailed discussion about the keystroke dynamics for e-learning is done in the next section.

KEYSTROKE BIOMETRICS FOR E-LEARNING

The keystroke biometrics user authentications are the technologies that are measurable on physiological or behavioral characteristics (Jagadamba G, Sharmila S P, Tejus Gowda, 2014). A behavioral biometric recognition system can run in two different modes; identification, and Verification. Here automated approaches will be defined for an identity for each user (human or process). The verification process realizes the data capture, the feature extraction and the comparison with the biometric model. However, keystroke dynamics/rhythm is a natural choice for security parameters, which is an act of recognizing an individual based on typing patterns. Keystroke dynamics/rhythm is not what we type, but how we type and how we use. Hence, the keystroke dynamics became one of the cost-free identification techniques among the biometric identification. Telegraphs and Morse code lead to the earliest examples of keystroke dynamics, and a behavioral biometric indicator which can be used to recognize an individual rather than the more common physiological measures (The Web CT, Simon Fraser University).

Keystroke Dynamics System

1. **Data Acquisition:** The acquisition process collects the raw keystroke data through various input devices such as computer keyboard, virtual keyboard, smart phone, cellular phone etc.
2. **Feature Extraction:** The feature extraction process obtains the raw data and are treated and stored as reference template for future usage. This step includes the threshold evaluation, feature selection, dimension reduction, outlier detection, etc.
3. **Matching:** Matching includes the identification of the extracted value with the threshold value to make further decision by using algorithms.
4. **Decision:** The obtained data is presented to the authentication system and compared with the reference threshold/template via algorithm and the respective decision will be taken as the user is legitimate or masquerade.
5. **Retraining:** The stored reference threshold/template is updated often to reflect the ongoing changes.

Keystroke dynamics/rhythm includes several different measurements like latency between consecutive keystrokes, duration of the keystroke, hold-time, etc., (Jagadamba G, Sharmila S P, Tejus Gowda, 2014), which can be detected when the user presses the keys in the keyboard. In this chapter, we propose an efficient safeguard system for recognizing specific unique and habitual patterns with a user's typing rhythm called as keystroke dynamics/rhythm. About Keystroke

In the process of authentication, the keystroke dynamics/rhythms of the user are measured to develop a unique biometric template of the user's typing pattern. Raw measurements available from each keyboard can be recorded to determine dwell time (the time a key is pressed) or flight time (the time between key down and the next key down and the time between key up and the next key up). After recording data is processed through the algorithm, which serves as the primary pattern for future comparison and analysis.

Different methodologies have been adopted depending upon dwell time and flight time. A summary of existing methodologies with remarks has been collected from [Fabian Monrose, Aviel D. Rubin, 2000; Jagadamba G, Sharmila S P, Tejus Gowda, 2014, Monrose, F., Reiter, M. K., & Wetzel, S., 2002; Ness, J., 2017; Marcus Karnan., Akila, M., Kalamani, 2009; Shanmugapriya, D., Padmavathi, G., 2011, Qinghai Gao, 2010; Lawrence O Gorman, 2003) are listed in Table 2.

Keystroke Metrics

The keystroke biometric technique includes the static and continuous keystroke dynamic extraction approaches. Static verification approach analyzes the keystroke characteristics at the specific times (for example- during the login sequence). The static approach provides more robust user verification than simple passwords, but do not provide adequate identification. These techniques are not capable of detect-

Table 2. Different keystroke methodologies used for training and testing

Ref.	Methodology	Remarks
(Jagadamba G, Sharmila S P, Tejus Gowda, 2014).	Hold key Timings	Total time periods and pressure was measured using Euclidian distance measure between two vectors of typed characters, stored as template.
(Lawrence O Gorman, 2003; Marcus Karnan, Krishnaraj, N, 1010)	Ant Colony Optimization Technique	Mean and Standard Deviation was used to extract the features from the Keystroke duration, latency and digraph.
(Monrose, F., Reiter, M. K., & Wetzel, S., 2002)	Keystroke Timings	Successive keystrokes were recorded and used for authentication and result with FAR 4% for seven users.
(Romain Giot, Bernadette Dorizziy, Christophe Rosenberger, 2011)	Standard and Measure.	The mean, standard deviation of keystroke latencies and digraph between reference profile and test data are compared. A result of 17% FAR and 30% FRR was obtained.
(Ramu, Dr.T. Arivoli, 2013)	Pattern Recognition and Neural Network	Fuzzy ARTMAP, RBFN, and LVQ neural network paradigms were used. BPSTF, potential function and Bayes' rule algorithms gave moderate performance.
(Shanmugapriya, D., Padmavathi, G., 2011)	Bio Password	This technique is fast, accurate, transparent and scalable.
(Yong Sheng, Vir. V. Phoha, Steven, M, Rovnyak, 2005)	Parallel Decision Trees (DTs) on keystroke patterns	By this method the average false reject rate was 9.62% and the average false accept rate was 0.88%.
(D. Stefan, & Yao, D. Oct. 2010)	Telling hUman and Bot Apart (TUBA)	Monitoring user's keystroke-dynamics patterns and identifying intruders. The Bot-generated keystroke sequences are detected with high true positive rates 93%.
(Jagadamba G, Sharmila S P, Tejus Gowda, 2014)	Virtual key force	This method improves the accuracy by 90.4% and reduces the training and testing time within 0.025sec

ing a substitution of the user after the initial check. Continuous verification, on the contrary, monitors the user's typing behavior throughout the interaction and acquires the correct typing rhythm.

Further, the keystroke dynamics include several different measurements which can be detected when the user presses keys on the keyboard. Below, we have identified the possible measurement of the keystroke, namely.

- Latency between consecutive keystrokes.
- Duration of the keystroke, hold-time.
- Overall typing speed.
- Frequency of errors (how often the user has to use backspace).
- The habit of using additional keys in the keyboard, for example, writing
- Numbers with the numpad.
- The order the user press keys when writing capital letters, is the shift or the letter key released first.
- The hitting force for keys while typing (requires a special keyboard).

In (Chaya Shree G and Jagadamba G, 2016) latency is used as the measuring technique to identify the user typing dynamics. The latency is defined as the difference in the timestamps between successive keystrokes, the key-down to key-up time. Consider, a two-character pair P and A, where t1 be the time when the character P is typed, and t2 be the time when character A is typed. The difference between t1 and t2 gives the latency between these two characters.

Most of the researchers have used two types of latencies - string latency and character latency. The string latency is the time difference between the first and the last key pressed, i.e., typing frequency of the string (t2-t1) as shown in Figure1.

In character latency, the time difference between the successive pair of keys pressed, i.e., if the password length is 8 then there will be 7 different key frequencies between pair of successive keys. The Figure2 shows the character latency{(t2- t1)+{(t3-t2)+{(t4-t3)+{(t5-t4)+{(t6-t5)+{(t7-t6)+{(t8-t7)}.

In (Alves, D. D., G. Cruz, and C. Vinhal, 2014) the authentication system is applied using string keystroke dynamics of equal size or greater than five characters. The restriction of equal size string which lead to an unambiguous statement for the user. Hence, the keystroke latency based on character is preferable for structured or unstructured passwords. In (Chaya Shree G and Jagadamba G, 2016), both the latency types are tested to have the experimental confirmation for the structure passwords.

One of the major hurdles that keystroke dynamics has encountered is typing variations substantially during tired or angry, switch computers, reposition their keyboard, or even talk on the phone or otherwise be distracted. These seemingly small shifts could affect the way that the computer interprets a user's

Figure 1. String latency

Figure 2. Character latency

keystroke dynamics. These variations will cause error rates, both false positives, and false negatives. A valid solution using keystroke dynamics must take these elements into account, and strive to decrease false positives and adverse actions. Hence, the keystroke dynamics efficiency is measured using the two performance metrics: False Acceptance Rate (FAR) and False Rejection Rate (FRR) (Monrose, F., Reiter, M. K., & Wetzel, S., 2002).

In the keystroke dynamics, the false acceptance rate (FAR) is the measure of the likelihood of the biometric system will incorrectly accept an access attempt by an unauthorized user. A system's FAR typically is stated as the ratio of the number of false acceptances divided by the number of identification attempts. The false recognition rate (FRR), is the measure of the likelihood of the biometric system will incorrectly reject an access attempt by an authorized user. A system's FRR typically is stated as the ratio of the number of false recognitions divided by the number of identification attempts. In any biometric security systems False Rejection Rate (FRR) and False Acceptance rate (FAR), should be expected to be as less as possible. We have Equal-Error Rate (EER) or Crossover Error that defines the overall accuracy as well as a comparative measurement against other system. In our chapter we have considered FAR and FRR as the evaluation metric as we are comparing with the three factor authentication system (https://www.bayometric.com).

SOLUTIONS AND RECOMMENDATIONS

We broadly identified the two categories (Abdullah Alshehri, Frans Coenen, Danushka Bollegala, 2018; Fabian Monrose, Aviel D. Rubin, 2000) of keystroke authentication and are as defined:

1. Keystroke static authentication (KSA): Authentication directed at static (predefined) text such as passwords, usernames, and pin numbers.
2. Keystroke continuous authentication (KCA): Authentication directed at dynamic (nonpredefined) text such as text, number and alphanumeric.

More research was required for KCA compared to KSA; hence in this section, we discuss the approach of adopting the KSA method for authenticating the user in the process of e-learning. Many approaches (Jagadamba G, Sharmila S P, Tejus Gowda, 2014; Yong Sheng, Vir. V. Phoha, Steven, M, Rovnyak, 2005; Ramu, Dr.T. Arivoli, 2013) are followed for user authentication using KSA. In this chapter, we discuss the KSA approach which consists of a client, server and database. The client acts as an end user with various types of activities. The client performs activities like login to system, transactions, view to his data, etc. Server includes the functions like registering the server with Remote Method Invocation (RMI) registry, establishing the communication to client, processing of all clients requests, running keystroke procedure to check for valid user, sending the successful key typing latencies to database, provides application services to authenticated clients, if any variation a third level security policies will be defined. The third level policies may include sending the random code to the user's mobile, user's email account or check security answer sent by the client and proceed further processing. The database is used for storing all the relevant keystroke latency data and also useful information. The database receives the queries from the server process, execute them and return the corresponding results to the server process. A database server can store user details. The secret information of a user, i.e., the password can be stored in an encrypted form.

Adaptive Learning Classification (ALC) Algorithm

ALC algorithm initiates the working through the user signup process. Whenever the user signs up, user typing samples are captured without the knowledge of the users. The samples are collected whenever the user attempts the login. The samples are collected from the first five consecutive logins and a threshold value will be evaluated for the individual. Once, the sample collection process is initiated for the first time; the keystroke biometrics authentication mechanism comes into operation. For the evaluation of threshold, an average of the recent five successful transactions typing frequency is considered and then used to cross check with the current typing frequency. In the process of authentication, the current typing frequency of the users is matched with the average threshold value of the individual. If the user current threshold value will be between the identified threshold values, the user will be authenticated; otherwise, a user will be unauthenticated and third level security policies will be presented. The following steps are followed in the ALC algorithm to decide the user is a genuine user or intruder.

ALC Algorithm

Step 1: Initialization: Capturing the values, x= d, from username= `Specific User'.
Step 2: Template Design: Mean and Standard deviation is calculated for the specific user and the corresponding threshold is calculated.

$$\mu = \frac{\sum_{i=1}^{n} x_i}{N}$$

$$\sigma = \sqrt{\frac{\sum_i (x_i - \mu)^2}{N}}$$

Upper Threshold= $\left(\mu + \sigma\right)$

Lower Threshold= $\left(\mu - \sigma\right)$

Step 3: Analysis: if (Lower Threshold ≤ μ ≤ Upper Threshold)-authenticator, else Intruder.

Step 4: Decision: if (Authenticator) Service is continued else (Intruder) Third Level Security policies will be introduced.

Results of the KSA Approach

The approach was tested based on the Password string latency, i.e., flight time between the first and the last key pressed and Password character latency, i.e., flight time between the successive pair of keys pressed. During the experiment, the delete operations and irrelevant shift key pressed were eliminated from the data. A user can login successfully only when the successful match is made between the stored original template and live samples. The valid user is identified by verifying the matched value. The false match is registered when a live sample exceeds the specified threshold, and then the user will not be authenticated. In the testing procedure of string latency, forty participants were allowed to log in to their account and data collected where analyzed. The Figure3 shows the frequency variations for ten attempts and Figure4 shows the experimental results where failure rates of FRR fall around 10%-40%. In the same way character latencies where collected for forty participants by allowing them to log in to a particular account and frequency variations are depicted in Figure5. The success rate of FAR falls between 0%-20% and success rate of FAR is almost 0.0025%, which is shown in Figure6. All the tests are done for an e-learning application and the response time was observed to be concerning 2-5 milliseconds with the internet connectivity of 10.0Mbps, with some 50% of traffic in the channel in the client-server architecture.

Figure 7 shows the intruder detection analysis done over the keystroke dynamics. When we increased the number of the intruders, the experimented results were able to detect the same percentage of an intruder. Thus, the experiment shows the validations of the work.

As keystroke dynamics are inexpensive with no burden for implementation as no extra hardware is required. The keystroke authentications alone are not so feasible, hence adopted with the two or three factor authentication techniques. In this regard, an attempt to optimize the keystroke authentication was done (Alves, D. D., G. Cruz, and C. Vinhal, 2014) based on the simple statistical methodologies. A work on similar approach was done in our earlier work (Jagadamba G, Sharmila S P, Tejus Gowda, 2014) and achieved good efficiency with adaptive learning algorithm. The work was useful concerning to two-factor authentication technique when compared to (Alves, D. D., G. Cruz, and C. Vinhal, 2014). A smart card authentication with the password was efficient for two-factor authentication (Das, Manik Lal, 2008). The approach was able to overcome with the many attacks, but not with the stolen attacks of smart card and eavesdropping attack. Hence, the three-factor authentication, based on the keystroke biometric which does not require the complicated procedures to extract biometric data is the solution to strengthen the authentication process.

Hence, the identification process is verified by taking the three factors i.e., password, smartcard and keystroke dynamics. The architecture consists of clients, application server and authentication server database. The client acts as an end user with various types of activities. Clients operating terminal includes an interface for entering the password and extracting the keystroke dynamics with smartcard code. The client side authentication data are sent through a secure channel to the application server. The application server includes the functions like registering the server with Remote Method Invocation (RMI) registry, establishing the communication to client, processing of all clients requests, running password, keystroke dynamics and smart card verifying procedure to check for the valid user. The application server sends the successful key typing latencies to database server before providing application services to authenti-

Figure 3. Frequency variations of string latency

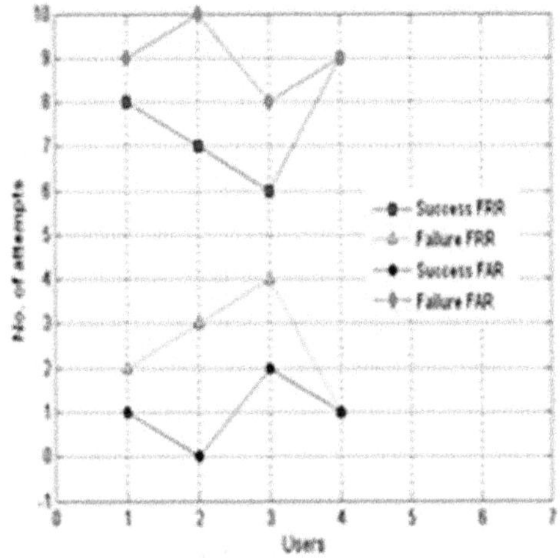

Figure 4. FRR and FAR for string latency

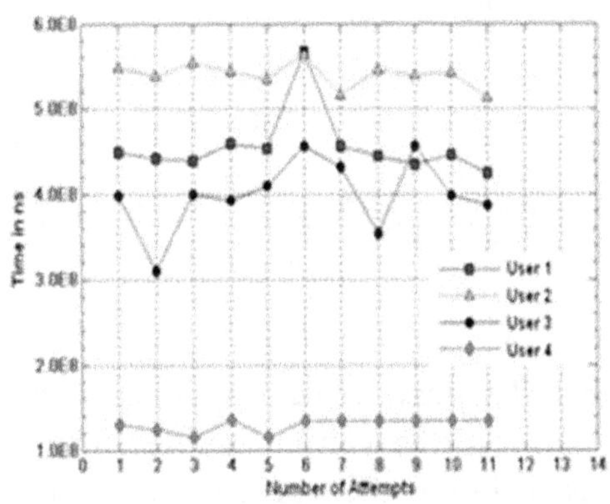

Figure 5. Frequency variations of character latency

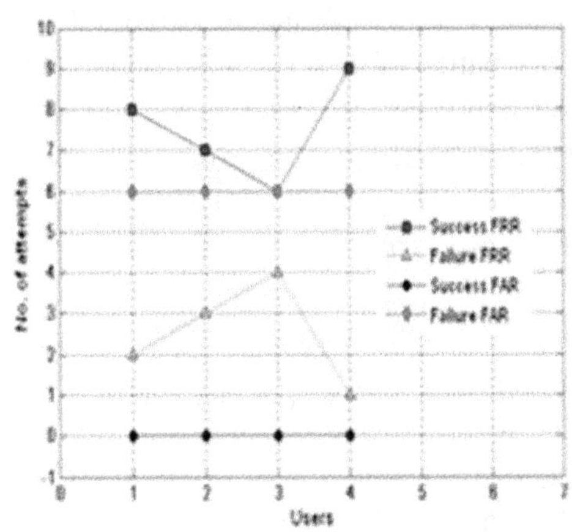

Figure 6. FRR and FAR for character latency

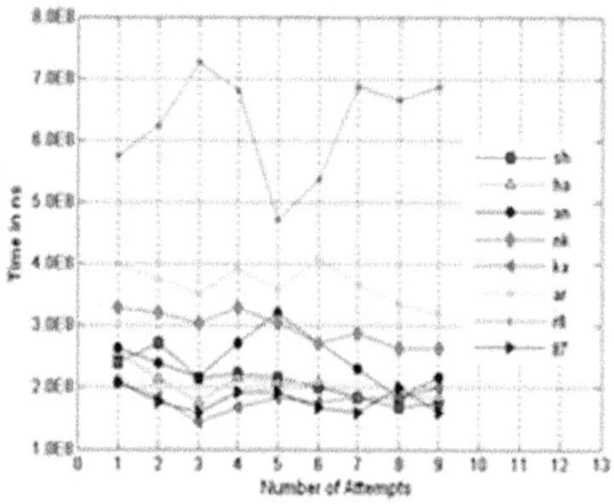

Figure 7. Overall analysis for intruder and valid user with keystroke analysis

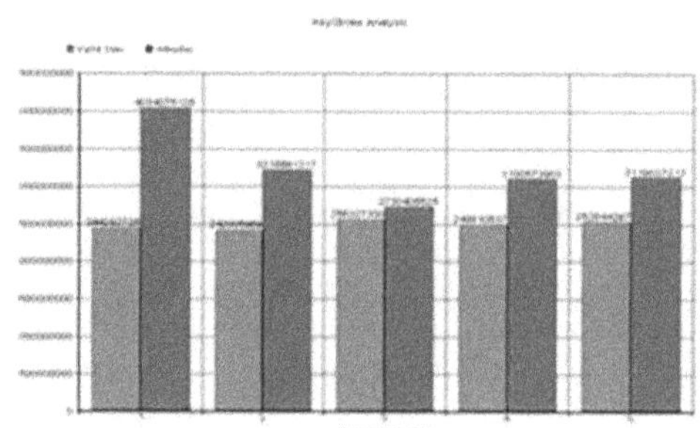

cated clients. If any variations in the keystroke dynamics, a third level security policies will be defined. The database is responsible for storing all the relevant user information I including the smart card hash code P and keystroke latency B data of the entire registered user. Once the request arrives from the client side with the authentication credentials (I, P, B), the server receives and verifies the exactness of the credentials before proceeding to provide services. Those above mentioned will receive the queries from the server process, execute them and return the corresponding results back to the server process. A database server includes information like user details, commodity prices, details of transactions, incoming and outgoing messages, and authentication details in different tables. The first degree of security is provided by storing the secret information of a user in an encrypted form.

Client: Basically clients are our end-users, used for sending and receiving requests. The various types of client activities like login, changing of account password, viewing of transaction details, viewing of account details and so on.

Server: Server is computer running a windows operating system as our Server System. This system is responsible for the following tasks

- Registering server with rmi regisrty.
- Establishing communication with clients.
- Processing of all clients requests.
- Running keystroke procedure to check for valid user.

- Sending the successful key typing latencies to database.
- Provides services to authenticated clients.
- If a variation occurs, send random code to user mobile and email account.
- Check security answer sent by the client.

Database: It is accountable for storing all the relevant keystroke latency data and also user information. It will receive the queries from the server process, execute them and return the corresponding results back to the server process. A database server is used to store various information required by the server process. Information like user details, commodity prices, details of transactions, incoming and outgoing messages are stored in the database in different tables.

Uid Ci (0<i<16)

A schema of character_frequency or string-frequency of all character and string latencies recorded during typing of the password is shown as a template for the reference. Here, only successful login values are stored in the character/string frequency table with the 16 attributes as shown in schema diagram. Where Uid is the username holder attribute, Ci ($0 < I < 16$) represents the time taken to type Ci+1 character after typing of Ci.

The proposed work verifies the authenticity of the user in the three stages of the test: (1) things that the user knows i.e., username and password, (2) things that the user has i.e., smartcard, (3) and things that the user is i.e., keystroke biometric. The system authenticates the user using the entire three tests without interruption. The first test and the second should be matched before proceeding to the third step. The third test- biometric- keeps the secret to the user by allowing the server to perform its authentication.

On the process, if the user fails to provide the requirements, the systems deny the service requested by the user. Hence, the proposal is physically two-step as the keystroke authentication is automatically verified without the knowledge of the user, using the Adaptive Learning Algorithm. The formal three-factor authentication scheme for event E is defined as follows:

A_1: The user U passes the checking of the password for event E.
A_2: The user U passes the checking of the smart card for event E.
A_3: The user U passes the checking of the keystroke biometric data for event E.

Security Phases in 3-Factor Authentication

This proposed three entity authentication security scheme is modeled in 5 phases, namely: User Enrolment Phase, User Login Phase, User Authentication Phase, Remote Server Authentication Phase, User Password Change phase.

Before proceeding to the individual phases of adaptation, let us consider some of the definition used in the security scheme. Where UE – User Entity, UL – User Login, UA – User Authentication, RSA – Remote Server Authentication, UP – User Password Change, SR – Remote Server, UR – Remote User and CS – Smart Card.

1. **User Enrolment Phase:** Each every user register him/herself before accessing any service in the system which is executed only once for every user.

UE1: The U_R chooses his Identifier I and Password P and sends to the R_S. The Keystroke pattern (B) is also extracted.

UE2: S_R has received the enrolment request from U_R and executes the following task.
UE3: Computes G as follows,

$$G = h\,(I \oplus P) \tag{1}$$

where, *h (.)* is One-way hash function, $\oplus$ is the bitwise XOR operation.

UE4: Computes K as follows

$$K = h\,(S) \oplus G \tag{2}$$

where, *S* secret key of $S_{R,}$

UE5: Stores the parameters {h (.), e, I, B, K and G} into C_S, where *e* is the secret number stored in each enrolled user`s Smart Card $C_{s.}$

UE6: Sends the C_S to the U_R through a private channel.

2. Login Phase:

The U_R inserts his C_S into a card reader and enters the required details.

UL1: The C_S computes L as follows,

$$L = K \oplus h\,(I \oplus P`) \tag{3}$$

where, I is Users Identifier and P Password

L = h(S). If it is equal, then C_S performs the following tasks otherwise rejects the Login request.

UL2: Computes O as follows,

$$O = G \oplus h\,(T) \tag{4}$$

where, *T* current time of User Login

UL3: Computes N as follows,

$$N = h(h(e) \oplus K \oplus h(T)) \quad (5)$$

UL4: Sends the Login request {O, N, T, B} to the S_R through a public channel.

3. User Authentication Phase:

This phase is executed after the Login request is sent to the S_R.

UA1: The S_R has received the Login request {O, N, T, B} at time T* and executes the following tasks.
UA1.1: S_R Checks the difference between T* and T. If the time difference is a valid transmission delay, then S_R performs the next task otherwise rejects the Login request.
UA1.2: Computes N` as follows,

$$N` = h(h(e) \oplus h(S) \oplus O) \quad (6)$$

where, *S* is secret key of S_R

UA1.3: The S_R checks whether N = N` and LB` < B < HB`, where LB` and HB` are Lower and upper thresholds respectively. If it holds S_R accepts the Login request of U_R otherwise it rejects.

4. Remote Server Authentication Phase:

Once the user is authenticated to the server, the next task is to authenticate the server. This phase validates the authenticity of S_R.

RSA1: S_R computes M as follows,

$$M = h(T^{**} \oplus h(T \oplus h(S) \oplus h(I \oplus P))) \quad (7)$$

where, T^{**} time stamp

RSA2: S_R sends {M, T**} to the U_R through a public channel.
RSA3: U_R receives {M, T**} at time T***. C_S checks the difference between T** and T*** whether is a valid transmission delay or not. If it is valid then C_S performs the next task.
RSA3.1: C_S computes M` as shown in Table 3.

$$M` = h(T^{**} \oplus h(T \oplus K)) \quad (8)$$

RSA3.2: The C_S checks whether $M = M`$. If it is holds the S_R is authenticated.

Table 3.

References	Communication cost	Computation cost	User privilege to choose password	Server side matching and decision	Security
(Fan, Chun-I., and Yi-Hui Lin, 2009)	5 messages	High	Yes	Yes	Proved
(A Bhargav Spantzel, et. al., 2007)	-----------	High	Yes	No	Not proved
(E. J. Yoon and K. Y. Yoo, 2005)	4 messages	High	Yes	No	Not proved
(Tan, Z. W, 2013)	-----------	Low	Yes	No	Dissatisfied
(Tan, Zuowen, 2014)	5 messages	High	Yes	No	Proved
(Das, A. K., 2011)	------------	Low	Yes	No	Dissatisfied
(Das, Manik Lal, 2008)	4 messages	High	Yes	Yes	Not proved
(Yu, J., Wang, G., Mu, Y., & Gao, 2014)	4 messages	Very High	yes	Yes	Proved
(Das, A. K. 2015)	3 messages	High	yes	yes	Proved
(Chaya Shree G and Jagadamba G, 2016)	3 messages	Low	Yes	Yes	Proved

5. **User Password Change Phase:**

The U_R inserts his C_S in a card reader, enters his identifier I and Password P. The C_S verifies the entered I and P with the stored value of I and P in the C_S. If all the verifications are passed successfully, then C_S executes the following tasks.

UP1: Ask the U_R to enter a new Password and the user chooses a new Password P\`.
UP2: Computes G\` as follows.

$$G` = h(I \oplus P`) \text{ and } K` = h(S) \oplus G` \quad (9)$$

UP3: P\`, G\`, and K\` are stored in the place of P, G and K respectively.

To verify the performance of the keystroke dynamics with other two authentication factors, the Communication and Computational cost was verified by including all the three factors i.e., password, keystroke and smartcard. The computation and communication cost metrics are evaluated during the registration, login, and authentication stages. But, we restrict ourselves to estimate for the login and authentication phase of the registration process, as this may vary concerning to the user and applications requirement.

The following notations were considered while evaluating the computational cost for login and authentication steps. Where, *treqc* is the time for the login request from the client to the base station, *treqb* is the time from the base station to the server, *th* is the hashing time, *trsae* is the encryption time, *tress* from the server to the base station and *tresb* is the time taken from the base station to the client. The each phase of computation is measured in milliseconds for each operation. The proposed scheme consumes the computational cost as of the following:

Login phase: (*treqc, reqb+th+trsad*)

Authentication phase: (*2treqc, tresb+th+tadl+ trsad*)

The computational cost for three factors and two factor authentications is as low as possible, but the communication cost is comparably more, i.e., 3 message length compared to two factor authentications. Detailed performance metrics can be found in Table.1 for various works with the verification of security parameter.

APPLICATIONS/PRODUCTS DEVELOPED BASED ON KEYSTROKE DYNAMICS

Real-time applications are studied to introduce our self to the technics and methodologies and applications which were using the keystroke dynamics. We are able to discuss a few applications here to give an insight (Yong Sheng, Vir. V. Phoha, Steven, M, Rovnyak, 2005, https://www.techopedia.com) of working:

- **Typing DNA:** It is the application developed on artificial intelligence engine that can match any two typing patterns with unprecedented accuracy. They mentioned the ease of using keystroke dynamics authentication for eLearning, fraud prevention, securing logins, detecting intruders and online biometric authentication, multifactor authentication, and user identification. This product provided a continuous authentication app, for Windows and Mac.
- **ID Control:** This product is developed by a Dutch company to provide an affordable authentication solution using keystroke dynamics. This product can be integrated with MS Windows logon, Citrix, VPN and many others.
- **BioTraker:** This software uses continuous authentication to verify the identity of a user. Plurilock Security Solutions developed this for cyber security technology. US defense department is utilizing this software to have the complete control over the corporate network by providing real-time surveillance of all network activities.
- **BehavioSec:** This product is developed by a Swedish company specialized in continuous authentication systems to monitor activities from the genuine account owner who is using the computer. This product uses mouse dynamics with keystroke dynamics.
- **KeyTrac:** This product collects the input entered by the user and creates a user profile by integrating on a web server. The only disadvantage of KeyTrac is the integration with CAPTCHA applications that can be read only by humans, not computers.

FUTURE RESEARCH DIRECTIONS

Thus keystroke biometrics exhibits an inexpensive user authentication system for e-learning and also not limited to applications like internet banking, online commerce, etc. Performance can be improved by minimizing the FAR by adapting relative keystrokes with proper fault tolerance methods.

In the recent days, we are looking at the research work concentrating on the KCA approach. Hence, in future KCA approach can be involved that does not require any predefined password storage. However, this needs stringent algorithms which include artificial intelligence.

From the study (Shauna Beaudin, Yair Levy, James Parrish and Theon Danet, 2016) we observed that to improve the authentication strength, two single-factor authentications can be combined into a two-factor authentication . At the same time, a two-factor authentication approach is the use of live-proctor authentication along with username/password or biometric authentication. From (O'Gorman, 2003; Ross, 2007; Ross, Nandakumar, & Jain, 2006) a multi-factor authentication, combining three factors, creates a very strong authentication and improves reliability against impersonation fraud. Hence, we strongly believe that, the research work can be extended by considering various combinations of authentication factors to make any security system strong enough to have full pledged efficiency. We have experiment the three factor authentication system for online banking also by considering the PIN, card and keystroke dynamics. The experiment results showed an expected efficiency; hence we are encouraged to do the future work to develop the security system by considering various combinations of authentication factors based on the type of applications we choose.

CONCLUSION

This chapter presented a framework for defining a quick and secure authentication approach that can be followed in e-learning process which doesn't require any specialized hardware except, the keyboard for the applications in the highly-connected world. This approach does not affect privacy since the behavior typing rhythm of an individual is considered. As individual behavior cannot be copied, it becomes an identification token for verification in any online learning process/examination.

REFERENCES

Adams, C., & Armstrong, A. A. (1998). A Web-based testing: A study in interesting. *World Wide Web (Bussum)*, *1*(4), 193–208. doi:10.1023/A:1019224017027

Alshehri, Coenen, & Bollegala. (2018). Iterative Keystroke Continuous Authentication: A Time Series Based Approach. In *Künstliche Intelligenz*. Springer. DOI: . doi:10.100713218-018-0526-z

Alves, D. D., Cruz, G., & Vinhal, C. (2014). Authentication system using behavioral biometrics through keystroke dynamics. *IEEE Symposium on Computational Intelligence in Biometrics and Identity Management (CIBIM)*, 181-184. 10.1109/CIBIM.2014.7015461

Alwi, N., & Fan, I. (2010). E-learning and information system management. *International Journal of Digital Society*, *1*(2), 148–156. doi:10.20533/ijds.2040.2570.2010.0019

Arivoli. (2013). A framework of secure biometric based online exam authentication: An alternative to traditional exam. *International Journal of Scientific & Engineering Research*, *4*(11).

Bailie, J. L., & Jortberg, M. A. (2009). Online learner authentication: Verifying the identity of online user's. *Journal of Online Learning and Teaching / MERLOT*, *5*(2), 197–207.

Bailie & Jortberg. (2009). Online Learner Authentication: Verifying the Identity of Online Users. *MERLOT Journal of Online Learning and Teaching*, *5*(2).

Beaudin, Levy, Parrish, & Danet. (2016). An empirical study of authentication methods to secure e-learning system activities against impersonation fraud. *Online Journal of Applied Knowledge Management, 4*(1).

Bhargav-Spantzel, A., Squicciarini, A. C., Bertino, E., Modi, S., Young, M., & Elliott, S. J. (2007). Privacy preserving multi-factor authentication with biometrics. *Journal of Computer Security*, *15*(5), 529–560. doi:10.3233/JCS-2007-15503

Boakye, O., & Marfo, Y. (2016). Utilizing Keystroke Dynamics as an Additional Security Measure to Password Security in Computer Web-based Applications - A Case Study of UEW. *International Journal of Computers and Applications*, *149*(5).

Chaya Shree, G., & Jagadamba, G. (2016). Keystroke Dynamics in Authentication Scheme. *International Journal of Computer Information Systems*, *12*(3).

Das, A. K. (2011). Analysis and improvement on an efficient biometric based remote user authentication scheme using smart cards. *IET Information Security*, *5*(3), 145–151. doi:10.1049/iet-ifs.2010.0125

Das, A. K. (2015). A Secure and Efficient User Anonymity-Preserving Three-Factor Authentication Protocol for Large-Scale Distributed Wireless Sensor Networks. *Wireless Personal Communications*, *82*(3), 1377–1404. doi:10.100711277-015-2288-3

Das, M. L. (2008). *A Flexible and Secure Remote Systems Authentication Scheme Using Smart Cards.* arXiv preprint arXiv:0801.0575.

Desouza, E., & Fleming, M. (2003). A Comparison of In-Class and Online Quizzes on Student Exam Performance. *Journal of Computing in Higher Education*, *14*(2), 121–134. doi:10.1007/BF02940941

Fan, C-I., & Lin, Y-H. (2009). Provably secure remote truly three-factor authentication scheme with privacy protection on biometrics. *Information Forensics and Security, IEEE Transactions on, 4*(4), 933-945.

Gao, Q. (2010). Online teaching: Do you know who is taking the final exam? In *Proceeding of Mid-Atlantic ASEE Conference*. Villanova University.

Giot, R., Dorizziy, B., & Rosenberger, C. (2011). Analysis of Template Update Strategies for Keystroke Dynamics. *Proc. IEEE Symposium Series on Computational Intelligence*, 21-28. 10.1109/CIBIM.2011.5949216

Gorman. (2003). Comparing Passwords, Tokens, and Biometrics for User Authentication. Proc. IEEE, 91(12), 2019-2040.

Gunetti, D., & Picardi, C. (2005). Keystroke analysis of free text. *ACM Transactions on Information and System Security*, *8*(3), 312–347. doi:10.1145/1085126.1085129

Ho, B., & Kimand, H. (2015). Proposal: A Design of E-learning learning User Authentication System. *International Journal of Security and Its Applications*, *9*(1), 45–50. doi:10.14257/ijsia.2015.9.1.05

Jagadamba & Sathish Babu. (2015). Security in Cyber-Physical Systems: Generalized Framework Based on Trust with Privacy. In *Cyber-Physical Systems: A Computational Perspective* (pp. 443-462). CRC Press, Taylor & Francis Group.

Jagadamba, G., & Sharmila, S. P. (2014). A Secured Authentication using Keystroke Dynamics. Lecture Notes in Electrical Engineering, 248, 253-260.

Joosten-Ten Brinke, D., Van Bruggen, J., Hermans, H., Burgers, J., Giesbers, B., Koper, R., & Latour, I. (2007). Modeling assessment for re-use of traditional and new types of assessment. *Computers in Human Behavior*, *23*(6), 2721–2741. doi:10.1016/j.chb.2006.08.009

Joyce, R., & Gupta, G. (1990). Identity authorization based on keystroke latencies. *Communications of the ACM*, *33*(2), 168–176. doi:10.1145/75577.75582

Karnan & Kalamani. (2009). A Feature Subset Selection in Keystroke Dynamics Using Ant Colony Optimization. *Journal of Engineering and Technology Research*, *1*(5), 72-80.

Karnan, M., & Krishnaraj, N. (2010). Bio Password-Keystroke dynamic Approach to secure Mobile Devices. *Proc. IEEE-2010, International Conference on Computational Intelligence and Computing Research.* DOI:10.1109/ICCIC

King, C. G., Guyette, R. W., & Piotrowski, C. (2009). Online exams and cheating: An empirical analysis of business students' views. *The Journal of Educators Online*, *6*(1). doi:10.9743/JEO.2009.1.5

Levy, Y. (2008). An empirical development of critical value factors (CVF) of online learning activities: An application of activity theory and cognitive value theory. *Computers & Education*, *51*(4), 1664–1675. doi:10.1016/j.compedu.2008.04.003

Monrose, F., Reiter, M. K., & Wetzel, S. (2002). Password hardening based on keystroke dynamics. *International Journal of Information Security*, *1*(2), 69–83. doi:10.1007102070100006

Monrose, F., & Rubin, A. D. (2000). Keystroke dynamics as a biometric for authentication. *Future Generation Computer Systems*, *16*(4), 351–359. doi:10.1016/S0167-739X(99)00059-X

Ness, J. (2017). *Presentation Attack and Detection in Keystroke Dynamics* (Master's thesis). NTNU.

O'Gorman, L. (2003). Comparing passwords, tokens, and biometrics for user authentication. *Proceedings of the IEEE*, *91*(12), 2019–2040. doi:10.1109/JPROC.2003.819605

Patricia, A., & Robert, I. (2009). Student Identity and Authentication in Distance Education: A Primer for Distance Learning Administrators. *Continuing Higher Education Review, 73.*

Peres, P., Lima, L., & Lima, V. (2014). B-learning quality: dimensions, criteria and pedagogical approach. *European Journal of Open, Distance and e-Learning, 17*(1), 56-75.

Rogers, C. (2006). Faculty perceptions about e-cheating during online testing. *Journal of Computing Sciences in Colleges*, *22*(2), 206–212.

Sarrayrih, M. A., & Ilyas, M. (2013). Challenges of online exam, performances and problems for online university exam. *International Journal of Computer Science Issues*, *10*(1), 439.

Shanmugapriya, D., & Padmavathi, G. (2011). An Efficient Feature Selection Technique for User Authentication using Keystroke Dynamics. *International Journal of Computer Science and Network Security*, *11*(10), 191–195.

Sheng, Phoha, & Rovnyak. (2005). A Parallel Decision Tree-Based Method for User Authentication Based on Keystroke Patterns. *IEEE Transactions on Systems, Man, and Cybernetics Part B: Cybernetics, 35*(4). Available: http://www.ucertify.com/certifications//ic3.html http://polinance.blogspot.in/2010/06/keystroke-dynamics-how-unique-keyboard.html

Shree, C. G., & Jagadamba, G. (2016). A Three Factor Authentication System Using Keystroke Dynamic. In *Proc. of National Conference on recent trends in communication and computing*. Davengere.

Stefan, D., & Yao, D. (2010). Keystroke-dynamics authentication against synthetic forgeries. In *2010 6th International Conference on In Collaborative Computing: Networking, Applications and Worksharing (CollaborateCom)* (pp. 1-8). IEEE.

Tan, Z. (2014). A user anonymity preserving three-factor authentication scheme for telecare medicine information systems. *Journal of Medical Systems, 38*(3), 1–9. doi:10.100710916-014-0016-2 PMID:24643750

Tan, Z. W. (2013). An efficient biometric-based authentication scheme for telecare medicine information systems. *Przegl. Elektrotech., 89*(5), 200–204.

Yoon, E. J., & Yoo, K. Y. (2005). A new efficient fingerprint-based remote user authentication scheme for multimedia systems. *9th Int. Conf. Knowledge-Based & Intelligent Information & Engineering Systems (KES 2005)*, 332–338. 10.1007/11553939_48

Yu, J., Wang, G., Mu, Y., & Gao, W. (2014). W. An efficient generic framework for three-factor authentication with provably secure instantiation. *IEEE Transactions on Information Forensics and Security, 9*(12), 2302–2313. doi:10.1109/TIFS.2014.2362979

Chapter 2
Keystroke Dynamics in E-Learning

Radha Sundararajan
Shrimathi Devkunvar Nanalal Bhatt Vaishnav College for Women, India

ABSTRACT

Enhanced authentication is the need of the hour in today's technology. Commonly used login and password are not enough as they may be guessed by imposters. Most of the websites adopt the traditional authentication as login and password. But they don't verify whether the same person is accessing their information continuously in the current session. This is of great concern in distance-based e-learning systems. The institutes offering the e-courses must verify whether it is the same student who enrolled, is accessing their materials, doing the assignments themselves, and completing the examination without any cheating. In this case, one of the techniques, behavioral biometrics-keystroke dynamics, plays a very important role. Along with other authentication methods, keystroke dynamics can be combined to provide a more secured system for the students in e-learning environments. In this chapter, the basics of keystroke dynamics and some of the applications that use them are discussed.

INTRODUCTION

The emergence of Massive Open Online Courses(MOOC) has brought a new attention to ensuring integrity in a global online classroom. The verification of student in online examinations and distance education programs are very important factors in the successful implementation of an e-learning program. After completion of the course, writing examination and giving of credits to the student who completed the given task is very important in successful running of the course and maintaining the reputation of the institution and running the course continuously. The traditional class room teaching is more equipped with these types of courses and those who are not able to join colleges and universities can benefit from these types of online distance learning programs.

Almost all the governments across the world implements the e-learning courses which can be taken by students of different levels, get certificates and enrich their knowledge in a particular course. Some of the online courses offered by various organizations are:

DOI: 10.4018/978-1-5225-7724-9.ch002

Coursera - founded by Stanford professors Andrew Ng and Daphne Koller offers online courses. Coursera is partnering with world top universities and offer courses online. It provides 800+ online free courses and works with world class universities and other organizations to offer online courses, specializations, and degrees in a variety of subjects such as engineering, humanities, medicine, biology, social sciences, mathematics, business, computer science, digital marketing, data science, and others.

Udemy.com is a global marketplace for learning and teaching online which offers nearly 65,000 courses taught by expert instructors.

Udacity is a for-profit educational organization founded by Sebastian Thrun, David Stavens, and Mike Sokolsky offering massive open online courses (MOOCs).

Edx.org provide various online courses with cutting edge technology in collaboration with various universities like Massachusetts Institute of Technology, Harvard University, Berkeley University of California, The Hongkong Polytechnic University, University of Texas, and so on.

Alison is an e-learning provider founded in Galway, Ireland in 2007 by serial entrepreneur, Mike Feerick. It is one of worlds' largest company offering free online education to enable everyone, anyone, at any time, to gain basic education and workplace skills. Alison is collaborating with American third level institutions such as MIT and Stanford University, Columbia University, Cambridge University, Yale University, and companies such as Microsoft and Google to provide high-class knowledge to their students

Khan Academy is a non-profit educational organization created in 2006 by educator Salman Khan with a goal of creating a set of online tools that help educate students. They release their courses as YouTube videos and the supplementary practice exercises and materials are available to their students. Other than English, they also provide the content in languages like Spanish, Portuguese, Hebrew, Italian, Russian, Chinese, Turkish, French, Bengali, Hindi, German.

MIT OpenCourseWare (MIT OCW) is an initiative of the Massachusetts Institute of Technology (MIT) to publish all of the educational materials from its undergraduate – and graduate-level courses online, freely and openly available to anyone, anywhere. The program was originally funded by the William and Flora Hewlett Foundation, the Andrew W. Mellon Foundation, and MIT.

SWAYAM is a programme initiated by Government of India and its main principle are access, equity and quality. The objective is that even most disadvantaged also will be benefited at maximum. The courses are prepared by more than 1000 specially chosen faculty and teachers across the country. All most all the courses are free of cost for studying. Only if the student want to get certificate then he/she has to pay a minimum amount and they are allowed to write online examination and the certificates and credits are awarded. The courses are developed in IT platform with rich multimedia, audio and video presentation which make the learning experience an enjoyable one. In order to ensure best quality content are produced and delivered, nine National Coordinators have been appointed: They are AICTE for self-paced courses, NPTEL for engineering, UGC for post-graduation education, CEC for under-graduate education, NCERT & NIOS for school education, IGNOU for out of the school students and, IIMB for management studies. UGC has already issued the UGC (Credit Framework for online learning courses through SWAYAM) Regulation 2016 advising the Universities to identify courses where credits can be transferred on to the academic record of the students for courses done on SWAYAM.

When the students are taking the classes, going through the videos, etc the course provider must make sure that only the registered students are going through their materials. Many fraud takes place even in writing online examination where the students make other people to use their identity and write the examination for them and try to cheat the examination system. That is why the Indian Government is adopting a very strict supervision on students who are writing NEET examination (National Eligibility cum Entrance Test).

After entering the user id and password in an online examination, the user may ask the expert to write the exam instead of writing the exam himself which is a very serious crime. So to prevent this there must be a technology to check whether the registered users are the one who are writing the examination and are still in the session. The technique that can be adopted for this is Keystroke dynamics which is based on the user typing pattern as security measure which can find out bogus person.

The genesis of keystroke dynamics dates back to World War II, when the military began using dots and dashes based on the syncopation of telegraph keys. In the 1980s, the National Science Foundation asked Rand to test keystroke dynamics for use as a computer security technology.

While the tests were successful, it wasn't until 2000 when keystroke dynamics passed the Financial Services Technology Consortium/International Biometric Group comparative testing program, making it a potential candidate for use in IT security and e-commerce.

E-commerce, mobile banking and other transaction-based applications are driving interest in keystroke dynamics to better protect digital identities, according to market research company Global Industry Analysts. GIA expects the size of the market to nearly double, to **$796.5 million by 2020.**

When it comes to authentication, behavioral biometrics are usually considered less reliable than physical biometrics because they are different in nature. With physical biometrics, one is dealing with absolutes – pass or fail. But behavioral biometrics are more about statistics and percentages.

The main objective of this chapter is outlined as Identity theft which explains how imposters use the students or user identity. After this the background research done in this area is discussed and then role of authentication and the use of devices for acquisition of data, feature selection for keystroke dynamics, methods used in this field, advantages and disadvantages of keystroke dynamics, and finally some of the applications using keyloggers are discussed.

IDENTITY THEFT

The growing use of internet gives a very big crime of exploitation of authorized people through identity theft. Identity theft is used by intruders to access web accounts through internet. Identity theft is a fraud, in which someone pretends to be someone else in order to steal money or get other benefits like money laundering, trafficking human beings, stock market manipulation, breaching of national security or terrorism, doing fraud in online examinations by using the student's identity etc. According to (Moskovitch R et al,2009) Identity theft can be broadly divided as:

- Financial identity theft (using some one's identity to stole goods and services) example bank fraud.
- Criminal identity theft (posing as another when apprehended for a crime)
- Identity cloning (using another's information to assume his or her identity in daily life)
- Business/ commercial identity theft (using another's business name to obtain credit)

There are online cheat-for-hire companies which openly advertise as they will assist the online course students in submitting assignments, completing the online classes in time, writing posts for discussion boards, taking exams and emailing the faculty posing as student. Some of the dangers involved in this are

1. They ask lot of personal information like birthdate, home or campus address, usernames, passwords, etc- basically the keys to one's personal information
2. Identity thieves might apply for or collect students' financial aid. An online student's personal information could be used to apply for scholarship or educational loan. The victim is responsible for repaying.
3. The companies may threaten the student for payment. They threat the student that they will report to the respective institution for academic dishonesty unless they pay additional fee.

Online students considering using these services seldom consider the additional nightmares of identity theft, financial aid fraud, extortion, failure or expulsion. It's understandable that busy online students can feel cheating is the only way out of a tough semester. It isn't so.

Keystroke Dynamics

In the mid of 19th century when telegraph was used for communication between two different places, it was observed that telegraph operators were able to identify the other operators by identifying their typing rhythms. During world war II, the methodology known as 'Fist of Sender' was used to identify the sender of the telegraph by using the rhythm of sender. Military troop movements could be tracked using this. (Shaffer L.H, 1978) showed that typing is a motor programmed skill that movements are organized prior to their actual execution stating that a persons typing pattern is a skill developed over a period of time which can not be lost of forgotten.

Biometrics Research Group, Inc defines biometrics as measurable physical and behavioural properties that make it possible to authenticate an individual person's identity. Keystroke dynamics uses the users typing pattern to develop a unique biometric template to identify the individual along with other security features like login id and password. Keystroke recognition or dynamics is the process of measuring and assessing the typing speed or rhythm of a person on digital devices like mobile, computer keyboards, touch screen etc. It refers to the timing about when a key is pressed and released as the person types. According to neuro-physiological factors that make written signatures of an individual unique, the typing pattern of a person may also be unique. Besides fingerprint recognition, iris recognition etc the keystroke dynamics is now considered as a behavioural biometric technique. Other behavioural biometric techniques are Voice recognition, gait (way of walking), signature, Lip Movement, mouse dynamics etc. Research on keystroke dynamics biometrics has been increasing, especially in the last decade. The main motivation behind this effort is due to the fact that keystroke dynamics biometrics is economical and can be easily integrated into the existing computer security systems with minimal alteration and user intervention. The main advantage is that it does not need any extra software or hardware, just the keyboard of the computer where the biometric analysis has to be performed is needed.

The research in this field is to determine the identity of a person who is producing the keystrokes. In this the features considered are along with key pressed, key released the other characteristics like seed-time (time taken to get to and depress a key), hold-time (time taken to hold down the key) and the errors a person make while typing also can be considered. Some of the common typing errors are "substitutions", "reversals", "drop outs", "double-strikes", "adjacent letter hits", "homonyms" and "hold-length-errors" (shift key held down too long or too short at a time). Even the pattern of errors made by a person is enough to distinguish two people. Digraphs, which are the time latencies between two successive keystroke of the user and trigraph, which are the time latencies between every three consecutive keys and similarly n-graphs are considered as features.

Keystroke dynamic systems involves two modes namely identification and verification. In the identification mode, a sizable quantity of keystroke data of users is collected and stored. A biometric template for each user is formed at the training phase. During verification mode, the users currently typing pattern are collected and matched against their respective template already created. If a match is found then they are said to be valid users. The matching is done by comparing with every template stored, resulting in a score or a distance illustrating the similarity between the pattern and the template. The one with the closest similarity to the biometric template is considered

BACKGROUND

(E.Yu and S.Cho, 2003) constructed a 4-layer auto associative multilayer perceptron (AaMLP) to improve the performance of a novelty detector and used SVM for keystroke dynamics identity verification. (M.V.S.Paula et al, 2005) proposed a mechanism which involves three techniques, conducted cluster analysis to segment the users in three groups. New user registration and existing user authentication were used f or authentication mechanism. They record the typing profile along with user name and password and then use them to analyse and store them for later authentication purpose. In this way a user is asked to type his username and password about 10 times. They tested the data with Artificial Neural Network as verification unit, multilayer perceptron (MLP) with back-propagation algorithm and learning-vector quantization (LVQ) and self-organizing map (SOM) and Support Vector Machine. MLP was implemented with 17 input layers, 25 hidden layers and 2 output layers. They conclude that MLP and LVQ gives better result.

(H.Ceker and S.Upadhyaya, 2015) use Gaussian Mixture Model (GMM) for representing digraphs. They work on the fact that digraphs do not necessarily exhibit pure Gaussian but a mixture of Gaussians. As the user keystroke don't remain constant and it varies based on various environmental issues, just a single Gaussian distribution for representing user pattern is not enough and go for GMM. In this they used GMM with 2 components that enhances the recognition by 93.8% over the commonly used pure Gaussian. This gives a reduced error rate from 1.3% EER down to 0.08% ERR.

(R.Giot and C.Rosenberger, 2012) paper proposed a novel framework to guess the gender of an individual through keyboard typing pattern identification. They use SVM for classifier for learning and recognition of male and female patterns. For this they have tested four different scenarios like 1. no gender information, 2. Manual labelling 3. Automatic labelling, 4. Semi-automatic labelling. They have obtained an accuracy of 91.63%. They suggest that finding information like this will help to automate personalising of web applications and fighting against paedophilia or social network monitoring.

(Moskovitch R et al, 2009)proposed multimodal framework which combines three biometric technologies: face, mouse dynamics and keystroke dynamics. They name their system as ExamShield platform and it is a virtual exam-center that integrates multiple heterogeneous services like multi-biometric authentication, video streaming and recording etc) . They developed it as web portal and deployed it in cloud.

(Deian Stefan and Danfeng Yah, 2010) in their work use keystroke dynamics as a tool to identify anomalous activities that may be due to malware. They discussed about synthetic and automatic attacks and forgeries. They worked on the main idea that an attacker may write a program that performs static manipulation and synthesis to produce keystroke sequences in order to spoof others which is a serious threat. They present a protocol in a client-server architecture using the X Windows System (X or X11). They used the email addresses, password and web addresses which they receive from users as input and found that performance vary according to the inputs received. For the evaluation of their system they used two bot programs namely GaussianBot and NoiseBot. These programs generate synthetic data i.e they mimic a particular user keystroke dynamic. They developed a model called TUBA (Telling Human and Bot Apart) which detects the patterns developed by the Bot programs.

(Saira Zahid et al., 2009) in their work provide a novel keystroke dynamic-based PIN (Personal Identification Number) verification mode to ensure information security in smart phones. For this they collected 25 diverse users keystroke form their smart phones and selected six distinguished keystroke features. They used fuzzy classifier as front-end and for optimization of this classifier they used Particle Swarm Optimization (PSO) and used Genetic Algorithm (GA) as back-end dynamic optimizers to adapt to variations in usage patterns.

(Yu Zhong et al., 2012) proposed a new distance metric which combines Manhattan distance and Mahalanobis distance in their work. In this they apply Mahalanobis distance to decorrelate and normalize the keystroke dynamics feature variables. In this the covariance matrix of the transformed feature vectors becomes an identity matrix. This new distance metric not only removes the undesirable correlation and scale variations, but also supress the influence of outliers the improve the performance. They used Nearest Neighbour classifier with the new distance metric to either ascertain a keystroke dynamics feature as originating from the genuine user when the distance to its nearest neighbour in the training is below a threshold value, or reject it as an imposter.

(Dharmendra Singh et al, 2017) used soft biometrics like facial recognition, gender recognition and clothes color recognition for employees who wear uniforms, mouse dynamics, Geo-location verification and a google authenticator for the purpose of unlocking a locked down mechanism. Their authentication system consists of the 5 components namely data acquisition, Feature extraction, Classification and Matching, Decision and Adaptation. The evaluation metrics they used are False Rejection Rate (FRR), False Acceptance Rate(FAR), Equal Error Rate(EER), Half-Total Error Rate(HTER).

AUTHENTICATION

Authentication is the process of verifying a person's legitimate right prior to the release of correct and secure resources. Most people are familiar with the three factors of authentication:

- **Something You Know:** A password, PIN Code
- **Something You Have:** Security token, credit card, and
- **Something You Are:** Fingerprint, keystroke dynamics, signature etc.
- But there's also a lesser-known fourth factor: Something you do.

User authentication can be categorized into three categories

1. **Knowledge-Based:** Something you know
2. **Object or Token-Based:** Something you have
3. **Biometric-Based:** Something you are

Traditional approaches like login and password has some disadvantage. One has to remember the password or write it in some secret place. If any body gets hold of this the security is lost. The passwords can be guessed. The smart cards can be lost and can be cloned. So these security features have their own drawback. In order to overcome these problems, a new form of authentication is needed. The keystroke dynamics combined with the password security measure will provide a high level of security to the system.

A person has to be identified before giving access to any resource and must be authenticated. Some of the biometric authentication methods are Physiological (static) like iris, fingerprints and retina etc. another type of biometric authentication method is behavioural (dynamic) like features that people have learned to do like Gait, signatures and keystroke dynamics.

The physiological methods offer a superior security protocol when compared with password-based approach, the infrastructure costs and their intrusive nature make them undesirable and impractical in many scenarios. An alternative is the keystroke dynamics method where the user typing pattern is analysed and identified. Some of the variables used for this technique are measuring of time period between key presses, and/or release of keys, or key-strike pressures, finger placement, overall typing speed etc. Keystroke dynamics promise a low-cost, additional layer of authentication beyond a simple user identification and password. By authenticating users based on their typing rhythms, organizations don't need to make large investments in new access control technology.

Two types of keystrokes – static and continuous are considered. In static verification, the keystrokes are analysed only at specific times e.g during login. Once the user enters the website there is no other mechanism to check whether he is the person who logged i.e the static method do not provide continuous security. They cannot detect the substitution of the user after the initial verification which is more important in e-learning where it has to be verified for the particular student.

Continuous verification monitors the user's typing behaviour throughout the session and can be used for detecting uncharacteristic typing rhythm. As the user is scrutinized ad infinitum, this ensures that the access to the system could be thwarted if there is any mismatch in the typing patters of the user uncovered through persistent observation.

Static authentication suffers with many drawbacks

- Passwords can be shared between users
- Passwords can be stolen in online by eavesdropper, hackers
- Entering of PIN can be trapped by key loggers
- It is easy to guess passwords as most of us use the date of birth, wedding date, spouse birth date etc.
- When a session is successfully authenticated most system have no approach to reauthenticate user. This leads to session hijacking and perimeter possession.

(Rick Joyce & Gopal Gupta, 1990) claimed that the problem of recognizing a given pattern as belonging to a particular person, either after exhaustive search through a large database, or by simply comparing the pattern with a single authentication template, could be formulated within a statistical decision theory. Their testing methodology based on a training set resulted in a classifier used for recognition, Euclidean Distance Measures– "similarity" based on the distance between the pattern vectors.

Further work suggested that median inter key latency of expert typists are approximately 96 ms, while that of a novice is near 825 ms, therefrom making digraph-specific measures would lead to measurable improvements in verification accuracy

In information security, the imposters detection or intrusion detection is the processes of monitoring events that are going on a computer or networks and analyse them to detect possible incidents which are violation of security policies discussed by (Scarfone K and Mell P, 2007). Mostly intrusion detection is done by a concept called user profiling. User profiling mainly deals with generating a model by observing and recording the user behaviour in normal circumstances. Then the currently observed events are compared to these stored models and possible deviations are classified as intrusions by (Wang L and Geng X, 2009)

- **Training:** Obtaining features from user profiles
- **Recognition:** Matching the observed feature with the models

Many classifiers are used for identification and verification of user. They are broadly divide in three categories namely statistical methods, neural networks and pattern recognition based techniques.

(Gaines et al,1980) applied t-test on digraph features to check the similarity of mean vectors and covariance matrices on two multivariate normal populations.

(Bergadano F et al., 2003) the authors define a function that returns a real number measuring the similarities and differences between two typing samples. The function returns a smaller value when it compares two samples from the same user and returns larger when it compares with different users sample. This function is used to classify a new sample among a set of users whose typing habits are known. They have used a mean distance measure to find the similarity of patterns.

KEYSTROKE DYNAMICS SYSTEM OVERVIEW

Generally keystroke dynamics system involves two phase.

Phase I: Enrolment

Enrolment: This involves the four different steps namely data acquisition, pre-processing, feature extraction and storage which is given in Figure 1.

1. **Data Acquisition:** Raw keystroke data are collected through various input devices like computer keyboard, virtual keyboard, smart phone, cellular phone etc.
2. **Pre-Processing:** Biometric data may be normalized before being used.
3. Feature Extraction: The raw data obtained are then processed and stored as reference template for future usage. These steps are feature selection, dimension reduction, and outlier detection.

Phase II: Authentication Steps

4. **Classification/ Matching:** In this data are categorized and discriminated for later use to make decision by using required classification algorithms available.
5. **Decision:** The obtained data is presented to the system and compared to the reference template via classification algorithm and the decision of whether the user is legitimate or otherwise is taken based on outcome.

Figure 1. Biometric enrolment process

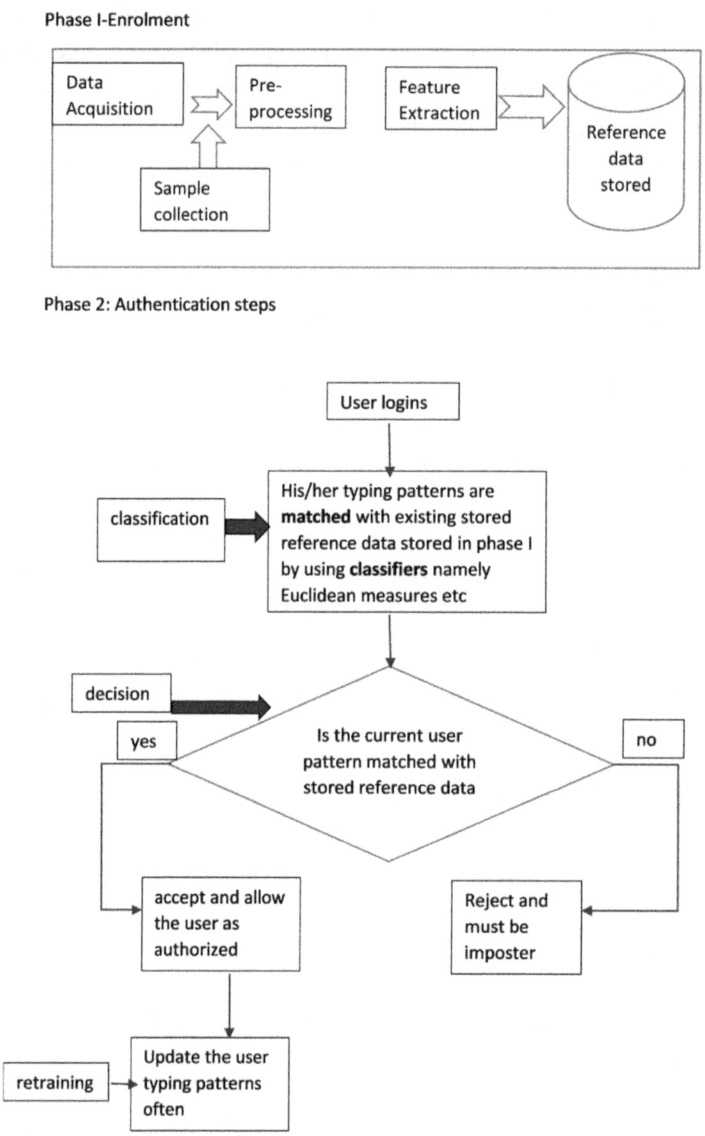

6. **Retraining:** The stored reference template should be updated often to reflect the ongoing changes.

KEYSTROKE DYNAMIC DATA ACQUISITION DEVICE

1. **Normal Hardware:** The main advantage of keystroke dynamics is that it requires already available device, the QWERTY keyboard which is available with every computer.
2. **Customized Hardware:** The normal keyboard can register timing vectors as feature data for analysis. But if another factor like pressure sequence while interacting with input devices is available than the analysis will give more accurate result and many researches tried to modify existing

devices to include pressure sensitive receivers. Another modification by authors (N.J.Grabham and N.M.White 2008) is a rubber membrane keypad that resembles an ATM machine input alignment, with the objective of improving security on a numeric PIN system.

3. **Mobile Devices:** Research works are performed on conventional numerical key pad cellular phone. But the cross-platform compatibility across diverse model of devices, is a big research issue.
4. **Other Hardware:** A less common equipment like four pairs of infrared proximity-sensing devices were used by (J.Mantyjarvi et al, 2002) to project a virtual numeric keyboard on a piece of white surface). In this the user has to keep his finger in 90 degrees on the keys for proper detection.

FEATURE SELECTION

A feature selection mechanism is applied to remove irrelevant features. This is to reduce the quantity of data and speed up the computation time and improve the performance. In general there are two kinds of feature extraction systems applied. They are:

- **Filter Approach**: This does not depend on the verification algorithm. The aim is to remove irrelevant features based on different measures (e.g the variance)
- **Wrapper Approach**: This depends on verification algorithm. Different feature subsets are generated and evaluated and the best one is considered

Keystroke activity generates hardware interrupt that can be time stamped and measured up to microseconds(ms) precision. After obtaining the information, like when the user presses the key and release them and timing between keystrokes, the obtained data is processed through a unique neural algorithm, which determines a primary pattern for future comparison. Some of the feature which has more significance in keystroke dynamics given by (Pin Shen The et al,2013) are:

Di-graph: Timing information of two consecutive key-strokes, known as di-graph. It is widely categorized as Dwell Time and Flight Time.

- **Dwell Time (DT):** Dwell Time is the amount of time between pressing and releasing a single key i.e how long a key was held pressing down.

It can be calculated by Pin Shen The et al,(2013)

$$DT_n = R_n - P_n \tag{1}$$

where R and P indicate the time stamp of release and press of a character respectively, n indicates the positions of the intended DT. The total number of timing vector of DT(V_{DT}) is given by

$$V_{DT} = \{DT_1, DT_2, \ldots DT_s\} \tag{2}$$

where s denotes the summation of characters in a string. Figure 2 shows the different keystroke events

Flight Time (FT): Flight time refers to the amount of time between pressing and releasing two successive keys. FT may exist in four different forms as in the below equation:

Figure 2 depicts the different keystroke events of two characters “J” and “Y”

Flight Time (FT): Flight time refers to the amount of time between pressing and releasing two successive keys. FT may exist in four different forms as in the below equation:

$$FT_{type1,n} = P_{n+1} - R_n$$

$$FT_{type2,n} = R_{n+1} - R_n \tag{3}$$

$$FT_{type3,n} = P_{n+1} - P_n$$

$$FT_{type4,n} = R_{n+1} - P_n$$

where *R* and *P* indicate the time stamp of release and press of a character, respectively, while *n* indicates the position of the intended FT. The total number of timing vector of FT (V_{FT}) can be generated as follows:

$$V_{FT} = \{FT_1, FT_2, \ldots FT_{s-1}\} \tag{4}$$

Figure 2.

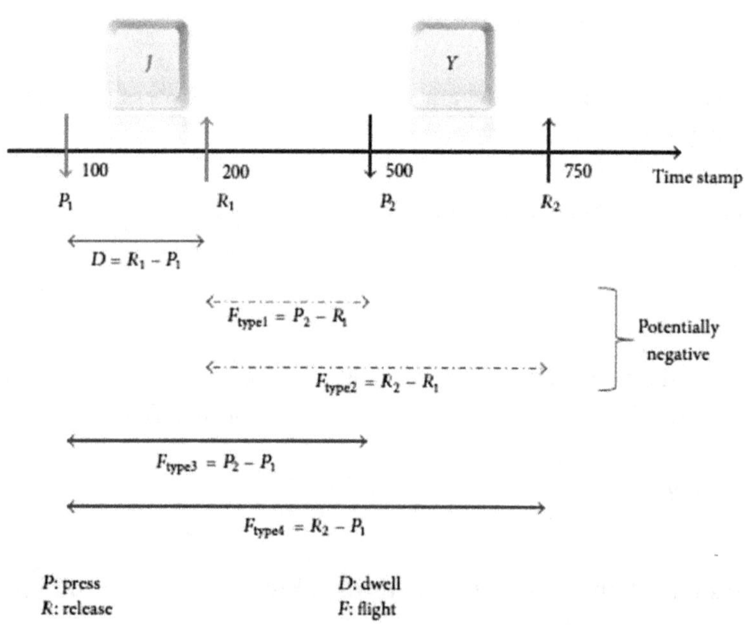

- **N-Graph:** N-Graph refers to the timing measurement between three or more consecutive keystroke events. It is known as elapse time(ET) between a key and the *n*th key event of a typing string. Consider

$$DT_k = P_{k+n} - P_k \tag{5}$$

where *P* indicates the time stamp of pressing a character, *n* denotes nth number of graphs employed, while *k* represents position of the intended elapse time. The total number of timing vector of ET exists in *n*-graph is given as

$$V_{ET} = \{ET_1, ET_2, ...ET_{s-n+1}\} \tag{6}$$

In addition to dwell and flight times the authors (Rybnik M et al.,2008) extracted

- Typing speed – average number of keystrokes per time interval
- Overlapping of specific keys combination ('shift', 'Caps Lock' keys)
- Amount of errors (pressing of backspace and delete keys)
- Method of error correcting
- Cursor navigation-specific keys (arrows, home, end, page up, page down)

(Flior .E and Kowalski.K, 2010) added two additional features to extract keystroke data:

- Keystroke seek time- different letters take a different amount of time for the user to locate and press.
- Examination of characteristic errors- holding the shift-key for too long, resulting in backspacing or simply common typographical errors.

(Monrose F and Rubin D.Aviel, 2000) extended the work of (Joyce and Gupta,1990). They collected the data over a period of 11 months and the participants down-loaded and executed the experiment on their local machines and the results were automatically encoded and electronically mailed back to them. Their clustering algorithm applied a heuristic approach to cluster the users in groups comprising of disjoint features sets in which the features in each set are pairwise correlated. They used factor analysis to determine feature sets. They used Euclidean distance measure to find the similarity between patterns, non-weighted probability to find the nearest neighbour person i.e to the person who maximizes the probability of the feature vector, weighted probability measure. They suggest the use of structured text instead of free text for better identification.

(Pragya Shukla &Rinky Solanki, 2013) in their work present a technique to recognize the emotional state of the user through analysing the keystroke patterns of the user. They consider the affective computing which focuses on giving computer systems the ability to recognize an individual's emotions and respond intelligently accordingly. This will be of very much useful for online tutoring systems that interact with humans and mimic human reactions. For this they collected the user data by asking them to enter some text in their application developed in Java. The user was asked to enter data at least one in a day which ensures to capture their different state of emotions. Every 15 minutes the software prompts six emotional states and asked the user to select one. So, the user can record his emotion at that time and this data along with key press, latency and all are collected. They extracted the following features

- **Session Time**: The total time spent by the user on the system
- **Keystroke Latency**: The time interval between the key release of the first keystroke and the key press of the second keystroke.
- **Hold Time (or Dwell Time)**: Is the time between key press and key release of the same key
- **Sequence**: Is the list of consecutive keystrokes.
- **Typing speed**: The total number of keystrokes or words per minute. This can be an indicator of emotions
- **Frequency of Error**: Use of backspace key and delete key. This provides the state of confusion of user.
- **Pause Rate**: The time the user spends responding to a question. Spending more time indicates the state of confusion or boredom.
- **Capitalization Rate**: Rate of using Shift + letter or caps locks key

The main advantage of the keystroke dynamics is that it is one of the easiest and cost-effective way of identifying a person as this is completely a software solution and no hardware is needed. As each keystroke is captured based on key press only, data can be transmitted over low bandwidth connections. The disadvantage is typing patterns can be erratic and inconsistent due to cramped muscles and sweat on hands which can change a person's typing pattern. Another disadvantage is the type of keyboard users are using. One cannot say all use the same brand.

TECHNIQUES USED IN KEYSTROKE DYNAMICS

Samples Collection

(Aleix Dorca Josa et al., 2017) in their work suggested that a set of users (may be 100) can be asked to type in the keyboard and their typed information can be collected in a database. The time interval between each key stroke can be collected and saved in the database for analysis. For every key event the information to be collected are: the user, his session identifier, the key event code, the type of event (key pressed or released), the timestamp of the moment the event had been recorded and data like user's keyboard type and location. These users must be undergoing the online courses. At least 4000 sessions must be recorded, that is for one person at least 100 sessions. The samples can be taken in such a way that the users age ranges between 15 to 70. Now the NPTEL online courses are released for 11th and 12th students also. So having considering that the age relaxation can be done. The users are definitely of various type like students, faculty, industrialists are working group who want to upgrade in certain topic and even those who are homemakers who want to do some self-work which will help them to earn more. The data needed to analyse is obtained by keystroke logging. **Keystroke logging**, often referred to as keylogging or keyboard capturing, is the action of recording (logging) the keys struck on a keyboard, typically covertly, so that the person using the keyboard is unaware that their actions are being monitored. Data can then be retrieved by the person operating the logging program. A **keylogger** can be either software or hardware. **Keylogging** can also be used to study human–computer interaction. Numerous **keylogging** methods exist: they range from hardware and software-based approaches to acoustic analysis.

Fixed Text vs. Free Text

The users are asked to type text which are either fixed text or free text.

- **Fixed Text**: In this the user is asked to type a predefined text a number of times. This text will be always same like entering username and password. This input is used to build a model with statistical methods or Machine learning techniques. Then the model is trained with samples belonging to authorized user and then compared with new sample of the user.
- **Free Text**: In this users type either long portions of text that simulate the idea of free text or, asked to type whatever they want to type without restriction. In this case the chosen algorithm has to extract relevant feature and build a model for each user. Later when the user login, they may type the same text or different text.

Interval Analysis

The Session of users has to be analysed for different Keydown (KD) and Keyup (KU) events to find the interval between them. This will lead to finding information of Press-Release (dwell time) and Release-Press(fly time) intervals for every pressed key. Then the process of detecting words can be done using delimiters like space key, comma key, the period key etc., and a maximum time interval of silence(300ms)

Data Structure

(Josa et al., 2017) developed a tree model that can be build to store the PR (Press Release), RP(Release Press),PP,and RR timing information. In the tree model, a node can have PR and RP timing intervals or not depending on whether the user has ever typed that particular whole word. The timing information is always stored in the node corresponding to the last letter of the word. If a word is detected more than once there will be a different PR and RP list for each instance found (i.e. ALL in the Figure 3). If a word is a sub-word of an already stored word, there will be PR and RP timing information in a non-leaf node (i.e. THE – THERE in the Figure 3) (Josa et al.,2017)

This tree model stores the information from the beginning to the end of each word. It is thus called a straight tree model. This means, for example, that for the word THIS the first node would contain the letter T, its first child node would contain the H, then the I and the leaf node, at depth 4, would finally have the S. The timing intervals would be stored on the S node

Session Evaluation

After the tree model is built, it is possible to compare new sessions against it and try to establish the author of a given session. The every word in new session is searched for matching in the tree model and the distance between origin word and the word matched in tree is calculated. If a word in tree model is present in more than one place then their mean is taken.

Figure 3. Straight tree model

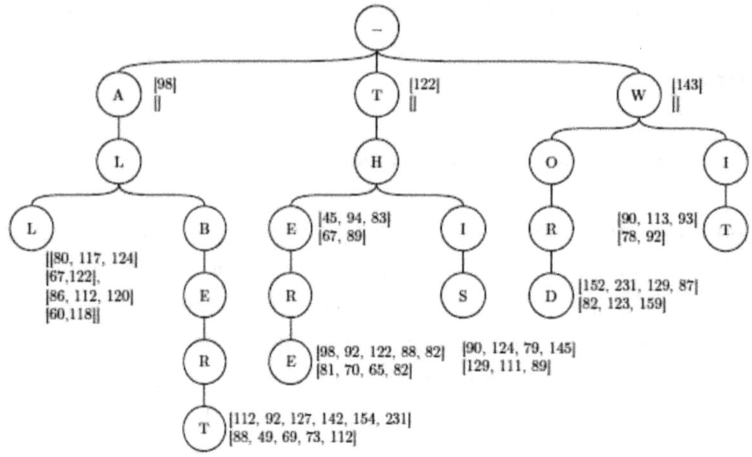

METHODS ADOPTED FOR KEYSTROKE DYNAMICS

1. **Classification:** According to (Pin Shen The et al.,2013) many classification methods can be applied to identify user patterns. Statistical approaches like mean, mode, median, distance measures, probabilistic modelling, cluster analysis can be applied to find similar patterns. Machine Learning approaches are widely used in pattern recognitions. Some of the Machine Learning methods are Neural Networks, Decision tree, Fuzzy Logic, Evolutionary computing, Support Vector Machines (SVM) and more.
2. **Retraining Module**: This refers to the recapture and refinement of users biometric template upon successful verification of their biometric credential. It is also known as incremental learning procedure. Unless the keystroke templates are keep on updated according to gradual shift in typing patterns of user, it is hard to keep the system accurate.
3. **Growing Window**: This method otherwise called as progressive mode where the latest user sample is appended to existing reference template. This may increase the reference template size and may cause storage overhead.
4. **Moving Window**: This adds the new user sample to template profile and subsequently release the earliest sample, and retain a constant template size.
5. **Intelligent mode**: This is the combination of progressive (growing window) and adaptive mode (moving Window). If the training vectors are added over a predetermined time of length, then adaptive mode is applied else progressive mode will be deployed.
6. **Retraining with Imposter Pattern**: The above methods involve in retraining with genuine users. This involves training samples with Imposters patterns.
7. **Outlier Handling**: Outliers in keystroke may be timing value of 3000ms
8. **Machine Learning**: Machine Learning is widely used in the pattern recognition domain. It is the ability of the algorithm to identify and classify pattern and make correct decision. Some of the machine learning algorithm adopts the technique of neural networks, fuzzy logic, evolutionary computing, support vector machine etc.

9. **Neural Network**: This mimics the biological neurons for processing of data. They consists of an input layer, hidden layer and output layer. Sample data is fed into network to produce some outputs based on the current state of its initial weights. The outputs are then compared to the true output and an error value is computed. This value is again fed as input and the weights are recalculated at each hidden layer to reduce error value. The sequence is reiterated until the overall error value falls below a predefined threshold.
10. **Statistical Model**: (Gaines R.S et al., 1980) assumes that a person (originator) who will later desire to gain access to a computer resources was asked to type predesigned text and the information regarding the keystroke-typing time are recorded. Later, another person (claimant) who wishes to access the computer and makes claim as originator was asked to type the predesigned text. His typing pattern also was recorded and both originator and claimant patterns are compared. If the pattern matches the claimant was considered to be originator else to be imposter. For this they used hypothesis H, stating that the claimant and originator are same and hypothesis A, will denote that claimant and originator are different persons. They used a test statistic U that reflects the difference in keystroke patterns between the originator and claimant. If the two individuals are same, U should be small and don't want to reject hypothesis H. If U value is large (significant), then hypothesis A should be accepted and H should be rejected. The statistical model which they develop involves testing the hypothesis that the mean vectors for two multivariate normal populations are or are not the same assuming the two populations have the same diagonal covariance matrix. They carried over a likelihood ratio test and found that the test statistic is a function of the corresponding student t-statistics for each of the digraphs.\

Some of the salient feature of the research work carried in keystroke dynamics is given in Table 1.

ADVANTAGES

1. **Uniqueness**: Since the Keystroke can be measured in milliseconds precision by software it is impractical to imitate one's keystroke pattern. This is very much needed as a student cannot just register and make another student to do his course as proxy.
2. **Low Implementation and Deployment Cost:** Traditional physiological biometric systems like fingerprint, iris recognition etc needs specific hardware device. But keystroke dynamics depends only on software which makes it ideal for implementation in remote authentication. This is very much needed as people from remote places access the online courses and need to be identifies as the same person taking the course through out.
3. **Increase Password Strength and Lifespan**: By using Keystroke dynamics biometrics, user can focus on creating a strong password whilst avoid being overwhelm by different sets of password.
4. **Replication Prevention and Additional Security**: Most security systems allow only limited number of erroneous input attempt before locking down the account. Hence it is hard to reproduce one's pattern '*n*' number of times till success. So guessing attack is impossible.
5. **Continuous Monitoring and Authentication**: Keystroke dynamics offer continuous validation of legitimate identity of user (student)

Table 1. Comparison between some of the research work done on keystroke dynamics

Sno	Author	FRR%	FAR%	ERR	Type of text	Technique used
1	(AleixDorcaJosa, et al, 2017)	0.0126	0.0002		Dynamic text	Straight tree model, interval analysis
2	Ceker H and Upadhyaya S (2015). Single Gaussian (1G) GMM with 2 components (2G) GMM with 3 components (3G) GMM with 4 components (4G) GMM with 5 components (5G)	 2.94 2.94 5.88 2.94 5.88	 0.61 0.09 4.58 4.93 5.44	 1.3% 0.08% 5.88% 3.36% 5.28%		Gaussian Mixture Model
3	D. Umphress and G. Williams. (1985)	12	6		Static text	Mean , standard deviation of digraph latencies
4	J. Leggett and G. Williams. (1988)	5.0	5.5		Static and dynamic text	Mean , standard deviation of digraph latencies
5	Gaines, R.S.et al,(1980)	4	0		Static text	Inter-key latencies
6	Rick Joyce & Gopal Gupta(1990)	16.67	0.25		Static text	Abs.distance
7	R. Giot and C. Rosenberger.(2012)				Static text	Support Vector Machine
8	M. V. S. Paula, E. A. Kinto, E. D. M. Hernandez, andT. Carvalho.(2005)				Static text	Multi layer perceptron, back-propagation network, learning-vector quantization, self-organizing Map and support vector machine
9	J.Mantyjarvi, J. Koivumaki, and P. Vuori, (2002)			78-99	Static text	KNN and multilayer perceptron
10	N. J. Grabham and N. M. White, (2008)	0	15	10	Static text	Euclidean distance

DISADVANTGES:

1. **Lower Accuracy**: Keystroke dynamics is inferior due to the variations in typing rhythm that are caused by factors like injury, fatigue or distraction.
2. **Lower Permanence**: typing pattern of a human may gradually change due to various reasons like fatigue, silence, laziness at the time of typing etc. To overcome this the database should be updated with latest keystroke profile.

EVALUATION CRITERIA

There are always error in recognition of user typing patterns. These error rates define the quality of the application of a keystroke dynamic system in distinguishing people from one another. The European Standard for Access Control (EN-50133) states that FAR should be less than 0.001% and FRR should be less than 1% for any commercially available authentication systems. Some of the metrics defined to calculate the accuracy and the efficiency of the biometric system are given below:

1. **False Rejection Rate (FRR) or Type 1 Error:** The percentage ratio between falsely denied genuine users against the total number of genuine users accessing the system. A lower FRR implies less rejection and easier. access by genuine user.
2. **False Acceptance Rate (FAR) or Type 2 Error:** the percentage ratio between falsely accepted unauthorized users against the total number of imposters accessing the system. A smaller FAR indicates less imposter accepted.
3. **Equal-Error Rate (EER) or Crossover Error:** The overall accuracy as well as a comparative measurement against other system

SECURITY

Biometric authentication systems have several drawbacks (El-Abed et al, 2011) proposed an extension of the (Ratha et al,2001) model to categorize the common threats and vulnerabilities of a generic biometric system. Their proposed model is divided into two sets as in figure 4.

The threats in the architecture in Figure 4 are:

1. Involves presenting a fake biometric data to the sensor.
2. And in a replay attack, an intercepted biometric data is submitted to the feature extractor or the matcher bypassing the sensor.
3. And the system components are replaced with a Trojan horse program that functions according to its designer specifications.
4. Involves attacks on the template database such as modifying or suppressing keystroke templates.
5. The keystroke templates can be altered or stolen during the transmission between the template database and the matcher.
6. The matcher result can be overridden by the attacker.
7. **Performance limitations:** (Doddington et al, 1998) assign users into four categories namely:
    a. **Sheep:** Users who are recognized easily (contribute to a low FRR)
    b. **Lambs:** Users who are easy to imitate (contribute to a high FAR)
    c. **Goats:** Users who are difficult to recognize (contribute to high FRR)
    d. **Wolves:** Users who have the capability to spoof the biometric characteristics of other users (contribute to a high FAR)

Figure 4. Vulnerability point in a general biometric system

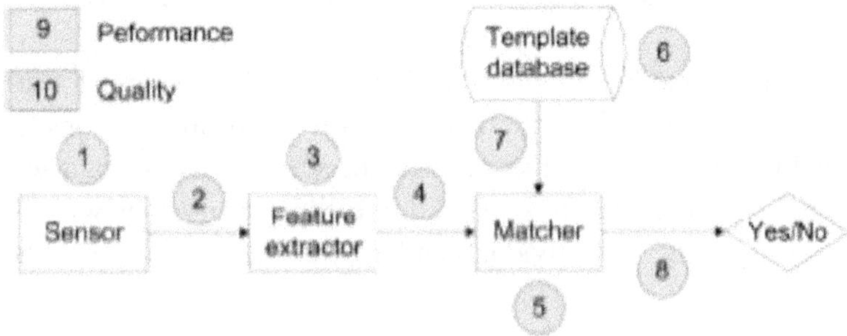

A poor biometric system in term of performance is easily attacked by lambs, goats and wolves users. Quality limitations during enrolment: the quality of samples is considered important for good performance. The absence of a quality test increases the possibility of enrolling authorized users with weak templates.

APPLICATIONS OF KEYSTROKE DYNAMICS

- **Identification and Authentication**: Keystroke dynamics are used for two purposes namely identification and authentication. During identification the user test sample is collected and is matched with all the templates stored in a database. By using similarity measures the profile which is close to the user is found. During authentication the identity of the user is verified. The user supplies his identity and the system verifies where the user is who he/she claims to be.
- **Intrusion Detection**: The continual authentication using free-text method is used for intrusion detection in a network.
- **Online Marketing**: Free-text keystroke dynamics can be used for identifying users over the internet. The websites use the visitors typing pattern to identify them as first time visitor or frequent visitor and send them notifications, emails etc., for marketing and advertising their product. They can find out their preferences and interests and apply their marketing strategies based on that.
- **Cybercrime:** Users typing patterns can be used by investigating agencies to identify anonymous users and cybercrime persons. The attacker profiles are collected by network forensics and when the attacker surfs the internet they can be identified to stop the occurrence of crime.
- **User's Emotion Detection**: Feelings like frustration, focus, anger, stress, relaxation, excitement were derived from the user's typing behaviour.
- **Bank Security**: As almost all banks adopt net banking and other sort of transaction only through internet the security is more important. Even the OTP (one Time Password) is not secured if the mobile is used by a disloyal family member or friend. In 2005, the Bank of Ireland implemented keystroke dynamics as a part of their authentication system in a bid to improve the security of their e-banking services. Ecuador bank deployed an Authenware solution to measure online behaviour and keystroke patterns. Authenware is a software that works by learning and understanding the nuances of a user's keystrokes and behaviours. Bank of Utah adopted a similar solution. Most of the bank uses BioPassword software which is a commercially available software which prevents the login of unauthorized users. It uses a neural algorithm to analyse data providing users with a cross-over error rate of 3%.

APPLICATIONS WHICH USES KEYSTROKE DYNAMICS ARE:

- **AdmitOneSecurity**: Formerly Bio Password is a patented commercial system which uses keystroke dynamics. It provides authentication and security software solutions for preventing fraudulent use of digital identities. With 12 keystrokes, the software can recognize a user. The software records 47 different measurements of keystroke entered by user. Along with dwell time, flight time it forms the typing signature for individuals and use them as models when they try again

- **Ardamax Keylogger**: It is an affordable, yet powerful keylogger for monitoring users activities. Operates silently in the background, this records every keystroke on the user's system and saves all input to a reliably encrypted log file accessible exclusively to the admin
- **AuthenWare**: This algorithms security capabilities are based on 'natural' fuzzy logic and it utilizes a number of unique and highly precise algorithms at its core. It incorporates the usual identification mechanisms (user id and password) along with keystroke dynamics. An additional capability of AuthenWare is its behavioral heuristics inference engine. AuthenWare utilizes this important engine to infer whether a person attempting to authenticate is the rightful user based on behaviors such as the use of the mouse, tab key, return key, preferred browser and browser version, preferred times of operation, number of mistakes they make, and many others behaviors. This robust level of sophistication helps AuthenWare to render useless any attempt to gain unauthorized access to a system or data even if the intruder is using accurate credentials. AuthenWare allows the security administrator to configure default security at the company, application and individual user levels. The system automatically warns the administrator if a user setting is "less secure" than that of the application or company levels.
- **bioChec:** Has a patented implementation which uses keystroke dynamics for ubiquitous web-based login as well as workstation authentication. It creates an online multifactor authentication interface. It facilitate the dynamic enrolment for seamless deployment and deliver secondary authentication using SaaSv (Software-as-a-Service)
- **BioTraker**: This is a software that uses continuous authentication to verify the identity of a user. Plurilock Security Solutions, a developer of network security products and digital identification solutions developed a software BioTracker for cyber security technology. It continuously authenticates network users based on their typing pattern and mouse movements. Plurilock work with U.S Department of Defense and developed this software which helps in identifying people by monitoring their use of computers, laptops and servers. After 20 minutes of tracking user's keystroke and speed, mouse use and other behaviours, the software builds a unique biometric profile. It gives complete control over the corporate network by providing real-time surveillance of all network activities.
- **BlackBox Express**: It can also take screen captures, record all the text that's typed, and record Web searches, e-mails, and chats. It runs invisibly, with not even an icon in the system tray that might alert someone to its presence. One of its most notable features is the ability to send e-mail and text message alerts when certain keywords or Web sites have been detected
- **DanuSoft**: is a simple and easy to use free tool to capture keystrokes. It can be configured to run as stealth during system startup. It doesn't have the ability to protect the program by setting a password. The reports are shown in a simple Notepad(.txt) file.
- **Free Keylogger**: This is able to log keystrokes, clipboard data and site addresses without the screen capture which is fine for people who don't want to capture screenshots. Parents can use this software to check what sites their children view and the head of the company can use it to check what employees are using the PCs for work.
- **G^3 ISam**: It runs hidden in the system and collects data from individual users who logs in. It allows the owner via a customized hot key combination which asks for password and then only allows to see the recorded data. It provides the facility of emailing the recorded information to a particular given email id. So even if the owner of system is somewhere outside still they are able to access information via their mobile phone.

- **Keystroke DNA**: This is not free. This can recognize the identity of a user whether they are on a keyboard or a touchscreen as individual typing pattern remain unique. This can be easily integrated into any web application with just a few lines of code. This requires only a keyboard or keypad making it the most convenient and affordable method of biometric authentication available. It is efficient in any language as it analyses only the typing patterns and what is actually being typed. Therefore, its algorithms are technically language independent.
- **KeyTrac:** This can be integrated on a web server. KeyTrac integrates with CAPTCHA applications that can be read only by humans, not computers. The KeyTrac collects the input entered by the user and create a user profile. The KeyTrac recording module is integrated to the existing form like registration form, a form for entering address and banking data, product descriptions, forum posts or some standalone applications like e-mail programs, office solutions etc.

The module records the way the user types and not what the user types, and turns the collected information into biometric keystroke data.

The best part is that the collected data does not contain any personal information of the user. It doesn't matter what language the user is typing and which keyboard the user users. Since each key is assigned a keycode, each time a key is pressed the keycode gets recorded which allows the solution to be used with any international keyboard layout.

Once the collected keystroke data is sent to the KeyTrac Core Engine, a user profile is calculated on the basis of several attributes that are extracted from it and stored in a database, to be used after in the identification process.

The identification process starts when an unknown user types in text into a form or an application. The keystroke data is collected again and sent to the KeyTrac Core Engine, which compares that information with the information in the user profiles collected in the database.

If the goal is to check if the registration is a duplicate – for example, if the user has forgot his login credentials and is creating a new account – the data will be compared to all the profiles in the database. If the aim is to detect an intruder with stolen credentials, the data will be compared only to the profile of the user whose account the fraudster is trying to hijack.

Among the things that KeyTrac can be used for is also online fraud prevention. Fraudsters could be recognized by their typing pattern, regardless of whether they are using stolen credentials or have opened a new account with bogus information. If a central database of online fraudsters and their typing behavior is set up, the data can also be compared against the profiles in that particular database, preventing, for example, the initiation of fraudulent transactions.

- **Myjad Keylogger**: This is an invisible monitoring software which allows to track the activities going on the computer. This record the clipboard, track the applications running, and records all the keystrokes the user is typing. It records the website visited by the users.
- **Real Free Keylogger**: Running it the first time will prompt the user to set a password to protect the program. It comes with a scheduler to automatically start logging and the log files can be exported as a HTML file. It is a freeware and works in Windows 8/Vista/XP. The advantage is it logs all keystrokes. The disadvantage is it records only keystrokes and visible to the user that they are being observed.

- **REFOG**: This can monitor keystrokes, visited websites, clipboard, ran programs and computer activities. This can keep track of chat room and instant messaging conversation. The drawback is the inability to run totally invisible and it shows tray icon in the notification area
- **Revealer Keylogger**: It the second most downloaded monitoring tool in CNET Download.com. It is very responsive and light on system usage. It can record keystroke, has multiple language support, startup settings, ability to run with elevated privileges, hotkey support and automatic log cleanup. It has automatic Screenshot capture, has remote monitoring feature as what's going on another computer, a tablet of a smart phone with delivery via email, Dropbox, FTP or LAN. The program can be protected with a password, and is invisible in the taskbar and in add/remove programs.
- **Spyrix Free Keylogger**: This software allows the user to monitor all activities that take place within the computer. It is a freeware and has the capability of taking screenshots of applications that are running at certain intervals; it also gives a periodic report of all programmes that are running on the computer. The main features are it runs in invisible mode, is undetected by antivirus software. If is a perfect monitoring software that includes monitoring of social networks like Twitter, Facebook, Myspace etc., and sends all reports the registered personal email id.
- **Spytech Keystroke Spy:** Is a keylogger and computer monitoring solution that allows easily and efficiently log what computer users are doing. It can log every keystroke entered by user and capture their activities through screenshots without their knowledge. It can run in total stealth, email the user when specific keywords are typed. With this one can spy in which websites the users are visiting, the emails they send, passwords they use and the applications with which they interact etc. It is a complete spy.

FUTURE RESEARCH OPPORTUNITIES

Introducing new detector or improved classification algorithm with high quality data the performance of keystroke dynamics recognition can be improved a lot. Training the classifiers with more number of samples increases the accuracy of classifiers,

In mobile platform and touch screen devices the introduction of special miniature hardware sensors such as multitouch screen, pressure sensitive panels, accelerometer, and gyroscope has the potential of unleashing new feature data. This improves the quality of data which leads to increase in performance of keystroke dynamics.

More research on Keystroke dynamics on the dynamic authentication of languages other than English are needed. The experimental platform should be accentuated on web browser-based authentication since the computer usage trend has been shifted from operating system, based application to browser-based cloud services. Thus, continuous authentication in online platform is on high demand.

The research community of keystroke dynamics should be encouraged to come up with a shared benchmark dataset. This dataset may be of three types: say with free text, fixed text and numerical data.

CONCLUSION

With the advent of technology and the boom of internet there is lot of identity theft cases, where people are using various technological fraudulent methods to make fake identities. Using a single authentication becomes a failure. Imposters may use the time, when the employees leave their workstation for few minutes for some refreshments. Even in that small gap they can enter into genuine employees login and do some unwanted thing which will bring great trouble to the genuine users. In these cases, a self sufficient, adaptive and impenetrable security system is needed. Such a system can be supplemented by keystroke dynamics along with other biometrics security features.

Keystroke dynamics could be paired with mouse dynamics to increase user authentication data and improve the identification authentication process. Implementing keystroke dynamics in online education is the next step. Each institution needs to be aware of its assessment goals and strategies and how keystroke dynamics can help to reach them.

In future the keystroke dynamics software could be embedded in the operating system itself in which case the users would not have to do anything special, and the OS could still keep the keystrokes away from the keylogger malware. Keystroke dynamics is probably not accurate enough to be used as the only identification and authentication method, but combining it with others might just prove to be the missing ingredient for a foolproof solution.

REFERENCES

Bergadano, F., Gunetti, D., & Picardi, C. (2003). Identity Verification through Dynamic Keystroke Analysis. *Journal Intelligent Data Analysis*, *7*(5), 469–496. doi:10.3233/IDA-2003-7506

Ceker, H., & Upadhyaya, S. (2015). Enhanced Recognition of Keystroke Dynamics using Gaussian Mixture Models. *Military Communications Conference, MILCOM2015-2015 IEEE*, 1305–1310. 10.1109/MILCOM.2015.7357625

Doddington, G., Liggett, W., Martin, A., Przybocki, M. & Reynolds, D. (1998). *Sheep, goats, lambs and wolves: A statistical analysis of speaker performance in the nist 1998 speaker recognition evaluation*. ICSLP98.

El-Abed, M., Giot, R., Hemery, B., Shwartzmann, J.-J., & Rosenberger, C. (2011). Towards the security evaluation of biometric authentication systems. *IEEE International Conference on Security and Technology (ICSST)*.

Flior, E., & Kowalski, K. (2010). Continuous biometric user authentication in online examinations. *ITNG2010 - 7th International Conference on Information Technology: New Generations*, 488–492. 10.1109/ITNG.2010.250

Giot, R., & Rosenberger, C. (2012). A new soft biometric approach for keystroke dynamics based on gender recognition. *International Journal of Information Technology and Management*, *11*(1/2), 35. doi:10.1504/IJITM.2012.044062

Grabham, N. J., & White, N. M. (2008). Use of a novel keypad biometric for enhanced user identity verification. *Proceedings of the IEEE International Instrumentation and Measurement Technology Conference (IMTC '08)*, 12–16. 10.1109/IMTC.2008.4546995

Josa, Pérez, & Moreno. (2017). *Using Keystroke Dynamics and context features to assess authorship in online learning environments*. Doi:10.21125/inted.2017.0819

Joyce & Gupta. (1990). Identity authentication based on keystroke latencies. *Communications of the ACM, 33*(2), 168-176.

Leggett, J., & Williams, G. (1988). Verifying identity via keystroke characteristics. *International Journal of Man-Machine Studies, 28*(1), 67–76. doi:10.1016/S0020-7373(88)80053-1

Mantyjarvi, J., Koivumaki, J., & Vuori, P. (2002). Keystroke recognition for virtual keyboard. *Proceedings of the IEEE International Conference on Multimedia and Expo (ICME '02)*, 2, 429–432. 10.1109/ICME.2002.1035630

Monrose & Aviel. (2000). Keystroke dynamics as a biometric for authentication. *Journal Future Generation Computer Systems, 15*(4), 351-359.

Moskovitch, R., Feher, C., Messerman, A., Kirschnick, N., Mustafic, T., Camtepe, A., . . . Elovici, Y. (2009). Identity theft, computers and behavioral biometrics. *IEEE International conference on intelligence and security informatics*, 155–160. 10.1109/ISI.2009.5137288

Paula, M. V. S., Kinto, E. A., Hernandez, E. D. M., & Carvalho, T. (2005). User Authentication based on Human Typing Pattern with Artificial Neural Networks and Support Vector Machine. Academic Press.

Ratha, N. K., Connell, J. H., & Bolle, R. M. (2001). *An analysis of minutiae matching strength*. Audio- and Video-Based Biometric Person Authentication. doi:10.1007/3-540-45344-X_32

Rybnik, M., Tabedzki, M., & Saeed, K. (2008). A keystroke dynamics based system for user identification. *2008 7th Computer Information Systems and Industrial Management Applications,* 225-230. doi. doi:10.1109/CSIM.2008.8

Scarfone, K., & Mell, P. (2007). *Guide to intrusion detection and prevention systems (IDPS)*. NIST special publication, 800(2007): 94.

Shaffer, L. H. (1978). Timing in the motor programming of typing. *The Quarterly Journal of Experimental Psychology, 30*(2), 333–345.

Shukla, P., & Solanki, R. (2013). Web Based Keystroke Dynamics Application for Identifying Emotional State. *International Journal of Advanced Research in Computer and Communication Engineering, 2*(11), 4489–4493.

Singh, Jaggi, Nayyar, & Kumar. (2017). Presskey-A Keystrokes Dynamics Based Authentication System. *International Journal of Advanced Research in Computer Science, 8*(5), 2665–2670.

Stefan, D., & Yao, D. (2010). Keystroke-dynamics authentication against synthetic forgeries. *6th International Conference on Collaborative Computing: Networking, Applications and Work sharing (CollaborateCom 2010)*, 1-8. doi: 10.4108/icst.collaboratecom.2010.16

Stockton, Lisowski, Press, & Shapiro. (1980). *Authentication by Keystroke Timing: Some Preliminary Results*. Santa Monica, CA: RAND Corporation. Retrieved from https://www.rand.org/pubs/reports/R2526.html

Teh, Teoh, & Yue. (2013). A Survey of Keystroke Dynamics Biometrics. *The Scientific World Journal*. doi:10.1155/2013/408280

Umphress, D., & Williams, G. (1985). Identity verification through keyboard characteristics. *International Journal of Man-Machine Studies*, *23*(3), 263–273. doi:10.1016/S0020-7373(85)80036-5

Wang, L., & Geng, X. (2009). *Behavioral biometrics for human identification, medical information science reference*. Hershey, PA: IGI Global.

Yu, E., & Cho, S. (2003). Novelty Detection Approach for Keystroke Dynamics Identity Verification. *Intelligent Data Engineering and Automated Learning*, *2690*, 1016–1023. doi:10.1007/978-3-540-45080-1_143

Zahid, S., Shahzad, M., Khayam, S. A., & Farooq, M. (2009). Keystroke-Based User Identification on Smart Phones. In E. Kirda, S. Jha, & D. Balzarotti (Eds.), Lecture Notes in Computer Science: Vol. 5758. *Recent Advances in Intrusion Detection. RAID 2009* (pp. 224–243). Berlin: Springer. doi:10.1007/978-3-642-04342-0_12

Zhong, Y., Deng, Y., & Jain, A. K. (2012). Keystroke dynamics for user authentication. In *IEEE Computer Society Conference on Computer Vision and Pattern Recognition Workshops* (pp. 117-123). IEEE. DOI: 10.1109/CVPRW.2012.6239225

Chapter 3
Mitigation of Cheating in Online Exams:
Strengths and Limitations of Biometric Authentication

Aparna Vegendla
Norwegian University of Science and Technology, Norway

Guttorm Sindre
Norwegian University of Science and Technology, Norway

ABSTRACT

E-exams have different cheating opportunities and mitigations than paper exams, and remote exams also have different cheating risks that on-site exams. It is important to understand these differences in risk and possible mitigations against them. Authenticating the candidate may be a bigger challenge for remote exams, and biometric authentication has emerged as a key solution. This chapter delivers a categorization of different types of high-stakes assessments, different ways of cheating, and what types of cheating are most relevant for what types of assessments. It further presents an analysis of which threats biometric authentication can be effective against and what types of threats biometric authentication is less effective against. Insecure aspects of various biometric authentication approaches also indicate that biometric authentication and surveillance should be combined with other types of approaches (e.g., how questions are asked, timing of the exam) to mitigate cheating.

INTRODUCTION

Cheating is a significant threat against high-stakes university exams (McCabe, Trevino, & Butterfield, 2001), whether the exams are conducted on-campus (Cizek, 1999) or online (Nixon, 2004; Rowe, 2004). Successful cheating may reduce the credibility of universities and their diploma, create unfair advantages for cheaters over the honest, hard-working students, and ultimately be detrimental to the university's learning culture (Davis, Drinan, & Gallant, 2011).

DOI: 10.4018/978-1-5225-7724-9.ch003

Impersonation (e.g., having somebody else sit a test for you) is not the most frequent way of cheating at university. Sheard et al. (Sheard, Dick, Markham, Macdonald, & Walsh, 2002) in a questionnaire survey of two Australian universities, found that only around 2% of students reported ever having hired a stand-in for an exam, and there were also a few who had hired somebody else to do graded homework assignments for them, while there were much higher percentages for various other types cheating. McCabe (McCabe, 2005), based on surveys with 80,000 North American students, indicated that the most common cheats for examinations were acquiring test questions ahead of time and copying answers among students during the test. For written assignments (e.g., take-home work), the most common cheats were collaboration (or Collusion) and usage of assistance on work supposed to be individual, as well as small fragment plagiarism.

Even if impersonation is not the most common way of cheating, it is important to protect against. Whereas the actual grade gain of some other cheating approaches is often limited, impersonation can boost somebody's grade all the way from F to A, given a sufficiently competent stand-in. Moreover, impersonation is difficult in a small class context where the staff administering the test knows the students personally, the chance of successful impersonation increases in large class settings, and even more so for distance education with remote exams.

This is further accentuated in online courses due to the availability of technology and internet in exams conducted in un-controlled remote environment (Agulla, Rifón, Alba Castro, & Mateo, 2008). Indeed, in the early days of online exams, distance education there were many arguments that students would have to travel to proctored locations for high-stakes tests, since it was just too easy to cheat if allowed to take the exam at home.

In many application domains faced with a threat of impersonation (e.g., somebody else claiming to be the legitimate user), a typical solution is to combine several factors of authentication for increased security. A factor can be something that the user (i) *knows* (example: password), (ii) *has* (example: access card, one-time pin-code generator), or (iii) *is*, i.e. biometric properties of the user (Firesmith, 2003). All the authentication methods have their own advantages and disadvantages (Idrus, Cherrier, Rosenberger, & Schwartzmann, 2013). As for biometric properties, these can be either Physiological (example: fingerprint, hand geometry, palm print, iris, face, DNA) or behavioral (example: gait, voice, lip movements, handwriting, keystroke dynamics). When authenticating students for an exam, factors of type (i) and (ii) are inherently insecure because potential cheaters may *want* to be impersonated. Hence, biometrics should be the preferred means of authentication for exam candidates, and if the combination of several factors is found necessary, this should be two different biometrics rather than combining with passwords or tokens. Traditionally, a kind of "manual biometrics" have been used, students having to show picture ID cards. However, this is vulnerable to stand-ins with similar looks, as well as to fake IDs, as for instance illustrated by the news article (Anderson & Applebome, 2011). Computer supported biometric authentication would be clearly preferable, both for increased security and saving manual authentication work by exam proctors. A major problem of biometric authentication is that it is not free from errors in the extraction of the human characteristics and comparison of biometric data. A typical solution is to combine more biometric features using multi biometric authentication which is more reliable compared to single biometric technology or unimodal (Asha & Chellappan, 2008).

The idea of using biometrics for authentication of remote exam takers is not new, there were proposals for this more than a decade ago (Hernández, Ortiz, Andaverde, & Burlak, 2008; Levy & Ramim, 2007; Ramim & Levy, 2007). However, technology for biometric authentication has made much progress recently, offering a wider range of measures to mitigate cheating. At the same time, cheating techniques will also become more sophisticated to adapt to improved mitigation techniques. Hence, it is important to be aware not only of the potential advantages of biometric authentication, but also of its limitations. For a quick example, having somebody else type the answer for you during an online exam might get caught both by keystroke dynamics and facial analysis of the person sitting at the keyboard. However, the slight variation of doing the typing yourself, but having an accomplice provide the answer before you type might have the same grade gain as the impersonation. Yet this would go beneath the radar of the mentioned mitigation techniques, as long as the accomplice stays out of view of the webcam which is a source for the face recognition.

The chapter addresses the following research questions:

RQ1: How do the main cheating risks for educational assessments (i.e., exams and other graded work) differ between online and on-campus tests, as well as between different types of assessments (e.g., oral exam, written exam, term paper / project work)?
RQ2: For which of these cheating risks will biometric authentication provide significant mitigation?
RQ3: For those risks where biometric authentication alone is not sufficient mitigation, how can biometric authentication be combined with other means to make the exam more secure?

The reminder of the paper is structured as follows. Section 2 describes the research methodology. Section 3 presents the research results followed by discussion of recommendations and conclusions in section 4.

RESEARCH METHODOLOGY

The chapter combines two research methods: literature review of academic publications about exam cheating techniques and how to mitigate them, and a risk analysis where threats and mitigations are systematized to consider various measures that could reduce the cheatability of exams in general and online exams in particular. The literature review takes as a starting point the book by Cizek (Cizek, 1999) and its taxonomy of cheating, and supplements this with several later publications about cheating in general and remote exam cheating in particular, e.g. (Nixon, 2004; Rowe, 2004). It also considers publications particularly investigating the mitigation potential of biometric authentication. These include continuous biometric authentication through mouse dynamics, keystroke dynamics, and facial scans (Traoré et al., 2017), biometric web authentication using facial recognition, biometric belt-braces that use combination of facial scan and voice recognition for authentication in distance education (Wiklund, Mozelius, Westin, & Norberg, 2016), automated online exam proctoring (Atoum, Chen, Liu, Hsu, & Liu, 2017), using the combination of live proctoring through a computer webcam and biometric-based proctoring that monitors student's mouse movement, and head and eye movements (Mitra & Gofman, 2016).

The risk analysis makes use of a modelling technique called Attack Defense Trees (Kordy, Mauw, Radomirović, & Schweitzer, 2014) and builds on some previous risk analyses (Sindre, 2015; Vegendla, Søgaard, & Sindre, 2016). An attack-defense tree (ADTree) is a node-labelled rooted tree describing the measures an attacker might take to attack a system and the defences that a defender can employ to protect the system. Our previous analyses focused on threats against on-campus digital exams and how they compared to on-campus pen-and-paper exams with respect to threats and mitigations, the new analysis is extended to focus on problems particular to remote online exams where there are no on-site proctors.

Moreover, the analysis is also extended to compare the cheatability of various modes of examination, e.g., oral exam, written exam (one sitting), and take-home exams / term papers (longer duration, e.g. several days)

RESULTS

RQ1: How do the main cheating risks for educational assessments (i.e., exams and other graded work) differ between online and on-campus tests, as well as between different types of assessments (e.g., oral exam, written exam, term paper / project work)?

The main motivation for using remote online exams is to avoid travel long distance by writing exams remotely scheduling exams in prior. They have both pros and cons.

CATEGORIZATION OF DIFFERENT TYPES OF HIGH-STAKES ASSESSMENTS WITH DIFFERENT METHODS OF CHEATING

Cheating in Written Exams

E-Exams on Site

For written school exams, there is an increasing tendency towards e-exams rather than old-fashioned pen-and-paper exams. A key approach for preventing cheating is the usage of e-exam software, for instance by means of lock-down browsers like Respondus[1] or SafeExamBrowser[2] for bigger end-to-end e-exam systems like Inspera[3] or WISEflow[4]. Key cheating protection mechanisms in such software is to prevent the candidate from using other applications (e.g., to go on the internet to search for answers or chat with accomplices; or to access forbidden documents stored on the computer in advance of the exam). These systems may also prevent printing, copying, screenshots, etc., especially relevant if questions must be kept secret for potential reuse in a later test. In addition, some systems may automatically capture screenshots of the candidates' PCs at random times during the test, for instance as suggested by (Migut, Koelma, Snoek, & Brouwer, 2018). This can provide defense in depth in case somebody somehow managed to beat the lockdown functionality.

In addition to this, the school exam will tend to have proctors who observe the candidates during the test – to discover cheating which does not go through the exam PC. For instance, candidates may try to use old fashioned cheat notes, look at others' answers, or use cell phones to search for information or get help from outside accomplices[5]. With alert proctors, the usage of phones in the exam room may have a high risk of being spotted, but they may still be easy to use during toilet breaks. Most universities have instructions that candidates shall be accompanied to the toilet by a proctor of the same gender, and some even state explicitly that the candidate shall always be in view of the proctor during the toilet visit. Nevertheless, in practice, there is a tendency – based on decency, embarrassment, and respect for privacy – for proctors to allow the candidate to close the door of the toilet booth. This can give a cheater opportunity to send and receive messages or access information with a smartphone with little risk. In the exam room, the sophisticated cheater would rather use concealed gadgets which are more difficult for a proctor to spot (Chugh, 2016) (Srikanth & Asmatulu, 2014). For instance, a micro camera may be camouflaged as a shirt button to capture exam questions from the computer screen. A small transmitter hidden under the clothes may send these images to an outside accomplice, who may then speak answers back to the candidate, to be heard through a tiny wireless earpiece. While such spy equipment was previously expensive and only known to people with special jobs, it is now cheap and easily available on numerous web shop. To sum up, while there is a lot of cheating on written school exams, many types of cheating can be mitigated:

- Cheating within the exam infrastructure can be prevented or mitigated by proper use of e-exam software. While no system is 100% secure (Dawson, 2015), the e-exam infrastructure may not be the weakest link (Fluck & Hillier, 2016).
- Cheating outside the e-exam structure (e.g., using cheat notes or other forbidden materials, collaboration, whispering and peeking among students) is mitigated by human proctors, to the extent that the cheating actions are visible and/or audible to the proctors (and not effectively concealed as legitimate actions).
- Toilet breaks pose a risk to cheating, even if the instructions say that an accompanying proctor should watch the student at all times, since in practice, proctors will tend to look away, or allow the candidate to close the door of the booth. This may enable communication or resource access through smart phone, as well as more traditional ways of cheating, such as using the toilet as mail box where one student hides in a previously agreed spot a note with answers for another student to pick up.
- Cheating which is outside the exam IT infrastructure, yet invisible and inaudible to the proctors – such as usage of concealed wireless communication equipment – is hard to mitigate if competently performed by the persons involved in the cheating. Such cheating might however be mitigated by extra technical measures, such as jamming wireless communication in the exam room (as long as this does not also destroy the legitimate communication between exam PCs and the server), or putting up equipment to detect suspicious communication and find the equipment involved, depending on whether such measures are legal in that jurisdiction.

As for the impersonation threat, the traditional approach in school exams is manual control of student IDs, but this is vulnerable both to faking and stand-ins with similar looks. Hence, biometrics could provide more effective authentication and at the same time relieve proctors of manual work with checking of IDs.

Written Exams Online

Most cheating threats (Ullah, Xiao, & Barker, 2016) become more difficult to handle when a written exam is done remotely (e.g., the student sitting at home) rather than in a proctored exam room. While many security challenges and technical solutions are discussed in literature on online exams (Castella-Roca, Herrera-Joancomarti, & Dorca-Josa, 2006; Kiennert et al., 2017), still there exist some procedural requirements that need to be implemented to mitigate cheating threats. If there is no proctoring (neither manual nor automated), any kind of cheating is easily accomplished. However, some level of proctoring may be achieved by having a webcam capturing the candidate during the exam, possibly recording both video and sound. This could then be used either by a human proctor monitoring the video and sound

Table 1. Comparison of cheating threats in On-site exams Vs Online exams

Type of exam	Cheating threats in on-site exams	Cheating threats in Online exams
Written exam	• Thorough checking can be done by proctor to find forbidden materials, • Getting assessment questions in advance, • grade changing is possible	• Forbidden materials cannot be seen very properly, • Getting assessment questions in advance, • grade changing is possible through instructor password theft
Oral exam	• Assistance/collaboration is impossible, • Use of forbidden materials is difficult, • Getting assessment questions in advance is easier, • Impersonation is difficult	• Assistance/collaboration can be possible if the assistant is in close range and behind the cameras • Use of forbidden material becomes somewhat easier, • Getting assessment question in advance can be reduced, • Impersonation can be possible
Term paper/ project work	• Cheating by using forbidden materials is non-existent, • Assistance/collaboration is easier, • Plagiarism of printed/internet sources	• Cheating by using forbidden materials is non-existent, • Assistance/collaboration is easier, • Plagiarism of printed/internet sources

from afar, or by an AI system trained to discover illegal behaviors. However, unless a huge amount of equipment is used, cheating is still much easier in this remote context than for a school exam:

- Impersonation is easier in online than on-site since exams are conducted in un-controlled environment. Several authentication solutions (Castella-Roca et al., 2006; Kiennert et al., 2017) are available including the authentication using challenge questions (Ullah, Xiao, Lilley, & Barker, 2013) in online exams to protect against impersonation.
- Usage of forbidden material is easier, if the candidate is able to place the material in view of himself, but outside the view of the webcam. The number of possible places for the forbidden material (e.g., textbook, smartphone to search the internet) can be reduced by requiring multiple cameras in the room where the candidate is sitting the test. For instance, in addition to the standard webcam setup getting a frontal image of the candidate's face and upper body, there could be cameras on all four walls to capture the candidate and larger parts of the room, as well as a camera on the candidate's head to always capture the direction of looking – and maybe even a camera carried by a hovering drone moving randomly around in the room. However, the more equipment, the more expensive, and still there might be dead spots that clever cheaters can easily utilize.
- Collaboration between candidates is easier, as long as candidates are far enough away from each other to avoid both being caught by the same webcam(s). There are other possible means to discover that two candidates are in the same place, such as tracking of IP addresses (Eplion & Keefe, 2005; Gao, 2012), however, candidates may fairly easily beat this by using different networks. By means of technology, two collaborating candidates need not be in the same room either. For instance, clever candidate A might have added some extra equipment to capture his screen and then share this with a device in use by less clever candidate C (another device than the one C uses for the exam, so not discoverable by screenshots of C's PC). C may then type along the same answers as A is typing and click the same multiple choice alternatives as A. The only requirement is to keep the device with the streamed screen video out of sight of any surveillance cameras.

- Getting assistance from an outsider is also easier, again the only requirement is that the accomplice must be able to stay out of view of the camera(s) and offer help in a way which is not heard on the captured audio. The accomplice need not be in the same room (if this is risky due to multiple cameras), but could be in the next room or elsewhere. Yet, the accomplice might see the same screen picture as the candidate sees (via a cable to an extra monitor, or via remote sharing), and could communicate answers back in a variety of ways. For multiple choice questions and other questions with a low communication burden, it would easily be achieved in ways that are neither audible nor visible. For instance, the candidate could have attached to the leg some equipment that is controlled by the accomplice (via cable from the next room, or remotely) to generate mild electric pulses, e.g., left toe for A, left ankle for B, right toe for C, right ankle for D.
- Toilet breaks, already a notable risk for school exams, is even more risky for exams done at home, since there is no proctor to accompany the candidate. Requiring the candidate to bring the webcam and film himself while in the restroom would likely appear too creepy to be enforced in practice. However, toilet breaks must be allowed, except maybe for exams of a very brief duration.
- If the same question set is reused through multiple runs of an online exam, there is also increased risk that students may get hold of questions beforehand. Although e-exam software may prevent candidates from taking screen shots, using the PrintScreen button, using copy-paste of exam questions, etc., it is way harder to prevent somebody from photographing their screen with some other device outside the exam infrastructure. Hence, while in a school exam students might have to rely on memorizing questions (though this already a substantial threat, especially if several students collaborate in a structured way), the online exam might easily leave candidates with a direct record of all the questions they got – which might then find its way to social media groups or similar. Related cheats may be to creating multiple accounts to take the test several times under different identities (Northcutt, Ho, & Chuang, 2015), or to changing the clock of the exam system (though the latter only possible with a flawed system design) to be able to retake the test. With this practice, students take tests several times to memorize answers and get good grades.

Cheating in Oral Exams

Oral Exams on Site

Traditionally, oral exams are done by having the candidate in a small room together with the examiners (typically two or more persons). Many ways of cheating are infeasible in such a setting:

- Collaboration or peeking at other students' answers is impossible, since you are the only candidate in the room.
- Usage of forbidden materials (e.g., cheat sheet) is difficult since you are in plain view of the examiners. Even if you were able to access materials without the examiners' notice (e.g., imagine a scenario where the candidate is seated behind a screen to avoid examiners being impacted by factors such as candidate skin color, charm, handsomeness, etc.) it would be hard to make much gain from materials due to the live conversation aspect of the oral exam. Unless questions address very straightforward facts easily looked up from a reference sheet, a candidate who has to peek in materials for every answer will come out as slow and fumbling, with a poor flow of discussion, and thus appear less knowledgeable than a candidate who does not need to consult material.

- Getting assistance from somebody else is difficult. The helper cannot be in the room, so the only way of communicating would be through small electronic devices such as hidden microphone, wireless earpiece and transmitter, receiving hints at answers from a more competent accomplice elsewhere.

On the other hand, one type of cheating that might have a higher risk for oral exams, is for students to find out about questions beforehand. This because candidates will normally be processed sequentially rather than in parallel. If, say, 15 candidates are examined during a workday, with half an hour per candidate, it is quite likely that the later candidates may have heard from the early ones about questions asked. This could yield unfair advantage if questions are reused. However, examiners are normally aware of this risk, thus preparing a pool of many questions and drawing randomly for each candidate. Oral exams also offer the opportunity for improvised follow-up questions, which could be used in cases where a student's response to the initial question appears overly rehearsed.

Oral exams do not scale well to huge numbers of candidates. Hence they tend to be used mainly in small classes where the teacher (who is often also one of the examiners) knows each student personally. If so, impersonation is difficult. If the examiners do not have such personal knowledge of the students, some secure way of identification will be needed, and here biometric authentication will of course be more secure than the traditional student ID card which may be fairly easily faked.

Oral Exams Online

Oral exams online might be done in much the same way as the on-site oral examination, the examiners questioning a number of candidates in sequence. The main difference would be that each candidate is not present in the room with the examiners, but instead seated elsewhere, for instance at home, and communicating through a video conferencing system or similar. Some cheating risks become bigger in this context, in particular:

- Use of forbidden material becomes somewhat easier; the candidate only needs to keep the material outside the camera angle. However, this does not necessarily present a big advantage to the cheater, as there is still the problem of the poor flow of answering if the candidate frequently has to look to the side to consult material.
- Having an accomplice in the room for assistance also becomes easier, as long as the accomplice stays outside the picture and avoids behavior which is audible to the examiners. However, alike the option of using a wireless earpiece for the on-site oral exam, the cheater will again easily appear as fumbling and have a poor flow of discussion if frequently depending on cues from an accomplice while answering, whether cues are in writing (or pointing to pre-written statements likely to be usable), sign language, whispering, or similar.

As for impersonation, the examiners typically have less personal knowledge of the candidates in the online course, so secure authentication will tend to be more important. On the other hand, the risk that candidates' examined later in the day find out questions from those examined earlier may be somewhat reduced, especially if candidates do not know each other. For the on-site oral, the next candidates will typically be waiting outside the exam room and see the former candidates leave. In the online setting, there is no such natural arena for information exchange. Nevertheless, it would be naïve of examiners to thus assume that this risk is gone, as students may have other ways of getting in touch. Hence, many alternative lines of questioning should be the rule also for online oral exams.

Cheating in Term Paper/Project Work (or Unproctored Coursework)

For take-home exams, term papers, projects and other types of coursework, there is little difference in cheating risk between online and campus-based courses, since even in campus-based courses students tend to do this work in uncontrolled settings, for instance at home. For such type of tests, usage of forbidden materials would be easy, however, they tend to be given with all materials allowed and questions fitting such a context, so this cheating risk is often non-existent. However, other cheating risks are bigger for these assessments both due to the unproctored situation and the longer time span for the task (Franklyn-Stokes & Newstead, 1995):

- Peer collaboration and assistance on supposedly individual tasks is much easier.
- Plagiarism of printed or internet sources, or having somebody else do the entire work for you (e.g., buying from a paper mill, or asking a favor from a more competent friend) is much easier for coursework than for a school exam.

Plagiarism checking software can discover such cheating to some extent, for instance, if the source is generally available on the internet or similar text has also been delivered by another student whose answer was also checked in the same plagiarism detection tool. However, it will not discover cases where the plagiarized text was not generally available (e.g., from an article behind a payment wall, not accessible to the university) or where ghost writer's text was a pure original, not also given or sold to any other student.

RQ2: For which of these cheating risks will biometric authentication provide significant mitigation?

Authentication has been used for security of any software in some phases e.g., start and end of its use. In online exams, authentication in the login phase verifies user personal information i.e. finger prints, iris, and facial image stored in database. In the next phases, if it is a proctored exam, proctor verifies exam environment. In this case, there can be serious threats to exam system if cheating occurs after initial login phase. To avoid cheating, during the last decade, continuous authentication has been used in online exams. Continuous authentication consists of verifying repeatedly the identity of a user throughout online session, with the purpose of preventing identity fraud (Traore, Traore, & Ahmed, 2011). Continuous authentication will be more secure than just end-point authentication (e.g., just at beginning and end of test). Continuous authentication with a multi-biometric system is discussed in (Fenu, Marras, & Boratto, 2017). Also, authentication embedded in the delivered piece of work will be more secure than authentication outside it. Some of the mitigations are provided based on the type of exams below.

Mitigations for Cheating Threats in Oral Exams

- **Impersonation:** The main usage of biometric authentication is to address the impersonation risk. Face and voice recognition can be used to check that it is indeed the right candidate partaking in the video conference. Of course, fingerprint, iris, etc. may also be used, but this may require extra equipment – the advantage of face and voice is that these are naturally present in the oral exam video conversation (e.g. Skype, web meeting etc.). During exam, Continuous Face Recognition tracks user facial movements. Same as continuous face recognition, continuous voice recognition records voice.

- **Assistance/collaboration**: To some extent, biometrics can also mitigate the risk of assistance from an accomplice. E.g., face recognition could detect if somebody else is sitting next to the candidate, and voice recognition if somebody else is speaking. However, biometrics will be ineffective if the accomplice stays outside the view of cameras and uses some form of silent communication.

The ADTree representing cheating in online oral exam scenario is shown in Figure 1.

Mitigations for Cheating Threats in Written Exams

- **Impersonation:** The main usage of biometric authentication is again to address the impersonation risk, and face and voice recognition can be used in the same way as for oral exams. Keystroke dynamics (Ahmed & Traore, 2014) and mouse dynamics (Monaro, Gamberini, & Sartori, 2017) for typing task exams and clicking task exams, respectively – the general idea should be to utilize as much as possible activities that students have to do anyway, thus reducing overhead for authentication.
- **Assistance/Collaboration:** Same as oral exam, biometrics might also detect if an accomplice sits next to the candidate, or supplies spoken hints. Using continuous authentication (Traoré et al., 2017), e.g. checking keystroke dynamics all the way through an essay exam rather than just authenticating the candidate at the start and end of the test, will prevent an accomplice from taking over the keyboard and writing parts of the answer. However, biometrics will be ineffective if the accomplice stays outside the view of cameras, refrains from typing, and uses some form of silent communication.

Figure 1. Cheating in online oral exam

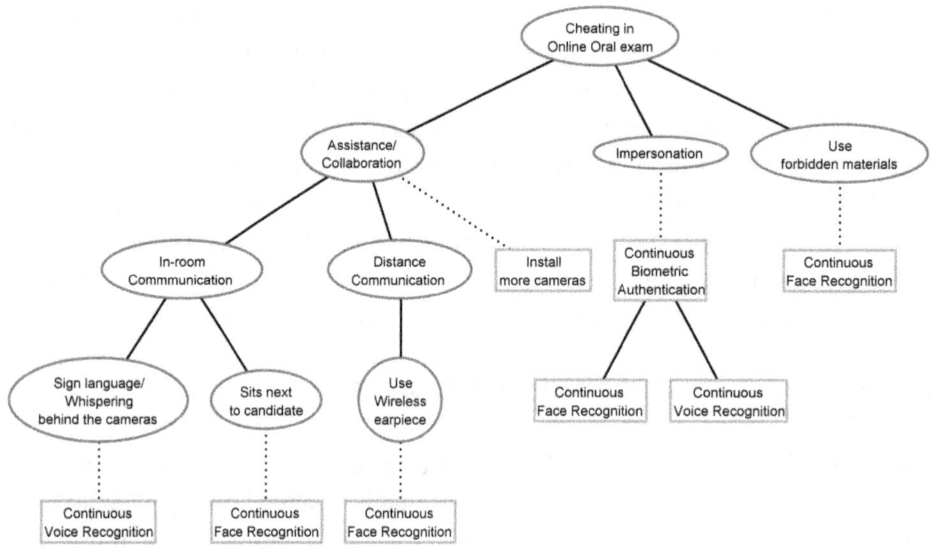

- **Unallowed Aids:** Biometrics have no use when it comes to detecting forbidden materials, e.g., a candidate looking up things in the textbook or searching for answers on the internet for an exam supposed to be no-aids.
- Biometrics (e.g. face recognition) could also be used to detect if the candidate has moved away from the view of the camera (i.e., no face is present), which might not be allowed expect for necessary toilet breaks.
- Considering risks related to reuse of questions, biometrics offers no help against students remembering questions and sharing them on social media. It might however help against the more specific threat of MOOC test takers registering multiple accounts to be able to retake a test and improve the grade. Biometrics might discover that two accounts are held by the same person (e.g., identical face, fingerprint, voice, …)

The ADTree representing cheating in on-site written exam scenario is shown in Figure 2.

The ADTree representing cheating in online written exam scenario is shown in Figure 3.

Mitigations for Cheating Threats in Coursework

The ADTree representing Cheating in Unproctored Course Work scenario is shown in Figure 4.

- **Assistance/Collaboration**: If candidates can only work on their take-home exams, project reports and term papers while logged into a system that has keystroke dynamics, it would be possible to check that these texts are indeed also typed by the candidates' themselves, rather than by somebody else. However, this will not protect against the scenario that the candidate has somebody else write the text, then retypes it. Hence, in this case usage of biometric authentication increases the hassle of cheating (the candidate has to type everything, rather than just getting a ready-to-deliver file from the accomplice) but does not prevent it.

Figure 2. Cheating in on-site written exam

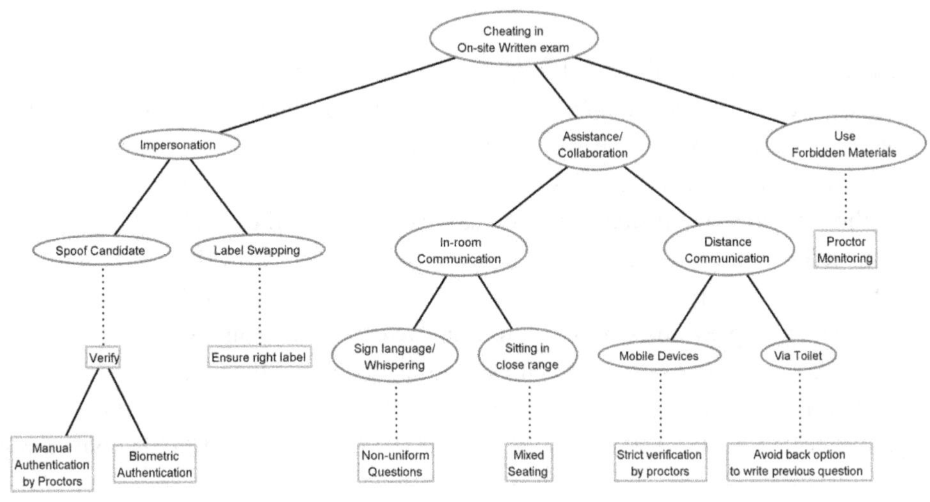

Figure 3. Cheating in online written exam

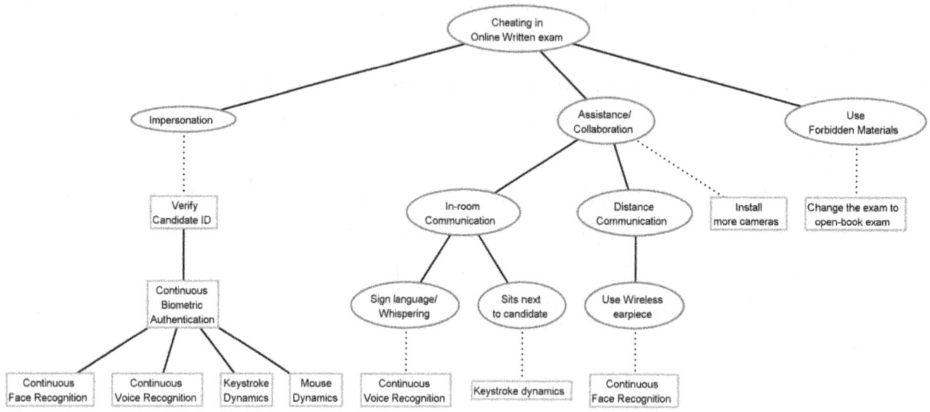

Figure 4. Cheating in Unproctored course work

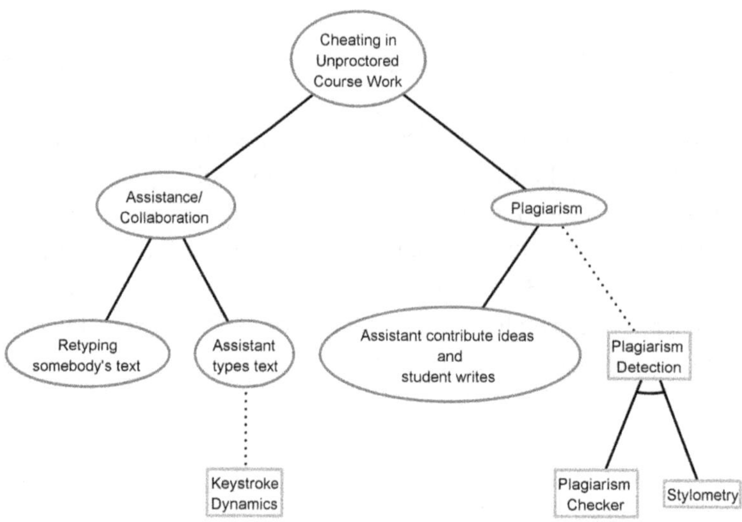

- **Plagiarism:** Stylometry / author profiling (Amigud, Arnedo-Moreno, Daradoumis, & Guerrero-Roldan, 2016; Ramnial, Panchoo, & Pudaruth, 2016) is an interesting alternative or supplement to plagiarism detection. A weakness of plagiarism detection is that it depends on similarity with known text written by others – thus bound to miss out on fully original text bought or gifted from an accomplice. Stylometry instead tries to establish whether the delivered text for some course-work (supposedly written by student X) is consistent in style with other text written by X. This requires that the university already has some collection of text positively known to be authored by X – for instance from previous school exams with stricter proctoring. Stylometry may not discover cases where the accomplice contributes ideas and feedback but the student writes the final answer in his own words. Still, it may at least increase the effort of cheating. Instead of a lazy delivery of a term paper written by somebody else, X instead has to formulate it in his own words. Also, the

Figure 5. Attacks against Biometric Authentication scenario for impersonation

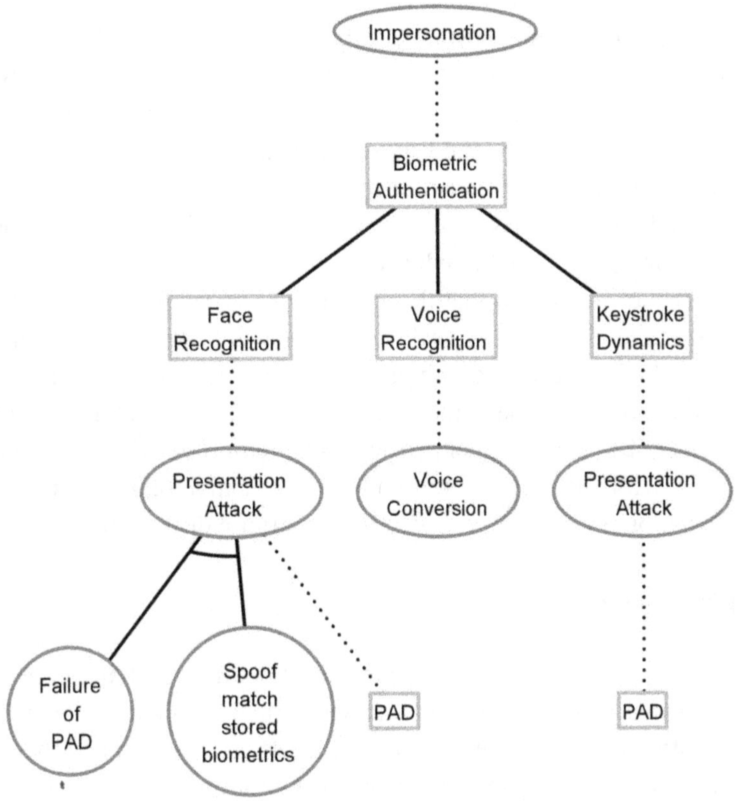

more knowledgeable accomplice will often have a much harder time coaching a C candidate to write an A paper than to simply self-author it.

Even the mitigations of cheating threats also prone to potential attacks. Some of the attacks against biometric authentication for impersonations are shown in the Figure 5. A presentation attack is defined as "presentation to the biometric capture subsystem with the goal of interfering with the operation of the biometric system" (Biometrics, 2015). A presentation attack detection (PAD) is an automated determination of a presentation attack. Several PAD algorithms for attacks against face recognition have been discussed in (Galbally, Marcel, & Fierrez, 2014; Ramachandra & Busch, 2017). In the literature, it is shown that biological biometrics are often prone to presentation attacks while few studies discussed on presentation attack in behavioral biometric systems. A presentation attack detection in keystroke dynamics is initiated in (Ness, 2017). Similarly, voice recognition can be attacked by voice conversion attack (Pal & Saha, 2015).

RQ3: For those risks where biometric authentication alone is not sufficient mitigation, how can biometric authentication be combined with other means to make the exam more secure?

For cases where biometric authentication alone is not enough, there are two main options.

Increase Surveillance

Increased surveillance during the exam, beyond just the authentication, having an exam system where an online proctor (human or AI) sees exactly what the candidate sees and hears exactly what the candidate hears. A typical setup for an oral exam is a candidate wearing a headset, whose microphone conveys to the examiners what the candidate says, and sitting in front of a PC with a webcam showing the face, upper body and close surroundings of the candidate. An accomplice may however be hiding elsewhere in the room, and the candidate may have hidden forbidden materials or devices out of sight from the camera. The possible hiding places can be reduced by adding more cameras, for instance one on top of the headset or in the form of glasses so that the examiners can see in another screen picture what the candidate sees, as well as more cameras on walls to see the candidate and PC from various angles, and possibly a small drone which hovers randomly around the room carrying a camera to detect anything suspicious. However, the more equipment, the bigger the cost. If the candidate is sitting at home, heavy surveillance may also feel like an invasion of privacy, as there may be other things going on in the house which has nothing to do with cheating but still feels very sensitive (e.g., the students' parents may be quarrelling loudly next door, a friend might come by and offer a drug deal, …).

Change the Type of Exam

Change the type of exam (e.g., mode, question genre) to something less vulnerable to cheating. For example, If an oral exam is performed in batch mode, where the student gets one big question at the start of the session, then is offered five minutes to think and take some notes, and then delivers the answer as a 20 minute uninterrupted monologue, having an assistant hiding in the room or communicating via wireless earpiece could be quite useful, as the assistant could give quite a lot of useful information in the five minute thinking break, and could keep on giving cues at natural points during the monologue. It is also quite easy to use hidden material placed out of view of the camera, since the candidate could use the thinking break partly to locate the relevant sections of the textbook or pages on the internet on which an answer could be based. On the other hand, if the candidate is supposed to start talking immediately, it is much more difficult for an assistant to give substantial input, and no time to look up relevant material. Frequent, quick follow-up questions, making a dialogue rather than monologue, will also make the cheating by assistant or material more difficult since attempts to use assistance or materials will make the candidate unnaturally slow in responding.

The traditional batch written exam will show the candidate the entire question set at the start of the exam, which means that a candidate in an unproctored setting can then easily forward the entire set to an accomplice in one operation. Then the candidate can start answering the questions he knows, while in parallel the accomplice can work on developing answers to other questions – which must then eventually be typed or clicked by the candidate himself to beat keystroke and mouse dynamics. If instead showing only one question at a time, and demanding this to be answered before the candidate can move on, there will be no opportunity for parallel work with the accomplice; the candidate will instead have to wait for cues from the accomplice on every uncertain question. If nothing else, this will slow the candidate down. As suggested by (Cluskey Jr, Ehlen, & Raiborn, 2011), if the exam is designed with high time pressure, so that A-students shall just be able to finish it, weaker students ending at various degrees of completion, it will hurt a candidate's performance if having to interpret input from an assistant or look up answers in a textbook rather than knowing the answer right away. Thus, the feasibility of cheating is reduced. A further idea could be to request candidates to think aloud during the exam, maybe not continuously (which would be strenuous) but when prompted for this at random points by the exam system. For instance at multiple choice exams, which otherwise have the highest potential for gain if there is a hidden assistant (since the assistant's communication can be very short, for instance just A, B, C, D, thus creating little delay for the candidate), the system might suddenly ask in an auto voice for some questions "May you please explain why you answered that?" – response expected immediately and recorded by the system for voice recognition and further analysis. If the candidate generally responds slowly and incoherently to such questions, or the prompt draws response from another voice than the expected one, this might lead to suspicion of cheating.

In this respect, interactive exams will have lower cheatability than batch exams, and oral exams will have lower cheatability than written exams. Hence, an interesting supplement to biometric authentication and surveillance in online exams may be the inclusion of interactive and oral elements, even for questions where answers are mainly written.

DISCUSSION AND CONCLUSION

A general observation is that while remote online exams do introduce several new cheating threats compared to on-site exams, they also enable several new types of mitigation against cheating that were either impossible, or practically infeasible, for on-site exams. Biometrics are just one of several potent mitigations here, others include for instance increase surveillance during the exam and change the type of exam. Moreover, while on-site exams have traditionally been believed to be quite secure compare to remote exams, this is not necessarily true – and their vulnerability towards cheating is actually increasing due to the emergence of cheap communication equipment (e.g., wireless earpieces) that can enable candidates to get substantial help from accomplices outside the proctored exam venue yet go unnoticed by invigilators. This increased availability of cheating technology means that the cheating threats of on-campus exams vs. remote exams – previously assumed to be quite different – are gradually converging. Hence, here we provide some recommendations for remote exams.

Recommendations for remote oral exams (assuming a typical video conferencing setup, capturing the candidate's face and voice)

- Use continuous face and voice recognition to protect against impersonation.
- Possibly also use voice recognition to alarm against potential assistance (e.g., if picking up other voices than that of the authenticated candidate). Of course, if somebody else (outside the view of the camera) speaks out loud to give hints to the candidate, this might also be discovered directly by the examiners. However, examiners might be fooled by an assistant impersonating the candidate's voice while the candidate is only pretending to speak (especially if sound and picture quality is poor, so that it is hard to see if lips are out of sync), while good quality voice recognition should be more robust against this. Also, an accomplice whispering cues to the candidate might be inaudible to the examiners, but could potentially be picked up by the computer system using amplification and filtering techniques.
- Conduct the exam as a flowing dialogue between candidate and examiners, rather than as one big question from examiners followed by a long, uninterrupted monologue answer from the candidate. This because the one big question will make it easier for an assistant hidden in the room with the

candidate or connected to the candidate through wireless earpiece to provide help or guidance. Even with a flowing dialogue, the candidate who has access to such assistance would have some unfair gain, but at least this gain would be reduced as the candidate appears slow and fumbling in the conversation when waiting for cues from the assistant.

- Request the candidate to begin responding immediately to every question and think while talking, rather than granting a thinking period of a couple of minutes before the response begins. This again because such a thinking period will make it much easier to receive help from a hidden assistant.
- Although the risk of late candidates learning about questions from early candidates may be smaller for dispersed online candidates than for an oral exam on campus, do not assume that this risk is zero. Whenever several candidates are examined in sequence, be sure to have alternative questions to ask, and some randomness in the selection, so that late candidates cannot know what to prepare for just before the interview.
- If a candidate's response seems overly rehearsed or the candidate talks as if following cues or reading of a manuscript, interrupt with some improvised follow-up questions to force the candidate to demonstrate independent reasoning abilities.
- Beware that "technical issues" may be used as a pretext for cheating. If video and/or audio connection with the candidate is lost for a couple of minutes during the exam, this may be accidental – but may also be deliberately engineered by the candidate to gain time to consult an accomplice. Such a communication breakdown would be particularly suspicious if the candidate starts an answer poorly, but then just after the connection reappears, turns out to be much more knowledgeable about the topic. Another case for suspicion is if communication tends to break just after questions have been asked. While it would be inappropriate to accuse a candidate of cheating without proof (since the breakdown might also be accidental), it is important to have a contingency plan for such cases, e.g. to shift to another equally difficult question afterwards so that the candidate does not gain any advantage from the break. Similarly, if a candidate needs a toilet break (which should not happen too often with an oral exam, typically lasting only 30 minutes or less), consider if a new question should be asked after the candidate returns, rather than continuing on the current one – to avoid any gain from consulting materials or assistance during the break.

Recommendations for remote written exams (assuming here a format similar to written school exams, typical duration 3-4 hours, but where the candidate takes the exam for instance at home rather than on campus or some other proctored test center).

- To protect against the impersonation, Implement secure protocols, for example, TLS (Transport Layer Security) for authenticating the users. TLS verifies the user based on x.509 certificate that is created when user register for exam. Additionally use continuous biometric authentication (e.g., face recognition of the person sitting in front of the PC, as well as keystroke and mouse dynamics).
- Since it is hard to control against the usage of materials in remote exams, a more secure approach is to allow materials (i.e., make the exam open book; then cheating by use of materials disappears because it is not cheating anymore). Of course, exam questions must then be adapted to that setting. This means that questions should focus more on higher levels of knowledge in the Bloom taxonomy (e.g., application of knowledge and beyond), rather than low level recall of facts – a change that may also be pedagogically beneficial. If some questions about recall of facts are

needed based on the learning goals for the course, these questions might be given with time pressure, so that candidates who need to look up everything in the textbook will not have the time to answer everything and thus get a poorer grade. This is in most cases more secure against cheating that trying to survey the room with a lot of cameras, which will still likely leave some blind spot and allow clever candidates to cheat.

- Multiple choice exams should be discouraged for online settings where no proctor is in the room with the candidate. This because it is way too easy to cheat effectively, and very difficult to prevent it, even with dozens of cameras in the room. It is straightforward for the candidate to have an accomplice who sees the questions as soon as the candidate receives them, either by (a) having the accomplice positioned with a view to the screen, though outside the view of any surveillance camera, or (b) having the accomplice seeing the questions on a second monitor in another room, and communicating back to the candidate in a way which cannot be seen or heard. Due to the low communication burden, there is little delay in communication with the assistant for such questions, so time pressure alone will be insufficient in mitigating cheating by assistance here.
- Similarly, short answer questions should also be discouraged for online settings, due to the limited communication burden and thus limited time loss for communication with the assistant (I.e., time pressure works against usage of materials, because the candidate has to spend time looking things up, reading the relevant paragraph in the textbook, and deciding what to answer – and the poor candidate will also be slower in looking things up than the clever candidate who has better mastery of course contents and is more familiar with the textbook; however, against continuous assistance from a more clever accomplice, time pressure is less effective as long as answers are short, since the clever accomplice might know what to answer right away and quickly communicate this to the candidate)
- Ideal properties of questions / tasks for a remote written exam are thus tasks which require longer answers from the candidate, possibly with a combination of many different types of input (e.g., writing, clicking, dragging, …) so as to increase the communication burden of a hidden accomplice. Following the observation that oral exams are harder to cheat on due to their immediacy, especially if done as rapidly flowing interactive dialogues rather than long monologues over few questions, written exams might try to mimic oral exams to some extent. For instance, candidates might be asked to write something (e.g., an essay, a mathematical proof or calculation, a piece of program code, …) and at the same time think aloud, explaining what they are doing, the oral explanation to be recorded together with the written answer. This would make it much harder for the candidate to make use of cues from an assistant. With spoken cues, it would be hard to talk and listen to cues at the same time. With written cues, it would be hard for the assistant to convey to the candidate both what to write, and what to say about it, at the same time. Hence, the candidate who tries to cheat by assistance would much more easily appear as incoherent and fumbling, for instance delivering spoken explanations which are inconsistent with what is being written. Thinking aloud through an entire exam might however be strenuous, so an alternative might be that this is required only for some tasks, or parts of some tasks – and an AI system could be implemented to prompt for think aloud questions at random times, or to ask follow-up questions for oral discussion based on what the candidate has written (e.g. for a programming exam, "Can you please explain the purpose of the variable count that you just declared?"; "Can you please describe the intended inner workings of the while-loop you are writing?")

- In spite of having questions that make it more difficult to cheat by assistance, do also use plagiarism detection for questions where this is appropriate – this can partly mitigate both copying from materials, collaboration between candidates, and assistance from outsiders (e.g., if several candidates are receiving remote assistance from the same competent accomplice)
- Similarly, consider author profiling for questions where this could be appropriate (e.g., essay questions). This might to some extent mitigate cases where a candidate gets help from a more competent accomplice even if there is no plagiarism from materials and no sharing of the text among several candidates. The text contributed by the accomplice will be in a very different (and typically much better) style than the candidate's normal output.
- For further mitigation of collaboration between candidates, consider not using the exact same question set for all candidates but having some variation, for instance drawing questions from a larger question base.

Recommendations for coursework (assuming here graded work which goes over longer time, from several days to the entire semester, such as term papers and projects) would be similar to those for written remote exams: Use continuous biometric authentication to assure that it is indeed the candidate who is writing, and use plagiarism detection and author profiling to ascertain that the text is really originated by the candidate and not somebody else. Allow materials and avoid multiple choice and short answers since it would be way too easy to cheat (and for longer term coursework, time pressure will not be a viable strategy to combat such cheating either), and consider having different questions for different candidates to mitigate collaboration. A key additional challenge with coursework, compared to the remote written exam, is the longer duration which means that the candidate will have plenty of off-hours (i.e., time periods when not seated in front of the PC / not logged in to the LMS, not working on the assignment). During these hours the candidate will be completely beyond any kind of control by the university, so it is impossible to avoid that the candidate might discuss the assignment with others, get feedback and advice, and even acquire text from others that could improve the work. While this cannot be prevented, one could at least make it somewhat more difficult to gain from it.

- Demand that the candidate does the work while logged in to the school's LMS / e-exam system, so that new iterations of the work (e.g., term paper, project report) can be continuously tracked. Hence, if the candidate receives text from somebody else, the candidate must at least retype it himself, as pasting in text from outside the e-exam system might either be prevented or discovered.
- Track the entire progress of the work, rather than just evaluating the end result. This way one could see that e.g. a term paper is developed in a natural way, e.g. starting with some bullet points gradually growing into a rough initial draft which is subsequently improved in several iterations. This may not entirely prevent a candidate from buying from a paper mill, but at least it will make it much more cumbersome. The candidate not only has to retype the term paper (to beat keystroke dynamics) but also fake over time the gradual growth of the work (or pay the ghost writer to fake that entire process, but that would be much more costly).
- Similar to remote written exams, try to mimic some aspects of oral exams to make it even more difficult to cheat. A traditional countermeasure for term papers and project reports is of course to have a kind of oral exam after the report has been delivered, discussing various aspects of it, and where a candidate who had bought the work or gotten lots of assistance, would do more poorly on the oral exam and then at least end up with a somewhat fairer grade. However, this approach may

not scale if the number of candidates is large. Another approach – similar to what is suggested above for remote written exams – could therefore be to automate some features mimicking an oral exam. E.g., at some random points while logged in and working on the term paper, the student may be prompted to think aloud or explain in some more detail the ideas behind the paragraph just being written, and these explanations could be recorded and listened to by the teacher or censor in parallel with reading the term paper for grading. In case a candidate appearing incoherent, providing an explanation which is of poor quality compared to the written text, this would certainly not be proof of cheating but could at least give reason to be suspicious. Possibly, candidates which give poor responses in these recordings might then specifically be invited for an extra oral exam, which would scale better than examining everybody. Such random prompts would make cheating by assistance more difficult, since for term papers and similar the easiest way to cheat is by asynchronous exchange between the candidate and assistant, i.e. the candidate receives the assignment from the teacher, then forwards this to the accomplice, then gets the answer (e.g. complete report) back after some time. Faking a natural growth process as suggested above already complicates this since the accomplice might have to send several iterations, and prompting for immediate response think-aloud about specific passages of the text complicates the cheating scheme even more, since the candidate would then need to have continuous access to the accomplice whenever being logged in and working (or pretending to work) with the assignment. This places a much bigger burden on the accomplice, making it harder to get that kind of assistance.

REFERENCES

Agulla, E. G., Rifón, L. A., Alba Castro, J. L., & Mateo, C. G. (2008). Is My Student At The Other Side? Applying Biometric Web Authentication To Elearning Environments. *Proceedings of the 8th IEEE International Conference On Advanced Learning Technologies, ICALT 2008.*

Ahmed, A. A., & Traore, I. (2014). Biometric Recognition Based On Free-Text Keystroke Dynamics. *IEEE Transactions on Cybernetics*, *44*(4), 458–472. doi:10.1109/TCYB.2013.2257745 PMID:23757560

Amigud, A., Arnedo-Moreno, J., Daradoumis, T., & Guerrero-Roldan, A.-E. (2016). *A Behavioral Biometrics Based And Machine Learning Aided Framework For Academic Integrity In E-Assessment.* Paper Presented At The Intelligent Networking And Collaborative Systems (Incos), 2016 International Conference On. 10.1109/INCoS.2016.16

Anderson, J., & Applebome, P. (2011). Exam Cheating On Long Island Hardly A Secret. *New York Times.*

Asha, S., & Chellappan, C. (2008). *Authentication Of E-Learners Using Multimodal Biometric Technology.* Paper Presented At The 2008 International Symposium On Biometrics And Security Technologies. 10.1109/ISBAST.2008.4547640

Atoum, Y., Chen, L., Liu, A. X., Hsu, S. D. H., & Liu, X. (2017). Automated Online Exam Proctoring. *IEEE Transactions on Multimedia*, *19*(7), 1609–1624. doi:10.1109/TMM.2017.2656064

Biometrics, I. I. J. S. (2015). *ISO/IEC DIS 30107–1. Information Technology - Biometric Presentation Attack Detection - Part 1: Framework.* International Organization For Standardization.

Castella-Roca, J., Herrera-Joancomarti, J., & Dorca-Josa, A. (2006). *A Secure E-Exam Management System.* Paper Presented At The The First International Conference On Availability, Reliability And Security, 2006. ARES 2006. 10.1109/ARES.2006.14

Chugh, R. (2016). *Students Are Using'smart'spy Technology To Cheat In Exams*. Academic Press.

Cizek, G. J. (1999). *Cheating On Tests: How To Do It, Detect It, And Prevent It*. Routledge.

Cluskey, G. Jr, Ehlen, C. R., & Raiborn, M. H. (2011). Thwarting Online Exam Cheating Without Proctor Supervision. *Journal Of Academic And Business Ethics*, *4*, 1.

Davis, S. F., Drinan, P. F., & Gallant, T. B. (2011). Cheating. In *School: What We Know And What We Can Do*. John Wiley & Sons.

Dawson, P. (2015). Five Ways To Hack And Cheat With Bring-Your-Own-Device Electronic Examinations. *British Journal of Educational Technology*.

Eplion, D., & Keefe, T. (2005). *Best Practices For Preventing Cheating On On-Line Exams.* Paper Presented At The E-Learn: World Conference On E-Learning In Corporate, Government, Healthcare, And Higher Education.

Fenu, G., Marras, M., & Boratto, L. (2017). A Multi-Biometric System For Continuous Student Authentication In E-Learning Platforms. *Pattern Recognition Letters*.

Firesmith, D. (2003). Engineering Security Requirements. *Journal of Object Technology*, *2*(1), 53–68. doi:10.5381/jot.2003.2.1.c6

Fluck, A., & Hillier, M. (2016). Innovative Assessment With Eexams. *Paper Presented At The Australian Council For Computers In Education Conference*, Brisbane, Australia.

Franklyn-Stokes, A., & Newstead, S. E. (1995). Undergraduate Cheating: Who Does What And Why? *Studies in Higher Education*, *20*(2), 159–172. doi:10.1080/03075079512331381673

Galbally, J., Marcel, S., & Fierrez, J. (2014). Biometric Antispoofing Methods: A Survey In Face Recognition. *IEEE Access: Practical Innovations, Open Solutions*, *2*, 1530–1552. doi:10.1109/ACCESS.2014.2381273

Gao, Q. (2012). Using IP Addresses As Assisting Tools To Identify Collusions. *International Journal Of Business. Human Technology*, *2*(1), 70–75.

Hernández, J., Ortiz, A. O., Andaverde, J., & Burlak, G. (2008). *Biometrics In Online Assessments: A Study Case In High School Students.* Paper Presented At The Electronics, Communications And Computers, 2008. CONIELECOMP 2008, 18th International Conference On. 10.1109/CONIELECOMP.2008.36

Idrus, S. Z. S., Cherrier, E., Rosenberger, C., & Schwartzmann, J.-J. (2013). A Review On Authentication Methods. *Australian Journal of Basic and Applied Sciences*, *7*(5), 95–107.

Kiennert, C., Rocher, P., Ivanova, M., Rozeva, A., Durcheva, M., & Garcia-Alfaro, J. (2017). *Security Challenges In E-Assessment And Technical Solutions.* Paper Presented At The 2017 21st International Conference Information Visualisation (IV). 10.1109/iV.2017.70

Kordy, B., Mauw, S., Radomirović, S., & Schweitzer, P. (2014). Attack–Defense Trees. *Journal of Logic and Computation, 24*(1), 55–87. doi:10.1093/logcom/exs029

Levy, Y., & Ramim, M. (2007). A Theoretical Approach For Biometrics Authentication Of E-Exams. *Paper Presented At The Chais Conference On Instructional Technologies Research, Raanana*, Israel.

Mccabe, D. L. (2005). Cheating Among College And University Students: A North American Perspective. *International Journal For Educational Integrity, 1*(1).

Mccabe, D. L., Trevino, L. K., & Butterfield, K. D. (2001). Cheating In Academic Institutions: A Decade Of Research. *Ethics & Behavior, 11*(3), 219–232. doi:10.1207/S15327019EB1103_2

Migut, G., Koelma, D., Snoek, C. G., & Brouwer, N. (2018). Cheat Me Not: Automated Proctoring Of Digital Exams On Bring-Your-Own-Device. *Proceedings Of The 23rd Annual ACM Conference On Innovation And Technology In Computer Science Education*. 10.1145/3197091.3205813

Mitra, S., & Gofman, M. I. (2016). *Towards Greater Integrity In Online Exams*. Academic Press.

Monaro, M., Gamberini, L., & Sartori, G. (2017). The Detection Of Faked Identity Using Unexpected Questions And Mouse Dynamics. *PLoS One, 12*(5). doi:10.1371/journal.pone.0177851 PMID:28542248

Ness, J. (2017). Presentation Attack And Detection. In *Keystroke Dynamics*. NTNU.

Nixon, M. A. (2004). Cheating In Cyberspace: Maintaining Quality In Online Education. *AACE Journal, 12*(1), 85–99.

Northcutt, C. G., Ho, A. D., & Chuang, I. L. (2015). *Detecting And Preventing "Multiple-Account"*. Cheating In Massive Open Online Courses.

Pal, M., & Saha, G. (2015). On Robustness Of Speech Based Biometric Systems Against Voice Conversion Attack. *Applied Soft Computing, 30*, 214–228. doi:10.1016/j.asoc.2015.01.036

Ramachandra, R., & Busch, C. (2017). Presentation Attack Detection Methods For Face Recognition Systems: A Comprehensive Survey. *ACM Computing Surveys, 50*(1), 1–37. doi:10.1145/3038924

Ramim, M. M., & Levy, Y. (2007). *Towards A Framework Of Biometrics Exam Authentication In E-Learning Environments*. Academic Press.

Ramnial, H., Panchoo, S., & Pudaruth, S. (2016). Authorship Attribution Using Stylometry And Machine Learning Techniques. *International Journal of Intelligent Systems Technologies and Applications*, 113–125.

Rowe, N. C. (2004). *Cheating In Online Student Assessment: Beyond Plagiarism*. Academic Press.

Sheard, J., Dick, M., Markham, S., Macdonald, I., & Walsh, M. (2002). *Cheating And Plagiarism: Perceptions And Practices Of First Year IT Students*. Paper Presented At The ACM SIGCSE Bulletin. 10.1145/544414.544468

Sindre, G., & Vegendla, A. (2015). E-Exam Versus Paper Exams: A Comparative Analysis Of Cheating Related Security Threats And Countermeasures. *Norwegian Information Security Conference (NISK)*, Ålesund, Norway.

Srikanth, M., & Asmatulu, R. (2014). Modern Cheating Techniques, Their Adverse Effects On Engineering Education And Preventions. *International Journal of Mechanical Engineering Education, 42*(2), 129–140. doi:10.7227/IJMEE.0005

Traoré, I., Nakkabi, Y., Saad, S., Sayed, B., Ardigo, J. D., & De Faria Quinan, P. M. (2017). *Ensuring Online Exam Integrity Through Continuous Biometric Authentication. In Information Security Practices: Emerging Threats And Perspectives* (pp. 73–81). Springer International Publishing.

Traore, I., Traore, I., & Ahmed, A. A. E. (2011). *Continuous Authentication Using Biometrics: Data, Models, And Metrics.* IGI Publishing.

Ullah, A., Xiao, H., & Barker, T. (2016). *A Classification Of Threats To Remote Online Examinations.* Paper Presented At The 2016 IEEE 7th Annual Information Technology, Electronics And Mobile Communication Conference (IEMCON). 10.1109/IEMCON.2016.7746085

Ullah, A., Xiao, H., Lilley, M., & Barker, T. (2013). *Design, Privacy And Authentication Of Challenge Questions In Online Examinations.* Paper Presented At The 2013 IEEE Conference On E-Learning, E-Management And E-Services. 10.1109/IC3e.2013.6735964

Vegendla, A., Søgaard, T. M., & Sindre, G. (2016). Extending HARM To Make Test Cases For Penetration Testing. In J. Krogstie, H. Mouratidis, & J. Su (Eds.), *Advanced Information Systems Engineering Workshops: Caise 2016 International Workshops* (pp. 254-265). Cham: Springer International Publishing. 10.1007/978-3-319-39564-7_24

Wiklund, M., Mozelius, P., Westin, T., & Norberg, L. (2016). Biometric Belt And Braces For Authentication In Distance Education. *Proceedings Of The European Conference On E-Learning, ECEL.*

ENDNOTES

1 https://www.respondus.com/products/lockdown-browser/

2 https://safeexambrowser.org/about_overview_en.html

3 http://www.inspera.com/

4 https://europe.wiseflow.net/; https://www.respondus.com/products/lockdown-browser/ ;

5 Many universities require candidates to put away phones in some designated place outside the exam room, but it is rare – probably due to both cost and inconvenience – to body search or scan candidates in case they might be carrying a second phone.

Section 2
Methodologies and Applications

Chapter 4
An Enhanced Integration of Voice-, Face-, and Signature-Based Authentication System for Learning Content Management System

Mukta Goyal
Jaypee Institute of Information Technology, India

Rajalakshmi Krishnamurthi
Jaypee Institute of Information Technology, India

ABSTRACT

This chapter explores a novel learning content management system. This chapter presents a novel system based on integration of voice authentication, face recognition technique, and signature of a person to recognize in e-learning system. Voice-based authentication, face recognition, and signature of a person is most widely used to authenticate human identity. The main concern in an e-learning system is to de-motivate unknown users from taking the examination in place of the learner. Different techniques have been introduced to stop this fraud if any unknown person wants to imitate person's identity. In order to avoid the fraudulent handling of e-learning systems, the authentication based on voice recognition is discussed as one of the efficient techniques in literature.

INTRODUCTION

In recent years, lot of research is been carried out to enhance the security of E-learning system. However, still there is lot of scope to develop a strong learning content management system (LCMS), that can authenticate and allow the right user to access the E-learning content and also to create materials in E-learning platform. In reality, the exponential growth in usage of smart hand held devices; need focus on building new enhanced LCMS to meet the requirement of mobility and versatile wireless environ-

DOI: 10.4018/978-1-5225-7724-9.ch004

ments. Further, security system of such LCMS should support mechanisms like data confidentiality, user authentication, data authorization and also user accountability.

The conventional methods of authentication like user id and password has several disadvantages. To specify, the user may forget or misplace the user id /password. Also, the misuse of user id /password may lead to voluntary damage to the E- learning content and data. The best alternate solutions are to use human physiological characteristics and next human behaviour characteristics. Under the human physiological characteristic, the biometric process of user authentication is performed. These biometric techniques prove to be more reliable and advantageous over conventional authentication mechanism. The biometric authentication involves using (i) speech signals (i) finger print patterns (ii) iris pattern of a user. In finger print biometric authentication, unique fingerprint patterns in individual is used to recognition the exact user. Similarly, in iris pattern mechanism, the unique furrows and ridges of the user iris is used. These techniques of authentication have several advantages over conventional password/ user id based authentication.

The human behaviour or characteristic based authentication system involves human brain activities and human eye movement activities. This system uses brain activity of user, measured through bio sensors attached to the users. Here, the brain waves are monitored using bio metric sensors. Then, the observed brain wave patterns are used for user authentication. Similarly, the eye movement patterns of users are observed and used as biometric sensor to authenticate the user.

The E-learning systems are designed in such a way that they able to mark the attendance of E-learners. The current multi model biometric systems perform identification, authentication and tracking of the user. Basically, this model involves behavioural biometric characteristics such as mouse movements, keystroke dynamics, El-Gayar, M. M., & Soliman, H., 2013; Anuradha.S.G, Kavya.B, Akshatha.S, Kothapalli Jyothi, Gudipati Ashalatha.,2016; Xiao, H., Ji, W., & Ullah, A.,2011) and physical characteristics such as face features (GUILLÉN-GÁMEZ, F. D.,2017; Beaudin, S.,2016; Obeidallah, R., Ahmad, A. A., Farouq, F., & Awad, S. 2015.; Penteado, B. E., & Marana, A. N. 2009).

In literature, different types of learning algorithms have been studied and the uses of these biometrics technologies have been demonstrated. Further, the experimental results of these studies have mainly focused the verification and attendance control processes. Also, it is observed that the existing solutions mostly need a lower level of students' collaboration. After the implementation of this model it can be added to the existing LMS and become compatible with it. Therefore, this model can be used to track the continuous attendance of the users in sensitive stages of E- learning. Next, the password authentication is one of the mechanisms that ensure the right student has login the web enabled learning courses. This mechanism has its own drawbacks such as if a learner do not want to attend a particular session or want to miss out any exam then learner can illegitimately share his login id or password to some other user. Overcome, this illegitimate sharing of passwords, there is need for biometric applications such as behavioural authentication or physical authentication mechanisms. Hence for secured authentication, popularization of low cost biometric enabled devices becomes essential. Physical authentication such as face recognition is another technique in which the image of a learner is captured online via webcam. Partial result shows success of confirmation of presence of user (Chutel, P. M., & Sakhare, A., 2014).

Due to technology advancement such as cloud based E-learning systems, it is possible to provide enhanced objectives such as availability, cost reduction, on demand, and self-service. Also, these systems are based on user's perspective or learner's provider perspective. Further, an E-learner should be ensured of the access control mechanism such as access of grades or other personal data should be accessed by him or another authorized users (like teachers/instructors). Further, the learner's behaviour during the learning session plays an important role to predict the performance of the learner in E-learning environment. An instructor who provides learning material in environment has to ensure that the exams are taken by the real user, not by another user. These authors have suggested the confidentiality of data is assured by homomorphism encryption, in which analysis is done over encrypted data, such that the output will be decrypted only by the user (Nita, S. L., & Mihailescu, M. I.,2018). The learning management system is mainly implemented on devices such as Smartphone's or tablets. The system provides service that has primary objective to analyse user's behaviour. The user's behaviour is monitored through biometric data patterns and then to determine if he/she is prepared for the course. Additionally, the fingerprint authentication method is used to avoid fraud situations, for example, situations like where final exam have not been actually performed by the real user.

However, the face recognition in biometric web authentication system increases the confidence level over password based systems (Ramu, T., & Arivoli, T.,2013; Mikolov, T., Sutskever, I., Chen, K., Corrado, G. S., & Dean, J., 2013). In such cases, intensive tracking is required to check the target student while access the learning material. This requires recording of information like student logging time, learning time, assessment result and validating it. It also helps to validate the student performance from session to session. Further, some of the researcher provided cognitive biometric application to secure the personal identification by combining both eye tracking and event relevant potential ERP of brain waves (Zhonghu, L., Lin, Z., Luling, W., Junhong, Y., & Jinming, W. 2017, ; Atlam, H. F., Attiya, G., & El-Fishawy, N.,2017).

Signature forgery involves if an unknown user counterfeits the signature of the legitimate exam taker. Thus signature verification is essential in day-to-day life. So, there is a need of automated signature verification software that projects against fraud. Since, all signatures are unique, relied upon for verification by educational organization, financial organization, business and government to authorize transaction and documents. The proposed automated signature works well in both signature from scanned documents and captured online on pads, tablets and Smartphone's. Face recognition is another technique which is most widely used and efficient system in biometric. Next section elaborates the background and requirement of these techniques.

BACKGROUND

This section elaborates the background of Learning Content Management System and biometric techniques used in mentioned.

Learning Content Management System (LCMS)

This section explores the various technologies which are being used to facilitate the LCMS system(Kang, B. H., & Kim, H., 2015). Most the LCMS system involves combination of two or more these technologies. These technologies are:

- **Audio:** Recent technologies allow audio streaming over the Internet for educational purpose. There are webcasts and podcasts which help students and teachers over the internet to download the learning material. Itunes are such example which allows student to free download.

- **Video:** Student and teacher can access the video lectures relevant to their study material instead of referring DVDs or VHS tapes. It is belief that a student can learn in a better way when he/she uses the learning material, numerical problems, examples, or any activity. This notion was well studied by many researchers such as Feldereman Silvermann, Kolb, Honey Mumford. According to Felderman Silvermann, visual learning is also one type of learning that tends to learn best by seeing the material rather than hearing or reading about it. There are some websites such as YouTube which are being used by the teachers as well as students. If the lectures are meant to be delivered in a systematic way, it considers as a promising media that impact student learning.
- **Blogging:** Blogging allows student and teachers to share their thoughts on some topic or it allows student to post their ideas and comment on some topic. It creates an interactive learning environment.
- **Screen Casting:** It allows user to share their screen directly from the browser and make the video available online so that the viewers can stream the video directly. Webinars' are conducted to deliver expert lectures through Skype, adobe connect etc.
- **Whiteboards:** Interactive whiteboards allow teacher and student to write on touch screen so that learning will be more interactive.
- **Virtual learning Environment:** Virtual classrooms and meeting are also known as virtual learning environment uses a mix of communication technologies. Web conferencing is one of the examples that enable student and instructors to communicate through webcam, microphone and real time chatting in a group.

It is to be noted that, these technologies have been used in LCMS system to deliver the right content/ learning material to student so that student can learn in a better way. Next, to know whether the student has acquired knowledge or not, student performance must be evaluated. Generally, performance of the student depends on the- E-exam, taken by the student. Hence learning system must be ensure that every time whenever the user has logged in, he/she is a legitimate user, or student who takes the exam is the one who enrolled for the course and exam. Hence, there is need to address advanced security concerns related to these learning systems. The next section explores various security concerns.

Security Concern

There are three major concerns regarding authorization of the user namely confidentiality of data, integrity of data and availability of data. It is necessary, to keep the personal information of the student as private and prevent from unauthorised use. Confidentiality of such data should be maintained and also such sensitive information should be protected from intrusions and threats. Data integrity is also another concern, in which, the objective is to protect the knowledge of change. It may be possible that someone deletes some of the legitimate names and then include illegitimate names instead, causing a lot of confusion and embarrassment to the party concerned. Data availability is another concern where the data should not be lost, theft or intentionally deleted by someone.

To protect the network from threats, including programs, equipments, information and data, there are many methods are currently available in literature. Methods are also available for protection of data carried out locally through the cloud which contains the applications of security on the site of network equipment or by remote communication. Viruses, threads, macros, and refusal of service are considered as intentional software attacks. While, software bugs, coding problems and gaps unknown are considered as other failure of errors. Similarly, any accidents and employee errors are considered as intentional faults or human failures. Gathering data through unauthorised means is another method of failure. Failure of hardware or business intentional information theft is another security concern. Also, the piracy and intellectual rights of authors and abuse is another security concern.

Authorization for a user in LCMS system is performed based on password authentication. In this method, various tasks are performed by the system namely access control on information, operations to be carried out, rights & authorizations grant to the user and services permissions applicable to the legitimate users. Only the authorized party can handle the confidentiality information. It should be ensure that there would be no modification or modifying it from people who are not authorized to access it. The data should not tampered by other users. The data can be retrieved for use whenever it needed. The right of access the data feature should also considered. As user interacts with the LCMS system, a wide variety of data such as their physical location, devices used, access patterns, learning progress, performance, etc. can be collected. Student course interaction feature can produce useful information. These data can be mined for integrity promotion purposes. For example, student-course interaction data can produce useful information about a student's level of engagement with the course, and generate low performance and/or procrastination warnings to steer at-risk students onto a path of success. These data can also be mined for integrity enforcement purposes. Unusual or suspicious activities (e.g., students who did not do their coursework and yet have a perfect score on an exam) can be identified from the data collected so that attention can be dedicated to investigate situations of significant integrity concerns. Predictive analytics, with its ability to extract information from data to predict trends and patterns of behaviour, will be well suited in this regard. This requires technologies to detect the right user to access the services. Next Section describes the biometric techniques for user identification.

Biometric Techniques for User Identification

This section explains the biometric techniques for user identification. There are two types of biometric techniques. One is the behavioural techniques and the other one is biological biometric techniques. Behaviour al techniques are as follows:

Behavioural Biometric

- Behavioural techniques such as signature matching, speech recognition, face recognition,
- Keystroke recognition, security and challenging question are explained further.

Signature Matching

Fraud detection can be done using signature matching and image detection. Signatures can be forged different types (Agulla, E. G., Rifón, L. A., Castro, J. L. A., & Mateo, C. G., 2008). Basically there are three types of signature forgery. First one is the blind forgery, in which the forger has no idea what the signature to be forged looks like. This is the easiest type of forgery to detect because it is usually not close to the appearance of a genuine signature. This type of forgery will sometimes allow an examiner to identify who made the forgery based on the handwriting habits that are present in the forged signature. In order to perform this type of examination, an examiner needs known cursive writing samples of the suspected forger in order to compare the handwriting habits in the questioned signature to the habits in the known cursive writing.

The second type of forgery is skilled, in which the forger has a sample of the signature to be forged. The quality of a simulation depends on how much the forger practices before attempting the actual forgery, the ability of the forger, and the forger's attention to detail in simulating the signature. This type of forgery is usually easily detectable by a trained document examiner when examining the original document or even a copy. Because the forger is trying to simulate the handwriting habits of someone else, the forger's own handwriting habits are masked. Therefore, it is often difficult to identify who forged a simulation.

The third type of forgery is a tracing. A tracing can be done by holding the model document and the questioned document up to light and using a pen to trace the lines of the model signature onto the questioned document. A tracing can also be done by using a blunt stylus on the questioned document to create an impression of the model signature in the paper. This impression is then filled in with a pen to create the appearance of the model signature. If the model signature used by the forger is not found, this type of forgery is sometimes difficult to detect from a photocopy. Figure 1 shows the different types of forgery in an examination hall.

Keystroke Dynamics

Keystroke dynamic is another method to improve the security in an online examination (Al Galib, A., & Safavi-Naini R, 2015; Dehnavi, M. K., & Fard, N. P., 2011; Agulla, E. G., Rifón, L. A., Castro, J. L. A., & Mateo, C. G. 2008, July; Srinivasan, H., Beal, M. J., & Srihari, S. N., 2005; Gupta, A., Khanna, A., Jagetia, A., Sharma, D., Alekh, S., & Choudhary, V., 2015). The dynamics of keystroke depends on the pattern of keystroke pressed by a user. It is defined as the relation between the key press times and the key release times. For this keystroke latencies and keystroke duration has to be measured to find the user pattern. The two feature vectors are key press duration and key press latency is defined as

Figure 1. Depicting different Type of Forgery in an Examination Hall

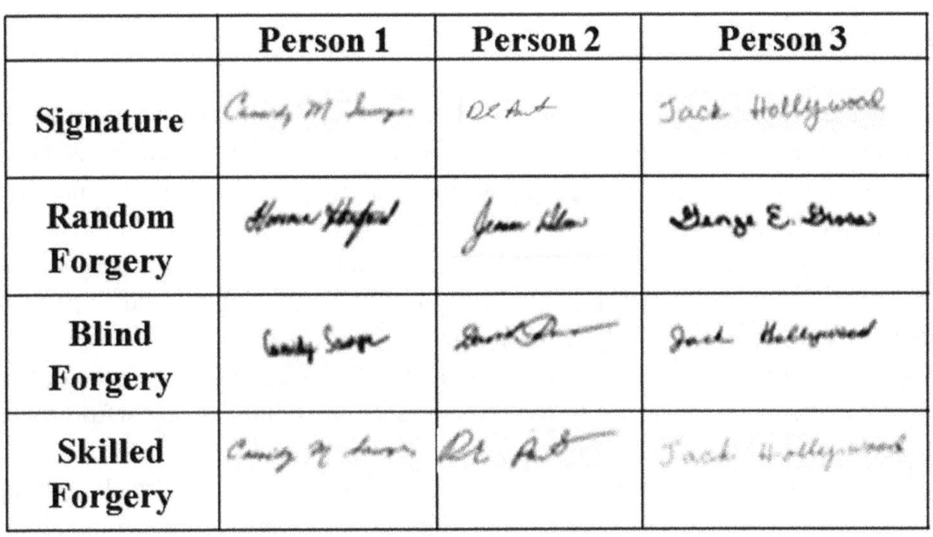

keyduration = releasetime- presstime

keylatency = presstime- releasetime

Where key duration time is represented as the time duration of any key pressed by the user and key, release time is the time at which the key was released, press time is the time at which the key was pressed and key latency is the time duration between the key releases of key press. The collected data is normalized to make it independent of the typing speeds that may vary depending on the condition of the user. Keystroke Patterns are heavily dependent on posture, device, keyboard layout, etc. The emotional state of the user also plays a determining factor on the typing speed. Similarly, the surface on which the computer is placed, type and brand of computer used could also have an impact on the efficiency with which a person can be accurately classified. There are various techniques such as fuzzy logic, neural network; statistical methods have been used as keystroke dynamics.

Security and Challenging Questions

In regard to the security concerns, different type of questions is being asked to student to judge his/her knowledge. From personal questions to password at entry level of system are the common methods for authentication (Agulla, E. G., Rifón, L. A., Castro, J. L. A., & Mateo, C. G., 2008, July; Song, K., Lee, S. M., & Nam, S. 2013, Khlifi, Y., & El-Sabagh, H. A., 2017). Today in real word applications are using login id or password system. But this scheme is more vulnerable. Now- a-days one time password (OTP) generation is another method where the application sent the authentication password to user mobile. If the user enters the correct OTP which is sent by the system, then user is able to enter into system. The examination scenario, web based learning environment, rely on these method are susceptible to collusion and malicious attack. Challenging questions or security based questions are the one which comes under the knowledge based questions.

Object based authentication is another method in the user has to identify the objects such as same objects from the several picture. It enhances the security due to the presence of the student as well as the objects. Again objects may be transferred or compromised then it may be threat to online examinations.

Brain Wave and Eye Movement

Brain waves have used to measure the event related potential which carries the information about the individual cognitive response (Dehnavi, M. K., Sharafi, S. M., & Nematbakhsh, N., 2011). ERP are to be considered as reliable personal identification due to its robustness and compact brain wave acquiring systems that were developed. ERP signals are related to the presentation of event and limited occurrence in the brain. These signal stimulated every 250 ~ 450ms which can be used as to explain the cognitive information process. After processing this cognitive information, it can be used as personal decision making, attention, uncertain solutions, and verification of stimuli.

One component of ERP signal is P300 (P3). The P300 wave is known to elicit the decision making process related to a person's reaction process to a given stimuli. The P300 is obtained from low-probability target items that are mixed with high-probability non-target (or "standard") items.

Research shows that boys are more curious to informational devices than girls. It is being noted that the difference in amplitude of P300 for gender were significant in prefrontal and frontal lobes. The prefrontal lobe is responsible for humanity, curiosity, decision-making along with reflection and the frontal lobe is responsible for concentration of attention and working memory.

Authentication Through Human Cognitive Ability

There are different type of cognitive tasks such as visual search, working memory & information processing and speed automatic processing & priming effect (Al Galib, A., & Safavi-Naini, R. (2015, January). In Visual search, performance of the user is depending on the time of the search. The search time is depending on the multiple factors such as the rate at which the user scans alternative. If he / she finds suitable alternative, her/his search stops. Working memory depend on the information stored the user mind and how he/she user to make the decision further while learning, searching. This varies from learner to learner. Sometimes it is required that information must be processed fast for better results. Automatic Processing process the behaviour of the learner, without information or disturbing the activity which is going underway at that time. These entire human cognitively has been handled through games.

Mouse Movement Verification

Mouse movement pattern are the pattern which does not require any special hardware to store the signature (Al Galib, A., & Safavi-Naini R, 2015, Dehnavi, M. K., & Fard, N. P.,2011). Mouse patterns can be stored while accessing a file, surfing the net during the class. Mouse features are depend on mouse speed, acceleration of mouse pointer, mouse direction, scroll wheel use, right and left click frequency and mouse idle time. The pointer of the mouse is responsible of the speed of a fixed period of time. The acceleration is calculated as the difference between the current speed and the speed measured during the previous time period. The data of mouse movement pattern for each user is depend on time spent to use mouse, calculated in milliseconds, can be processed and displayed in various graphs. These mouse movements can be recorded in eight directions, numbered from one to eight. These eight directions are considered with a gap of 45 degrees. If the user moves the mouse in first sector which is from 0 to 45 degree, average speed for the traversed distance is calculated.

Human Speech Based Authentication

Bonneau.J et. al., 2012 addresses, the human speech based biometric authentication system. Automatic Speech based authentication system consists of three main components namely client, server and connectivity as depicted in Figure 2. The end device of a client in an authentication system consists of smart phones, wireless devices, desktop computers, tablets etc. The client hardware system must be capable to capture quality audio input and should be able to transmit the audio input across the network through Internet, or a dedicated computing server. The client software application residing on the client device should be capable of executing the audio recording mechanism and also initiating APIs and network interfaces to connect to the server through Internet.

Once the recorded audio inputs of user are received by the server, it should perform the specialized authentication mechanism appropriately. For this purpose, the server should have the efficient computing capability to overcome the computing delays and other technical constraints. The server will then perform analyze of audio input, synthesize the fingerprints out of input and then determine the authentication of user as legitimate or an attacker. The features of the audio input are stored in the server system for

Figure 2. Architecture of Basic Human Speech Based Authentication System, where FX indicates the voice features of respective users

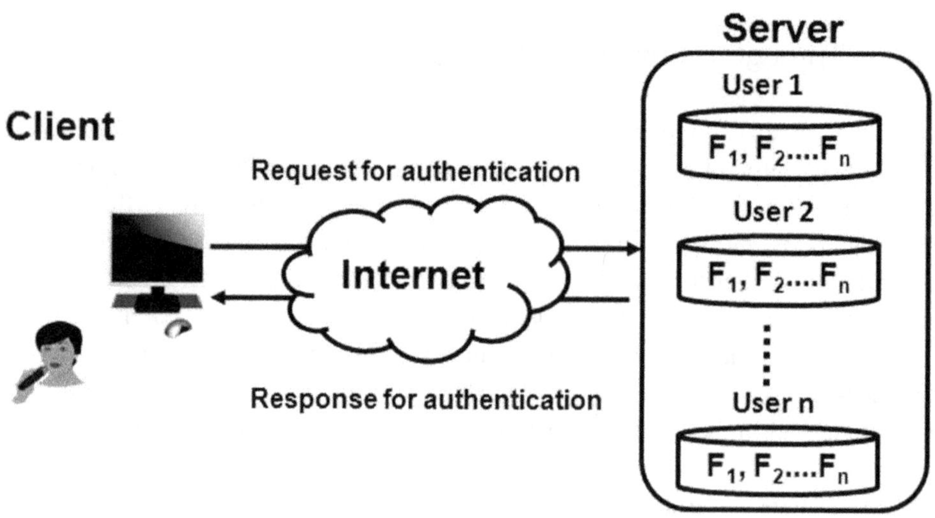

further authentication purposes. However, in case of defective audio fingerprints, the authentication mechanism will perform multiple sampling of user inputs. The features of the audio input are stored in the server system for further authentication purposes. The authentication based on the human speech as biometric input is classified into two types namely audio fingerprint based authentication and voice based authentication.

Audio Fingerprint Based Authentication

In this method the audio fingerprint is generated on the used audio input. Then these audio fingerprints are matched with the audio samples.

Fast Fourier Transformation: First, the user input is fetched as audio sample. Then Fast Fourier Transformation is performed on the audio sample. The sample is created using overlapping windows for the entire length of the input audio sample.

Spectrogram: The 2d array of spectrogram is created based on the FFT. Here, the amplitude is represented as the function that has time and frequency as its parameters.

- **Hashing:** Based on spectrogram, the peak of the two adjacent amplitudes is measured in terms of time and frequency is estimated. The frequency of the peaks and the difference in time between peaks are stored as hashed audio fingerprints.
- **Audio Based Recognition:** Whenever the input audio samples are given for authentication, the input is verified with all existing hashed audio fingerprints at the server. If there exists any match with the audio fingerprints, then the input user is authenticated and allowed to use the application services.

Voice Based Authentication Algorithm

The voice-based authentication system consist of six major components namely voice input, voice activity detection, long-term spectral divergence, Mel Frequency Cepstral Coefficient, Gaussian Mixture Model, User authentication as shown in Figure 3. The Voice based authentication system consists of two processes namely Registration and authentication.

Registration Process

The first step is the Registration process. In registration process, the user speech input is recorded, analyzed, perform voice activity detection, extract MFCC features, train Gaussian mixture model and then to store the records of user specific voice features in the server.

Authentication Process

The second step is authentication process. In authentication process, the steps involved are recording user voice input, analyze the voice, perform voice activity detection, extract MFCC feature, test through Gaussian mixture model, obtain the best likelihood between user input and stored records of voice features in the server, and finally authenticate the user accordingly.

- **Voice Activity Detector:** One of the important aspects of voice based authentication system is to detect the user voice activity exactly (J. Ramirez et al., 2004). There should be mechanism to avoid silence or noise during the user voice input. For this purpose, the Voice Activity Detector library is available in Python. The VAD determines the difference between the energy levels in different speech band. Then it estimates the total energy in each speech band. Then, the speech band having more energy level is evaluated and verified for voice activity being happened.
- **Long Term Spectral Divergence:** (J. Ramirez et al., 2004) has accomplished the concept of LTSD for speech recognition. The LTSD detects the boundaries of those speech regions and finds out where exactly the user is speaking the required random generated words. This process basically eliminates the noise regions in the speech band. It also helps to detect the silences within the user speech.

Figure 3. Block diagram of Voice Based Authentication Algorithm

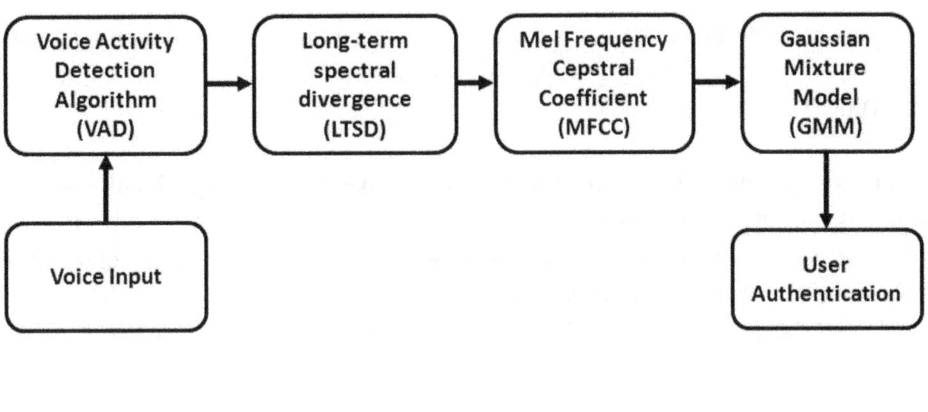

- **Mel – Frequency Cepstal Coefficient:** (X. Zhou et. al.,) has discussed the important role of MFCC in speech processing for an Automated Speech Recognition System. To being with, the speech signal is pre emphasized using filter. Next, the speech signal is sliced into frames using window function. Then, on each of the sliced frames are transformed using Fourier Transformed. Using Fourier Transformation, the power spectrum of each sliced frames are estimated. Further, filter banks are computed. Once, the filter banks are obtained, Discrete Cosine Transformation is applied on each of the filter banks. These filter banks also the system for retain only the candidate Mel – Frequency Cepstal Coefficients for speech recognition. The remaining filter banks are discarded. Then normalization is performed on each of the achieved coefficients.

Gaussian Mixture Model: (Reynolds et. al. 2000), discusses the advantages of Gaussian Mixture Model (GMM) for speech recognition over other machine learning techniques. The GMM considers each feature vectors as independent and generates log optimal output of different Gaussian Distributions. During Registration process, the user is asked to speak Radom generated words. For the audio input, GMM generates different the feature vectors and are stored in server. Upon authentication process, the user is again asked to speak random generated words. The authentication input audio is again model using GMM. Finally, the feature vectors of the registered features and the input features are compared and the best likelihood is achieved.

Gait

The motion of the human body while walking is a very popular way of analysis the surveillance camera captures for forensic investigation of the suspect when no face or other biometric features are available. The gait is affected by many aspects of the human body actual muscles and skeleton conditions as injuries, weakness, illness, etc.

Gesture

The gestures of the head/face and hands are usually captured by the camera (could be depth- or 3D-camera for better results) for the perceptual human machine interface design, using accelerometer or gyroscope the hands motions could also be captured. Using the predefined gesture signs or usual habit during the operation the verification or identification of the user can be realized.

Physiological Biometric Features

The subsection describes the physiological features of a human which is used as an authentication scheme.

Face Recognition

The facial features (eyes, nose, lips, and chin) of human have played a significant role in the recognition of individuals for long time. Depending on the unique shaping of each person's face, it has been a popular method in biometric recognition area. For authorization of a valid user, an enrolment is required for face verification. Enrolment requires a set of samples his/her face (Dehnavi, M. K., & Fard, N. P. 2011; Nita, S. L., & Mihailescu, M. I. 2018; Agulla, E. G., Rifón, L. A., Castro, J. L. A., & Mateo,

C. G., 2008, July; Gupta, A., Khanna, A., Jagetia, A., Sharma, D., Alekh, S., & Choudhary, V., 2015, Ketab, S. S., Clarke, N. L., & Dowland, P. S, 2017, GUILLÉN-GÁMEZ, F. D. (2017). To capture the image of a user' face video stream is processed and stored as verification modules. Though face features offers valuable information thus location of face, eyes, nose and mouth and their geometrical relationship allows selecting ("shot selection") a good candidate for a frontal face image to be used during the verification phase. The algorithms Viola and Jones are used for detection of structured and deformable objects. Frontal and profile face detection and also detection of internal features are performed using Lienhart's implementation included with the OpenCVlibrary. These images are stored at client side. At server side, the images are matched against the stored images. Decision on these stored images is on normalized scalar product. Statistical testing is required to compare the matching value which claim who is the actual user. The advantages of these systems are that it easy to use and the cost are also low. The drawback of these system are that the overall performance of biometric based face recognition system depend on number of factors such quality or resolution of collected photos for each individual, light conditions, angles of face rotation, etc. Also, an emotion also imposes some challenges to detect the identity of people.

Finger Print Recognition

Fingerprint based biometric authentication systems is one of the most popular authentication techniques, used by many organisation/system. It is more successful among other biometrics security methods for both of identification and verification processes of one's' identity. In this technique, fingerprints has to be enrolled in E-learning server database and biometric server database (Agulla, E. G., Rifón, L. A., Castro, J. L. A., & Mateo, C. G., 2008). Avoiding any malfunctioning in database, fingerprint scan are saved in an encrypted form. In an E-learning, environment student are assigned a student ID with an IP address. An intelligent agent is required to extract the fingerprint scans from the form the hardware device. It has to be ensuring that minimal noise should be present in the images. Later on, these enrolled scans will be matched by the intelligent agents of the system for the appropriate student. Figure 4 shows the fingerprint detection for E-learning systems.

Figure 4. Finger print detection in E-learning System
(Alotaibi, S.,2010)

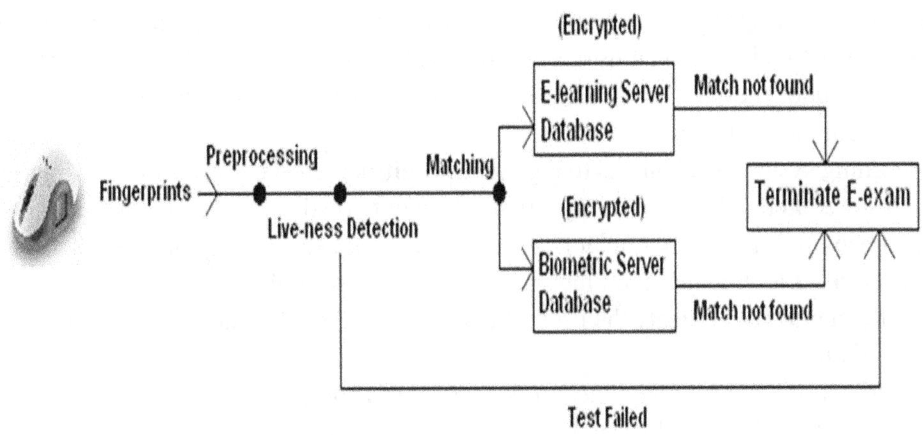

Though it is required a additional device to scan the fingerprint but scanning of fingerprint is required minimum time so that no other student can take the exam. Also, the scanning of students fingerprint is extract while working on the keyboard or the mouse, the pace of his work will not be affected. Moreover, due to low cost of optical sensor, it is easy to implement it. Fingerprint-based authentication system does not require much power as today implemented in many smart phones. The disadvantage of this technique is that it requires high quality of images of finger prints. Due to the issues of dirty, cuts, tear and wear that can easily affect the ridges and minutiae of fingertip. Finger nail plate is an emerging authentication technology has been introduced as biometric solution.

Palm Vein

The composition of vein in the palm of the right hand is a very accurate biometric feature. The main weakness of this feature is that scanner prices are very high and due to non-availability (Pleva, M., Bours, P., Hladek, D., & Juhar, J., 2016).

Hand Geometry

This biometric method is introduced and installed in late 60's in various security sectors. In this method, hand geometry is used as an automated technique for the identification or verification of authorized user. The measurement of the hand geometry is captured by specific scanner. For accurate data acquisition, a three-dimensional scanning takes different angles of the hand for better feature extraction. This method is widely accepted by user as it is friendlier authentication system than other biometric systems. It requires a special three dimensional scanner which acquire the full information of the palm. Moreover, it requires large space to store the information. Thus it is an expensive biometric authentication system. The future improvement is to make the device simplifier so that overall cost can be reduced.

Iris

Iris is described as the colored ring which is around the pupil of every human. It is different for every person. Student's iris of the eye is scanned for authentication. It is similar of taking the photograph away from 10-12 cm (Bal, A., & Acharya, A., 2011; Kaur, N., Prasad, P. W. C., Alsadoon, A., Pham, L., & Elchouemi, A. 2016, November). The eyeglasses do not affect much iris scan. Today Indian government is using iris and finger print authentication for making Aadhar card which can further using by any government verification. It exhibits a distinctive pattern that forms randomly in uterus, in a process called chaotic morphogenesis. The iris is a muscle that regulates the size of the pupil, controlling the amount of light that enters the eye.

In an E-learning system, student has to register himself/herself with iris pattern. All iris patterns are generated and stored in the database for a particular student. Testing has been done that no two patterns are same. It means that for every student there is unique pattern contain in the database. Hamming distance value is used to compare the uniqueness of different templates of different eyes. An image is captured of a learner to authenticate the user which helps to track the authentication of the learner during the examination.

For iris verification, aim is to find the boundaries of iris and pupil which are circular in shape. The extraction of iris pattern is simplified by knowing the boundaries of the pattern.

For iris detection Haugh circular transform algorithm is used on the outer edge of the iris. The Canny operator is used for a particular threshold value. A circle is defined as triplets (x,y,r) where x and y are the center coordinates and r represent the radius of the circle. Testing can be done using all possible radius and center coordinates. The circle with largest radius is considered to as iris boundary. The center and radius of the iris in the original image are determined by rescaling the obtained results. After locating the outer edge the inner one is located. The advantage of this method is that it is an optimal method as iris scanners spread in different security sectors for the automatic authentication of individuals. It is easy to use and flexible operating of scanning devices that make this technique in an increased demand. The disadvantage of this method is that if person wearing a glasses, eye lenses etc., it decreases the accuracy of iris recognition. Many researchers have proposed three dimensional cameras for iris scanning to get more accurate results.

Retina

This is another method similar to iris where a student eye is used for authentication. Retina is complicated network of neural cells at the back of the eye. The blood vessel pattern of retina is a distinctive pattern for each single retina of the same person. Identical twin does not have the same patterns of their retina. Retinal scan-based authentication system uses a specific sensor where the person is required to peep through this sensor. This sensor directs on a particular spot in the visual field in which the capturing of characteristics of blood vessel pattern is performed. This technique has a high level of accuracy of samples that can be obtained from the recognition of individuals. The disadvantage of this method is that it creates a little bit discomfort to the contributors while capturing their retina vessels. The development of this technology is that high resolution sensors have come sensor for capturing more accurate images of blood vessel sample (Israa.A.M., 2015).

Ear Recognition

Ear recognition is another biometric authentication technique based on the recognition of the unique shape and appearance of human being ear. A person is born with a visual shape of his/her ears and it is not subjected to change as person grows or aging. The features of ear extraction have the same process while face recognition. The specific measurement of outlook of the ear is taken and a mathematical model is created for providing a unique template. The method is more comfortable and friendly method in terms of user contribution than iris and retina recognition. The disadvantage of ear authentication scheme is that it do not consider as the most reliable authentication scheme. Many researchers are working on pattern recognition of the ear.

The major concerns of using the biometric solutions are universality, uniqueness, permanence, collectability, performance, and acceptability. Universality accommodates the features of the user. If user feature are missing then they need to be accommodated. Features should be unique. There should be no change in the characteristic of the user. User's authentication data can be easily available and measurable. The processing of these data must deliver a result. The user must approve the usage of Biometrics. Impersonation is one of the risks with current biometric solutions. A special device is required to get match from the result of fingerprint scanning. This may or may not possible to have a finger print device attached with every LCMS system due to the cost. Further it is also impossible to match with damaged or cut finger using fingerprint scanners. The other issue in biometric solution is that currently it is used for e-exams. The authentication of interaction between student and lectures has not been considered so far. Other situations such as E-commerce, online meetings, participation in forums should also be supported to other types of user interaction in E-learning.

Hence this chapter proposes a hybrid approach for authentication of interaction between student and learning content while using the LCMS system. A speech recognition, object identification, and signature verification and face recognition approach is used to verify the learner.

METHODOLOGY

This section explains the authentication of a learner/student during learning phase through LCMS. In this proposed system there are five basic phases namely (i) voice based authentication phase, (ii) Voice to text conversion phase (iii) Text based Image Retrieval phase (iv) Handwritten Signature based Authentication Phase (v) Face Recognition phase. The details of each of the modules are explained in this section. Figure 5 given below depicts the proposed methodology, various phases and also individual modules of each phase. The proposed system is implemented using the Python programming platform. Additionally, supporting libraries, API and external modules are included as and when required to python.

Phase 1: Voice Based Authentication

During both registration and authentication processes, the users are required to give their voice input. At this stage, it is essential to develop mechanism to handle the vulnerability imposed by attackers using stored voice of legitimate user as sources of input. First technique, to overcome this vulnerability is that the user is required to speak out the random generated alpha numeric words. The user interface will display the random generated alphanumeric words based on English Language. The user has to repeat those words character by character. The length of the string and number of characters used per word can be set according to the system requirements. In order to balance, the recording time and processing time, in this work, the number of random generated words is limited to 2 and the number of characters used is limited to 4 per word, total 8 alphanumeric characters are used. The complexity of random generated alphanumeric words can be further improved by varying the number of characters, types of characters (small letter alphabets, capital letter alphabets), numerical digits, special characters, punctuations. Further, the local language based characters can also be used (Roman alphabets, Hindi alphabets, etc.). Python libraries like string.ascii (), string.digits () generated the random string words for users as per the stem requirement. Upon user voice input the six other modules namely voice activity detection, long-term spectral divergence, Mel Frequency Cepstral Coefficient, Gaussian Mixture Model are carried out in sequence.

Phase 2: Voice to Text Conversion

To convert the speech to text, the Google API for speech recognition is used in this proposed system, For this purpose, additional modules like, pyaudio, pythondev, libportaudio, pip3 Speech recognition are also installed along with Python software. The recorded speech of user is fed into Google Speech Recognition API, the function recognize_google(audio) output the text for the inputted audio samples.

Figure 5. Methodology for the proposed authentication system for E-learning System

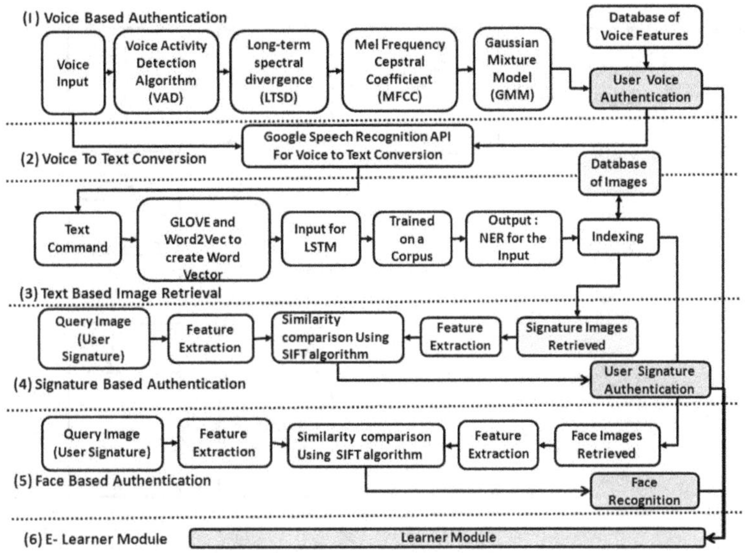

The generated text is given as input to the voice based image retrieval phases, to fetch relevant images from the stored Database.

Phase 3: Voice Based Image Retrieval

Once, the voice command is converted to text and then this string input is converted to a word array. On this word array the Machine Learning module for Natural Language Processing Toolkit (NLTK) is applied. This module is used to extract the 'location' tokens from the input text by the application of NER. Then the location tokens are sent to our image database where similarity coefficient between the name of the image and location token is found out and the most similar image is generated as output. This forms the query image for our later processes. Figure 6. shows the query image generated using voice command.

Phase 4: Handwritten Signature Based Authentication

The handwritten signature based authentication is based image retrieval process. The main objective of content based image retrieval is to search through the images of database and retrieve those images that are most suitable for the user query. Following are the steps of proposed CBIR system. Figure 7 shows this step.

Step 1: In this first step, the image descriptor is defined as a 3D color histogram in the HSV color space (Hue, Saturation, value). Typically, images are represented as a 3-tuple of Red, Green, and Blue (RGB). Image descriptors represent homogeneous features in an image or sub-image. A feature is a metric or some quantifiable value which is used to describe an image at high level perspective. The first step is to detect corners are contained in an image. After that we detect repeatability, able to detect the same physical interest points under different viewing conditions, followed by the

Figure 6. Architecture for Voice Command to Query Image generation

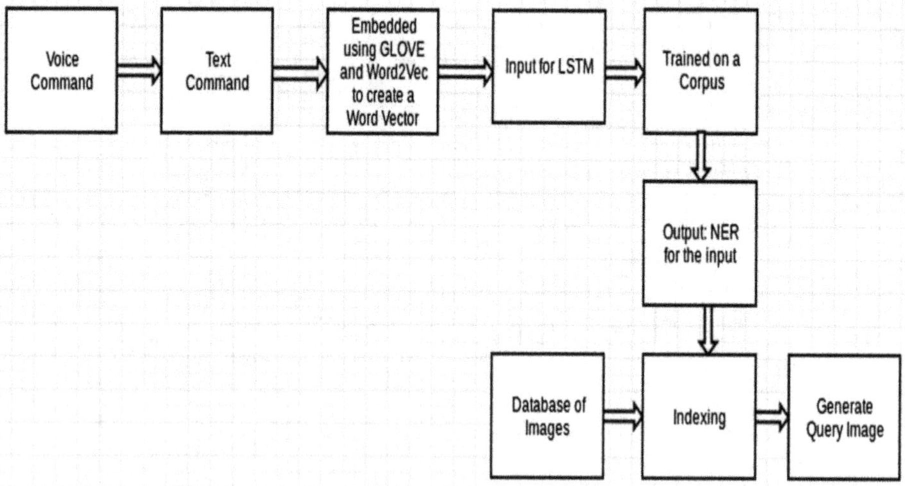

description calculation of the interest points. The feature needs to be unique i.e. if similar point is being described in two or more images then that point should have similar description and it should be of proper dimensions, a large descriptor will makes the computation longer. But if the descriptor is small then it may discard some useful information.

Step 2: Indexing of Dataset: After defining image descriptor, apply image descriptor in dataset and extract features from these images (i.e. color histograms) in dataset. The process of extracting features and storing them on persistent storage is commonly called "indexing". And, write the features to storage (ex. CSV file, RDBMS, Redis, etc.) so that they can be later compared for similarity.

Step 3: Defining Similarity Metric: In this step, feature vectors are compared to create a class that defines actual similarity metric between two images. Images that have a chi-squared similarity of 0 will be deemed to be identical to each other. As the chi-squared similarity value increases, the images are considered to be less similar to each other with the appropriate choices include the Euclidean distance, Cosine distance, and chi-squared distance, but the actual choice is highly dependent on data set and the type of features extracted.

Step 4: Searching: The final step is to perform an actual search. A user will submit a query image to the proposed system from an upload form or via a mobile app, for instance, and this system should able to

1. Extract features from this query image.
2. Apply similarity function to compare the query features to the features already indexed. And, the system will return top most relevant results.

Phase 5: Face Based Authentication

A pixel based classification is used to verify the signature in E-learning system. Classification is done on a per pixel level, using only the spectral information available for that individual pixel (i.e. values of

Figure 7. Architecture for extracting images based on query image

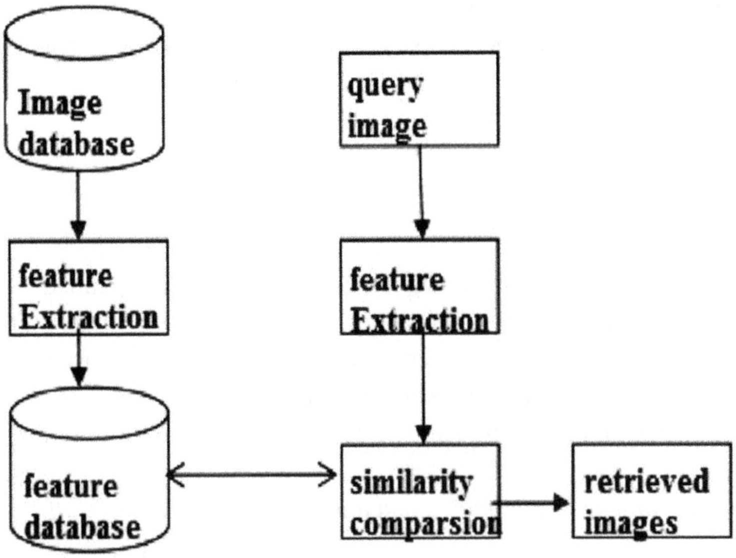

pixels within the locality are ignored). In this sense each pixel would represent a training example for a classification algorithm and this training example would be in the form of an n-dimensional vector, where n was the number of spectral bands in the signature image data. For image matching we have used SIFT(Scale invariant feature transform) Algorithm-For any object there are many features, interesting points on the object that can be extracted to provide a "feature" description of the object. This description can then be used when attempting to locate the object in an image containing many other objects. SIFT image features provide a set of features of an object that are not affected by many of the complications experienced in other methods, such as object scaling and rotation. Following are the steps for SIFT algorithm.

Step 1: Scale-Space Extrema Detection- This can be efficiently achieved using spacefunction is defined by-

$$L(x, y, \sigma) = G(x, y, \sigma) * I(x, y), \quad (1)$$

where * is the convolution operator, G(x,y,σ) is a variable-scale Gaussian and I(x, y) is the input image. Difference of Gaussians is one such technique, locating scale-spaceextrema is $D(x, y, \sigma)$ by computing the difference between two images, one with s scale k times the other. $D(x, y, \sigma)$ is then given by:

$$D(x, y, \sigma) = L(x, y, k\sigma) - L(x, y, \sigma) \quad (2)$$

To detect the local maxima and minima of D(x, y, σ) each point is compared with its 8 neighbors at the same scale, and its 9 neighbors up and down one scale. If the value is the minimum or maximum of all these points is an extrema.

Step 2: Key point Localization-This stage attempts to eliminate more points from the list of key points by finding those that have low contrast or are poorly localized on an edge. This is achieved by calculating the Laplacian, value for each key point found in stage 1. The location of extremum, z, is given by:

$$Z = -\frac{\delta^2 D^{-1}}{\delta X^2}\frac{\delta D}{\delta X} \tag{3}$$

If the function value at z is below a threshold value then this point is excluded. This removes extrema with low contrast. To eliminate extrema based on poor localisation it is noted that in these cases there is a large principle curvature across the edge but a small curvature in the perpendicular direction in the difference of Gaussian function. If this difference is below the ratio of largest to smallest eigenvector, from the 2x2 Hessian matrix at the location and scale of the key point, the key point is rejected.

Step 3: Orientation Assignment-The approach taken to find an orientation is:

Use the key points scale to select the Gaussian smoothed image L, from above
Compute gradient magnitude, *m*

$$m\left(x,y\right) = \sqrt{(L\left(x+1,y\right) - L(x-1,y))^2 + (L\left(x,y+1\right) - l\left(x,y-1\right))^2} \tag{4}$$

Compute orientation, θ

$$\theta\left(x,y\right) = \tan\left(\left(L\left(x,y+1\right)\right) - \frac{L\left(x,y-1\right))}{L\left(x+1,y\right) - L\left(x-1,y\right)}\right) \tag{5}$$

Step 4: Key point Descriptor-The local gradient data, used above, is also used to create keypoint descriptors. The gradient information is rotated to line up with the orientation of the key point and then weighted by a Gaussian with variance of 1.5 * key point scale. This data is then used to create a set of histograms over a window centered on the key point. Key point descriptors typically uses a set of 16 histograms, aligned in a 4x4 grid, each with 8 orientation bins, one for each of the main compass directions and one for each of the mid-points of these directions. This result in a feature vector containing 128 elements.

EXPERIMENTAL RESULT AND DISCUSSION

Data has been collected from Google and the dataset has 800 Images. Then, input is taken- voice as the input command then it is converted into text. A query image is generated from the dataset. A color histogram and SIFT algorithm is applied to extract the feature on the query image. Feature values are stored in the table. Now, these features are compared with the images stored in the database. A Chi square distance is used for similarly comparison. Based on the minimum value images has been displayed.

Voice Based Authentication and Image Retrieval

Figure 8 shows the input command of query 'Mukta Goyal'. Features are extracted on the base of RGB color index from 0 to 256. Based on this Histogram algorithm is applied to find out the probability of each color pixel. For a voice input command given below in Figure 8, output is generated in Figure 9 which is query Image for our Image Retrieval System.

And the output for the Image Retrieval is depicted in figure 9.

Test Cases for Fraud Detection Application

Using SIFT algorithm, we are extracting the feature of query image and other images present in the given database. Then, evaluating the distance between the two same features of query image and other images. Then compare the distance of query image with all the images one by one. And, if the distance is less than 0.1 these are matched features and else all other features are non-matched. And, ratio is calculated of matched and non-matched features and compare it with the threshold value. In this, we are taking signatures from different persons and calculating the similarity score between them by comparing with IMAGE 1. The score between Figure 9(a) and Figure 9(c) is 40.67% and the score between Figure 11(a) and Figure 11(c) is 53.36%. Figure 10 show the original input image.

Figure 8. Input Voice Command for Query Image Generation

Please say your username: Mukta Goyal
Please wait, processing....
Success, your are logged in!!

Figure 9. Output for Voice Command-Query Image

a) Image.1 (b) Image 2 (c) Image3

Figure 10. Input image

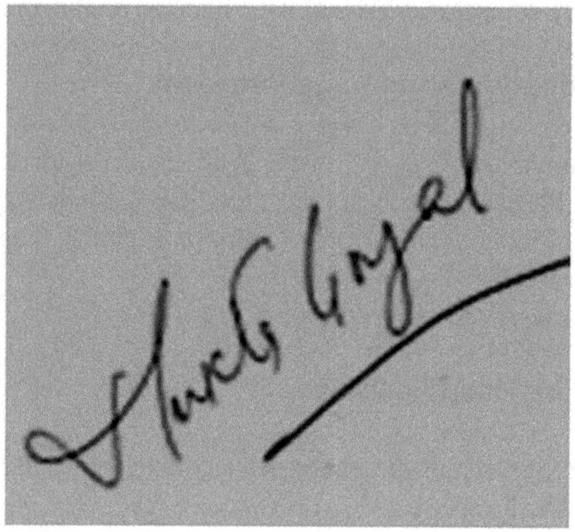

Figure 11. Signature images

a)Image 5 **b) Image 6** **c)Image 7**

Image Matching Using SIFT Classifies

Using SIFT Classifiers, for input query Image (Figure 12) shown below is matched with figure 13 even having contrast difference in both Images.

Similarly, Figure 14 and Figure 15 shown above are unmatched based on SIFT classifiers in our Image Retrieval System.

Hence the proposed methodology can enhance the authentication of a learner for learning system. Next section discusses the future research directions.

FUTURE RESEARCH DIRECTION

Future research in this area should focus upon the development of a complete e-invigilation system, with optimized result. Further, it is proposed to integrate and classify the techniques and add low level

Figure 12. Input query image

Figure 13. Output image retrieved from database

Figure 14. Input query image

Figure 15. Unmatched image retrieved from database

features, like texture, shape to the system. And to develop more efficient model when problem with false rejection when people change their hairstyle, grow or shave a beard or wear glasses and Identical twins for facial detection. The research should focus on testing the accuracy of the algorithms in detecting a single object within a scene. Future scope of this application emphasize in various field like Fraud detection has its advantage in Banking System, Passport Verification System.

Several performance factors also depend on the storage requirements and management of database to minimize the cost. Thus efficient database management algorithms should be proposed. Further impact of using the proposed biometric techniques on the increasing or decreasing the grades of the students, time period of the exam, controlling unethical behavior of students on the overall grades performance. Learners with special need to access LCMS should be authenticated.

Biometric solutions can be incorporated in E-exam in such a way that students who are taking e-exam using the fingerprints biometric solution will have lower grades or taking longer time to complete their e-exam than their counterparts. Results of such study can address the problem of unethical conduct in an E-learning exam taking. If the results of the proposed study show that students who are taking longer time to complete their exam or having lower grade are the ones who took the exams using the fingerprints biometric solutions, thus have a lower cheating rate.

To improve the performance of the LCMS system, one can design a model in such a way that the student's physical and behavioral information and also mental state should be monitored and processed.

CONCLUSION

The primarily this chapter addresses on review of existing methodology and how to enhance the existing authentication functionality of LCMS. Further, the second objective is to provide solution to the constrained requirements of mobility and versatile wireless environment based LCMS. Third objective is to propose solution to security system of such LCMS by providing support mechanism like data confidentiality, user authentication, data authorization and also user accountability (Adetoba, B. T., Awodele, O., & Kuyoro, S. O., 2016; Qwaider, W. Q., 2017).

In this chapter, as proposed solution to enhance the authentication of LCMS various techniques available authentication like voice based authentication, face recognition and signature based methods are discussed. Also, image retrieval mechanism required signature and face based authentication has been discussed. This image retrieval system mainly aims at retrieving the exact as well similar images from web based on a query image. Evaluation of three algorithms - COLOR HISTOGRAM, SIFT, and pixel-based algorithm has been done. Further, it is proposed to integrate and classify the techniques and add low level features, like texture, shape to the system. And to develop more efficient model when problem with false rejection when people change their hairstyle, grow or shave a beard or wear glasses and Identical twins for facial detection.

REFERENCES

AbdelRahman, S. M., & Sabir, A. A. (2017). Strategies and Contributions of Biometrics in E-Learning. *International Journal of Computer and Information Technology, 6*(2).

Adetoba, B. T., Awodele, O., & Kuyoro, S. O. (2016). E-learning security issues and challenges: A review. *Journal of Scientific Research and Studies*, *3*(5), 96–100.

Jain, A., Kulkarni, G., & Shah, V. (2018). Natural Language Processing. *International Journal on Computer Science and Engineering*, *6*(1), 161–167.

Agulla, E. G., Rifón, L. A., Castro, J. L. A., & Mateo, C. G. (2008, July). Is my student at the other side? Applying biometric web authentication to E-learning environments. In *Advanced Learning Technologies, 2008. ICALT'08. Eighth IEEE International Conference on* (pp. 551-553). IEEE.

Al Galib, A., & Safavi-Naini, R. (2015, January). User authentication using human cognitive abilities. In *International Conference on Financial Cryptography and Data Security* (pp. 254-271). Springer. 10.1007/978-3-662-47854-7_16

Alotaibi, S. (2010). Using Biometrics Authentication via Fingerprint Recognition in E-exams in E-learning Environment. In *The 4th Saudi International Conference*. The University of Manchester.

Anuradha, S.G., Kavya, B., Akshatha, S., Jyothi, K., & Ashalatha, G. (2016). Automated Face Detection & Recognition for Detecting Impersonation of Candidate in Examination System. *International Journal of Scientific & Engineering Research, 7*(3).

Athavale, V., Bharadwaj, S., Pamecha, M., Prabhu, A., & Shrivastava, M. (2016). *Towards deep learning in hindi ner: An approach to tackle the labelled data scarcity.* arXiv preprint arXiv:1610.09756

Atlam, H. F., Attiya, G., & El-Fishawy, N. (2017). Integration of Color and Texture Features in CBIR System. *Integration (Tokyo, Japan)*, *164*(3).

Bal, A., & Acharya, A. (2011, December). Biometric authentication and tracking system for online examination system. In *Recent Trends in Information Systems (ReTIS), 2011 International Conference on* (pp. 209-213). IEEE. 10.1109/ReTIS.2011.6146869

Bonneau, J. (2012). The Quest To Replace Passwords. *IEEE Symposium on Security and Privacy*, 553-567.

Cano, P., Batlle, E., Kalker, T., & Haitsma, J. (2005). A Review of Audio Fingerprinting. *The Journal of VLSI Signal Processing*, *41*(3), 271–284. doi:10.100711265-005-4151-3

Chowdhury, G. G. (2003). Natural language processing. *Annual Review of Information Science & Technology*, *37*(1), 51–89. doi:10.1002/aris.1440370103

Chutel, P. M., & Sakhare, A. (2014). Reverse Image Search Engine using Compact Composite Descriptor. *International Journal (Toronto, Ont.)*, *2*(1).

Chiasson, S., Biddle, S., & van Oorschot, P. (2012). A Second Look at the Usability of Click-Based Graphical Passwords. *Proc. ACM Symp. Usable Privacy and Security (SOUPS).*

Dehnavi, M. K., & Fard, N. P. (2011). Presenting a multimodal biometric model for tracking the students in virtual classes. *Procedia: Social and Behavioral Sciences*, *15*, 3456–3462. doi:10.1016/j.sbspro.2011.04.318

Dehnavi, M. K., Sharafi, S. M., & Nematbakhsh, N. (2011). Developing a E-learning Model for Tracking the Continuous Attendance of the Students. *Journal of Theoretical & Applied Information Technology, 24*(1).

El-Gayar, M. M., Soliman, H., & meky, N. (2013). A comparative study of image low level feature extraction algorithms. *Egyptian Informatics Journal, 14*(2), 175–181. doi:10.1016/j.eij.2013.06.003

Gonzalez-Agulla, E., Alba-Castro, J. L., Argones-Rua, E., & Anido-Rifon, L. (2010, January). Realistic measurement of student attendance in LMS using biometrics. *Proc. of the Int. Symposium on Engineering Education and Educational Technologies (EEET'09).*

Gupta, A., Khanna, A., Jagetia, A., Sharma, D., Alekh, S., & Choudhary, V. (2015, October). Combining Keystroke Dynamics and Face Recognition for User Verification. In *Computational Science and Engineering (CSE), 2015 IEEE 18th International Conference on* (pp. 294-299). IEEE. 10.1109/CSE.2015.37

Guillén-Gámez, F. D. (2017). Biometrics and education: A review about facial authentication software for the identification and verification of students who use virtual learning platform (LMS). *Advances in Educational Technology and Psychology*, *1*(1), 1–8.

Israa, A.M. (2015). Physiological Biometric Authentication Systems. *International Journal of Scientific & Technology Research, 4*(12).

Kang, B. H., & Kim, H. (2015). Proposal: A design of e-learning user authentication system. *International Journal of Security and Its Applications*, *9*(1), 45–50. doi:10.14257/ijsia.2015.9.1.05

Kaur, N., Prasad, P. W. C., Alsadoon, A., Pham, L., & Elchouemi, A. (2016, November). An enhanced model of biometric authentication in E-Learning: Using a combination of biometric features to access E-Learning environments. In *Advances in Electrical, Electronic and Systems Engineering (ICAEES), International Conference on* (pp. 138-143). IEEE.

Ketab, S. S., Clarke, N. L., & Dowland, P. S (2017). *A Robust E-Invigilation System Employing Multimodal Biometric Authentication*. Academic Press.

Khlifi, Y., & El-Sabagh, H. A. (2017). A Novel Authentication Scheme for E-assessments Based on Student Behavior over E-learning Platform. *International Journal of Emerging Technologies in Learning*, *12*(04), 62–89. doi:10.3991/ijet.v12i04.6478

Levy, Y., & Ramim, M. (2007). *A theoretical approach for biometrics authentication of e-exams*. Nova Southeastern University.

Mikolov, T., Sutskever, I., Chen, K., Corrado, G. S., & Dean, J. (2013). Distributed representations of words and phrases and their compositionality. In Advances in neural information processing systems (pp. 3111-3119). Academic Press.

Miller, G. (1998). *WordNet: An electronic lexical database*. MIT Press.

Moini, A., & Madni, A. M. (2009). Leveraging biometrics for user authentication in online learning: A systems perspective. *IEEE Systems Journal*, *3*(4), 469–476. doi:10.1109/JSYST.2009.2038957

Nita, S. L., & Mihailescu, M. I. (2018). Proposing a Secure E-learning System Based on Biometric Authentication and Homomorphic Encryption. *eLearning & Software for Education, 4*.

Obeidallah, R., Ahmad, A. A., Farouq, F., & Awad, S. (2015). Students authentication in e-assessment sessions: A theoretical biometric model for smartphone devices. *International Journal of Business Information Systems*, *19*(4), 450–464. doi:10.1504/IJBIS.2015.070204

Penteado, B. E., & Marana, A. N. (2009, May). A video-based biometric authentication for E-learning web applications. In *International Conference on Enterprise Information Systems* (pp. 770-779). Springer. 10.1007/978-3-642-01347-8_64

Pleva, M., Bours, P., Hladek, D., & Juhar, J. (2016). Using current biometrics technologies for authentication in e-learning assessment. In *Emerging eLearning Technologies and Applications (ICETA), 2016 International Conference on* (pp. 269-274). IEEE. 10.1109/ICETA.2016.7802043

Qwaider, W. Q. (2017). Information Security and Learning Content Management System (LCMS). *International Journal of Advanced Computer Science and Applications*, *8*(11), 588–593.

Ramu, T., & Arivoli, T. (2013). A framework of secure biometric based online exam authentication: An alternative to traditional exam. *Int J Sci Eng Res*, *4*(11), 52–60.

Ramirez, J., Segura, J. C., Benítez, C., de la Torre, Á., & Rubio, A. (2004). Efficient voice activity detection algorithms using long-term speech information. *Speech Communication*, *2004*(3-4), 271–287. doi:10.1016/j.specom.2003.10.002

Rejito, J., Abdullahi, A. S., Setiana, D., & Ruchjana, B. N. (2017). Image indexing using color histogram and k-means clustering for optimization CBIR in image database. *Journal of Physics: Conference Series*, *893*(1), 012055. doi:10.1088/1742-6596/893/1/012055

Reynolds, D., Quatieri, T. F., & Dunn, R. B. (2000). Speaker verification using adapted Gaussian Mixture models. *Digital Signal Processing*, *10*(1-3), 19–41. doi:10.1006/dspr.1999.0361

Song, K., Lee, S. M., & Nam, S. (2013). Combined biometrics for E-learning security. ISA 2-13. *ASTL*, *21*, 247–251.

Srinivasan, H., Beal, M. J., & Srihari, S. N. (2005, May). Machine learning approaches for person identification and verification. In Sensors, and Command, Control, Communications, and Intelligence (C3I) Technologies for Homeland Security and Homeland Defense IV (Vol. 5778, pp. 574-587). International Society for Optics and Photonics. doi:10.1117/12.601987

Xiao, H., Ji, W., & Ullah, A. (2011). Authentication of Students and Students' Work in E-learning. *Report for the Development Bid of Academic Year, 2010*, 11.

Zhonghu, L., Lin, Z., Luling, W., Junhong, Y., & Jinming, W. (2017, October). Research on adaptive SIFT algorithm in image matching of pipeline inner surface visual measurement. In *Electronic Measurement & Instruments (ICEMI), 2017 13th IEEE International Conference on* (pp. 305-309). IEEE.

Chiasson, S., Biddle, S., & van Oorschot, P. (2012). A Second Look at the Usability of Click-Based Graphical Passwords. *Proc. ACM Symp. Usable Privacy and Security (SOUPS).* doi:10.1109/ICEMI.2017.8265799

Chapter 5
Applications of Keystroke Dynamics Biometrics in Online Learning Environments:
A Selective Study

Anandhavalli Muniasamy
King Khalid University, Saudi Arabia

ABSTRACT

The biometric authentication in the online learning environment (OLE) is still exploratory and, despite an increase in keystroke dynamics biometrics research, many challenges remain in designing this authentication system due to the fact that it is economical and easily integrated with existing computer security system in OLE. Existing research in keystroke dynamics tends to focus on finding how keystroke dynamics of users can support non-intrusive authentications of users in OLEs. However, there is little evidence that researchers have approached the issue of unauthenticated users to take the role of authenticated users and perform tasks in the OLE with the intent of building models based on the keystroke dynamics of users. In a nutshell, the aim of this chapter is to provide an overview of the existing applications of keystroke dynamics as biometric authentication in the OLE, keystroke dynamics framework being designed for the OLE, advantages and disadvantages of a keystroke dynamics biometrics approach, as well as offering suggestions and possible future research directions.

INTRODUCTION

The prevalence of online education has increased dramatically in the past decade as it offers several significant benefits. (U.S. Higher Education System, 2016; Freidman, 2016; Newton, 2015). Among those benefits are availability, accessibility, affordability, updatability, and economical (Li & Irby, 2008; Ruiz, 2017). However, along with the benefits of online learning come several challenges, one of which is academic dishonesty among students (Grijalva, Nowell, & Kerkvliet, 2006; King, Guyette, & Piotrowski, 2009). Unethical behavior in schools is rampant (King, Guyette, & Piotrowski, 2009).

DOI: 10.4018/978-1-5225-7724-9.ch005

While cheating in school is not new, it has taken on new forms and is becoming easier to undertake due to increased use of technology to facilitate OLEs (King, Guyette, & Piotrowski, 2009; Sewell, Frith, & Colvin, 2010). The main challenge that the providers of online education facing is to assure a student who wants to take an exam is really the individual who is expected to attend to that test. Cyber cheating is far more widespread than originally believed because it is nearly impossible to verify the identity of an individual being assessed online (Moini & Madni, 2009).

To solve this cyber cheating issue, US government passed the Higher Education Opportunity Act (HEOA 2008) which requires that institutions offering online education have to make greater access control efforts for the purposes of assuring that students of record are those actually accessing the systems and taking online exams by adopting identification technologies as they become more ubiquitous (Monaco, Stewart, Cha, & Tappert, 2013). To comply with the requirements of this law, institutes of higher education who provide online courses have begun exploring the use of biometrics to authenticate students.

Biometric methods verify and authenticate the identity of a user based on a physiological or behavioral characteristic (Miller, 1994). One frequently used authentication mechanism is using usernames and passwords (Bours, 2012); thus "once only" authentication. However, in the case of online assessments and exams, there is a requirement to monitor the identity of users throughout the course of an entire assessment, thus continuous authentication. This includes how science and technology can be utilized to identify physiological or behavioral attributes that are unique to individual students (Karnan, Akila & Krishnaraj, 2011).

Commonly used biometric indicators (e.g., finger or palm prints, iris scans, facial and voice recognition) are effective because technology can accurately authenticate a user's identity by comparing samples of unique physiological characteristics (Karnan, Akila & Krishnaraj, 2011). The main concern with using biometrics in the authentication process is systems needed to capture and compare these metrics can be prohibitively expensive to implement (Panchumarthy, Subramanian, & Sarkar, 2012). They can also be somewhat intrusive (e.g., taking facial recognition videos while working on a course). These biometrics are good gatekeeper measures; however, they do not serve well as in-system verification tools. It promotes high recognition accuracy due to its permanence and high uniqueness. However, it possibly suffers from low public acceptance due to invasiveness (iris scanning) and high implementation cost (DNA analysis) in large-scale deployment.

Behavioral traits such as handwriting, signatures, keystroke dynamics, and mouse dynamics can be used like physiological characteristics to identify individuals (Karnan, Akila & Krishnaraj, 2011). Metrics using keystroke and mouse dynamics can examine the behavior of those admitted into the learning system to verify the person completing the coursework is the person signed up to take the course. These metrics can be somewhat less accurate than physiological characteristics as they often change slightly depending on circumstances, but they are less obtrusive and obtained during the process of an individual completing work in the course rather than merely at the beginning of a work session (Marsters, 2009).

Keystroke dynamics is a behavioral biometric characteristic based on the assumption that different people type in a unique manner on digital devices. (Joyce & Gupta, 1990) portrayed that keystroke dynamics in online assessment could lead to increased security when a learner is completing a test unsupervised at a distance using technology. Keystroke dynamics record and analyze the way a user types, based on habitual typing patterns (Monrose & Rubin, 2000). Like a fingerprint or a signature, the users can be identified using keystroke dynamics to create what is called a keyprint that represents the user's typing behavior. (Monrose & Rubin,1997) suggest that the use of keystroke dynamics to create a typing signature for individuals is the most reasonable and cost-effective approach for institutions of higher-education looking to improve their online assessment security. Multiple studies verify that keystroke dynamics work in free text and password scenarios and using multiple analysis techniques (Giot, El-Abed, & Rosenberger, 2011; Flior & Kowalski, 2010; Killourhy & Maxion, 2009). These research efforts in keystroke dynamics can provide a unique feature or identifier or signature for authenticating an individual's identity. The implications of these situations highlight the need to consider the applications of keystroke dynamics as biometrics authentication in the OLE.

This chapter first briefly highlights the basics of keystroke dynamics and further provides some aspects of the literature on the prevalence and prevention of online cheating based on keystroke analysis in combination with other techniques. It then sets out the keystroke dynamics framework being designed for the OLE and outlines the advantages and disadvantages of a keystroke dynamics framework approach. Then continue with an overview of how the various existing applications such as MOOC uses the concept of keystroke dynamics in the OLE and discussion on the keystroke dynamics issues in the OLE with solutions and recommendations based on previous keystroke dynamics in the OLE research. Finally, the chapter concludes with a summary of the contribution of keystroke dynamics applications in the OLE, and a roadmap for future directions.

BACKGROUND

Typing as a behavioral biometric authentication has been used for several years. By analyzing users typing patterns, several studies (Janakiraman & Sim, 2007; Gunetti & Picardi, 2005; Dowland, Furnell, & Papadaki, 2002; Kang, Hwang & Cho, 2007) conclude that the keystroke rhythm has features that represent consistent patterns of user typing behavior. It has been widely discussed that this rhythm tends to be unique for each person and that it can be a valid method of identifying, authenticating, constantly monitoring, or even classifying them. It can be implemented to quietly capture user typing during normal operation, thus making a user unaware of the process. It can be easily incorporated into the existing password-based authentication system and no additional device requirement increase the potential of keystroke technique to be deployed in a real-world application, such as Massive open online courses (MOOC).

Online Learning Environment

Online learning is a type of delivery method used in distance education that allows the synchronous and asynchronous exchange of resources over a communication network (Khan, 1998). OLE refers to the e-learning environment used for knowledge acquisition within computer mediated digital systems. The OLE is also the system surrounding the learner and the teacher in terms of technical and social aspects, which promote communication and collaboration among students and instructors (Khan, 2000). The literature on online learning is abundant of comparisons of distance education to the traditional teaching and learning contexts. The critical components of an effective learning environment are learner characteristics, course content, skills, learner support, resources, and assessment. Online learning tools refer to the online classroom, app, and assistive technology that can be accessed via an Internet connection and enhance a teacher's ability to present information and a student's ability to access that information. The new technologies support as to develop new learning environments and the instructors now have a rare chance to create a culture which are important for today's learners. As the Online learning continues to gain prevalent attention and succeed as a legitimate alternative to classroom instruction, so the educational institutions and online instructors face the challenge of sustaining student's authentication in OLEs.

Online Assessment

Online assessment or e-assessment is a procedure by which specific abilities or characteristics can be evaluated via the Internet. Online assessment is the application of formative and summative assessments within learning online and blended settings where the teacher and learners are separated by time and space and where a substantial proportion of learning and teaching activities are conducted through web-based information communication technology [ICT] (Gikandi, Morrow, & Davis, 2011). The systematic process of documenting and using practical data on the knowledge and skills in the form of e-assessment is very important in the educational sector. The most important components of e-assessments such as questionnaires and ability tests can be completed at home or in a secure OLE that is not accessible to others. The use of e-assessments saves educational institutions a lot of time and money. As all required knowledge is available in the online assessment, without the help of supervisors, multiple students can complete their assessments at the same time. Online identity verification is the confirmation of a focal person identity using supporting evidence. Online assessment security is the process of decreasing the actual and probable occurrences of cheating in an online course context.

Keystroke Dynamics

Keystroke dynamics (Keystroke Biometrics or Typing Dynamics) is the process of analyzing the way a user types to identify individuals based on typing patterns (Monrose & Rubin, 2000). Keystrokes, unlike other biometric features, convey an unstructured and very small amount of information. When keystroke dynamics has been used as a biometric technique, many different features have been extracted from a user's typing rhythm. Both the National Science Foundation (NSF) and National Institute of Standards and Technology (NIST), United States of America have conducted studies establishing that typing patterns are unique to the typist. According to (Jin, Ke, Manuel, & Wilkerson, 2004; Leggett &Williams,1988), measuring typing characteristics in keystroke biometric systems believed to be unique to an individual and difficult to duplicate. The typing dynamics or timing pattern has been measured and used in both verification and identification applications (Leggett, Williams, Usnick, & Longnecker,1991). Keystroke dynamics has been used to strengthen password-based user authentication systems by considering the typing characteristics of legitimate users.

The advantages of keystroke dynamics are obvious in the computer environment as it provides a simple natural method for increased computer security. Keystroke dynamics approaches can be characterized by the selection of features and the classification methods employed. Over the years, researchers have evaluated different features/attributes, feature extraction, feature subset selection and classification methods to improve the recognition capabilities of keystroke biometrics. Static keystroke analysis is performed on typing samples produced using predetermined text for all the individuals under observation. Dynamic keystroke analysis implies a continuous or periodic monitoring of issued keystrokes. It is performed during the log-in session and continues after the session.

The keystroke biometric is attracting interest because keyboards are readily available and keystroke data capture is not invasive. Products (BioPassword® and PsyLock®) currently exist which exploit keystroke biometric technology for enhanced security and increased confidence in user identification over standard password authentication methods.

COMMON FEATURES IN KEYSTROKE DYNAMICS

Timing features

Typically, the study of keystroke dynamics is concerned with the timing information generated from key presses (KD) and releases (KU). While typing, the computer can record the key pressed and the time at which it was pressed. In addition to this, it can also record the time at which the same key was released and for how long it was pressed. The time duration between key press and key release convert into security parameters. Each key event will be associated with a timestamp and this timestamp is the core component of the template generation process. Keystroke timing information shown in Figure1 in terms of hold or dwell time (HT), flight time (FT), difficulties of typing phrase, pressure of keystroke, typing rate, linguistic style, the sound of typing frequency of word errors can be extracted while a user is typing the data.

Dwell time (D_i) is the latency between the pressing (P_i – KDi) and releasing (R_i – KU_i) of a key. This is the most basic feature that can be extracted. The period of silence between successive keystrokes is Flight time (F_i). The delay between any n number of key press and key release or a combination of both events is known as an n-graph. Digraphs (Di_1) and trigraphs (Tri_1) are examples of this timing feature. This timing information is utilized to construct distinctive typing biometric templates for both enrolment and verification purposes. In the context of keystroke authentication process, such templates are also referred to as signatures (Ahmed & Traore, 2014) or reference profiles (Dowland, Furnell, & Papadaki, 2002).

Figure 1. Keystroke timing information

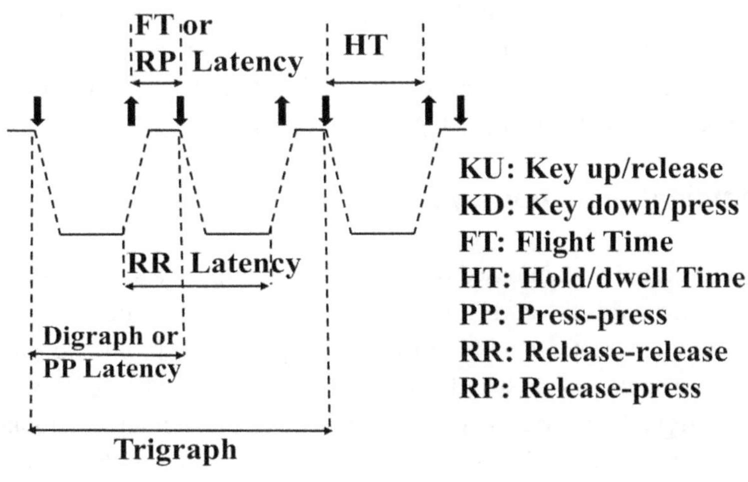

Contextual Information

Contextual features (Bartlow, 2005) can range from characteristics directly related to keystroke dynamics, like for instance, the position of the letters in a word, or be somewhat related to user behavior, like for example, the frequency a word is typed, or the age and gender of the user.

Fixed Text vs. Free Text

There are two main fields of research when dealing with keystroke dynamics.

1. Keystroke static authentication directed at static (predefined) text such as passwords, usernames, and pin numbers.
2. Keystroke continuous authentication directed at continuous (free) text.

In keystroke static authentication, the user is asked to type a predefined text (like a password) several times. Most of the literature (Leggett, Williams, Usnick, & Longnecker,1991; Joyce & Gupta,1990; Bleha, Slivinsky, & Hussien,1990; Brown & Rogers,1993; Obaidat & Sadoun,1997; Bergadano, Gunetti & Picardi 2002) on keystroke analysis falls within the category. The intended application of static analysis is at login time, in combination with other traditional authentication methods. Studies have shown that static approaches are believed to be more robust for user verification than simple passwords but do not provide continuous security to authenticate the person against the claimed entity after the initial verification. It should be clear that the analysis of free text is dynamic since even the length of the text is free.

To the best of our knowledge, the first research into keystroke dynamics analysis with long and freely typed text was conducted by (Monrose & Rubin,1997). They collected the keystroke profiles of short and fixed text and long and freely typed text from 42 participants over seven weeks. The average down–down time (DD-time) of the top N most frequent digraphs or tri-graphs. Authentication was performed by computing the Euclidean distance between the feature vectors of the valid user and a new trial, which resulted in a 23% classification error rate in their experiment. In keystroke continuous authentication, users type either long portions of text that simulate the idea of free text or, in other cases, type whatever they want whenever they want, without restrictions. It involves a continuous monitoring of keystroke typing and is intended to be executed during the entire session, after the initial authentication step. The works of literature (Dowland, Furnell & Papadaki, 2002; Gunetti & Picardi, 2005) shows that continuous authentication has great success to distinguish users effectively.

Keystroke Dynamic Verification System

A basic biometric verification system based on keystroke dynamic is a system that checks whether the user trying to be authenticated is the correct user or not, as shown in Figure 2.

The input device module captures and digitizes a sample of the biometric signal. A user tries to login gives his or her username and password, the typing features are extracted from the typed password and or username. Keystroke biometric samples are reduced to mathematical templates and the system stores only this "enrolled template". A template is a vector comprised of the extracted unique features of the biometric data. The template is generated during the process of feature extraction, which frees the raw data coming from the biometric sensor from irrelevant information. In this way, both the storage

Figure 2. Keystroke dynamic verification system

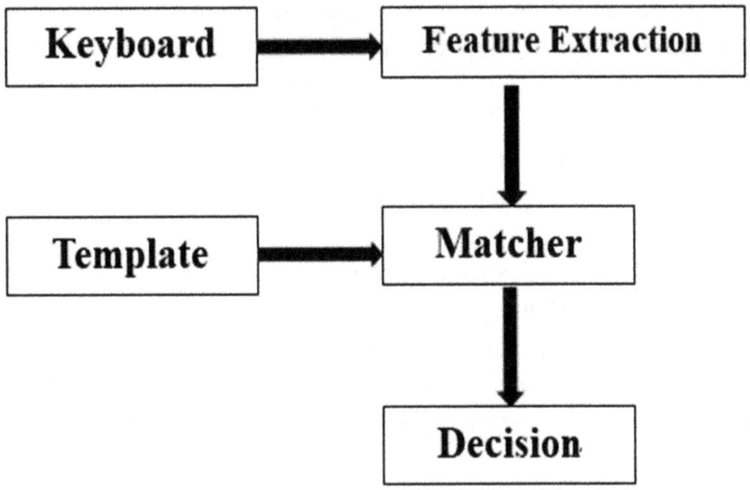

requirements and the matching expense are reduced. The matcher gets the saved reference template for the specific user and compares the collected typing features with the stored template in a distance metric. The distance metric score is compared with a threshold, and the user is accepted or denied access. Keystroke dynamic can easily strengthen existing password-based systems, by also checking how the password was typed and not only if it is the correct password. To authenticate a user against a claimed ID, the corresponding stored template is retrieved and matched against a new template derived from the current input. The output produced is where a match exists or "Yes/No" output.

The most commonly adopted metrics False Reject Rate (FRR) and the False Accept Rate (FAR) are used to evaluate a biometric system's authentication accuracy. (Leggett & Williams,1988) showed that keystroke digraph latencies had a potential for a static identity verifier at login time, as well as a dynamic identity verifier throughout a computer session. (Brown & Rogers, 1993) and (Obaidat & Sadoun, 1997) used short name strings for user verification. (Leggett, Williams, Usnick, & Longnecker, 1991) conducted similar experiments, reporting 5.0% FAR and 5.5% FRR on a long string of 537 characters.

Applications of Keystroke Dynamics

The applications of keystroke dynamics are many. A summary of the most common applications where the use of keystroke dynamics can be relevant is provided in this section.

- **Authentication:** Using keystroke dynamics users can be authenticated. The combination of a login, a password, and the biometric signature provides a means of accessing protected resources.
- **Verification:** Keystroke dynamics allows the possibility of continuously verifying a user using a computer system by constantly checking their way of typing against a template.
- **Identification:** Even if there has not been a proper previous authentication, a user could still be identified by doing a one-to-many check against a database of templates.
- **User Change Control:** By constantly monitoring the input of a user, it can be determined at which moment the user has been supplanted by another (Bartlow, 2005).

- **Exam Control:** Keystroke dynamics could be used to detect abrupt changes in the template of a user and conclude that another user is taking the exam.
- **Password Hardening:** Protecting access to highly sensitive resources can be achieved by adding a keystroke dynamics signature to the password (Antal & Nemes, 2016; Hussain & Al-Hassan, 2013).
- **Authoring:** When users submit information that has been controlled using a keystroke dynamics technique, it is possible to determine who wrote the entry by doing an identification check on the data.
- **Emotion Detection:** Some studies have proven that using keystroke dynamics, it is possible to detect some emotions (Giot, Hemery, & Rosenberger, 2010).
- **Attacks:** Keystroke analysis can lead to sophisticated timing attacks. By listening to a user typing, either via a network sniffer or using a key-logger, it is possible to learn how the user types or, even identify what users are typing over encrypted channels (Killourhy & Maxion, 2009).

Advantages of Keystroke Dynamics

There are many potential advantages of using keystroke dynamics in data analysis (Revett et. al., 2007; Checco, 2006; Gunetti & Picardi, 2005), which are as follows:

- It is believed that some characteristics of keystroke information, which form a unique biometric template of the user's typing pattern, are as individual as a signature.
- Keystroke event can be measured at a high resolution of milliseconds precision by software. Thus, it is difficult to imitate other people's keystroke pattern without lots of effort.
- Almost every workstation has a keyboard; thus, the process of recognition is all done based on software only.
- It is neither obtrusive nor intrusive, as a user will be using computer keyboard anyway.
- It allows data to be captured continuously over a length of time.
- It adds an extra layer of protection and increases the lifespan of a password as a user can focus on creating a strong password whilst avoid being overwhelmed by different sets of passwords.
- It is relatively inexpensive to implement as it does not require no extra equipment such as a scanner or camera.
- The technology can be embedded in any in-house software application.
- This technology does not require changes in existing network access policies.

Disadvantages of Keystroke Dynamics

- By using keystroke dynamics, it is possible to mitigate many password problems, like user sharing their passwords, weak passwords, brute force attacks, etc.
- Since the user's typing pattern change over time and their typing is not as unique as fingerprints, there will always be False Acceptance (FA) and False Rejections (FR).
- The variations in typing rhythm that caused by external factors such as injury, fatigue, adaptation to input devices or distraction can lower authentication accuracy of keystroke dynamics.

FRAMEWORK OF KEYSTROKE DYNAMICS

Keystroke dynamics is a non-intrusive and transparent technique to the user, but this does not mean that it is exempt from ethical and privacy issues. When performing authentication processes, the capture phase is limited to the login screen. At this point, when using keystroke dynamics, users know they are being monitored. Once the authentication phase is over, capturing also stops. Users do not have to be conscious of the fact that their actions may be analyzed locally or sent to a remote server for evaluation. When continuously monitoring a user working on a computer, in verification or identification schemes, the capturing phase of the natural rhythm of a user never stops. This means that all, or most, keystrokes are evaluated, and it would be very easy for a user to stop thinking about this. In fact, this should be a premise to allow users to show their real natural rhythm. Once the user has stopped thinking about a process of continuously monitoring their actions, without being aware that someone is watching. While the user is performing these tasks the keystroke dynamics module is recording or analyzing the information.

In the context of online learning environment where students and teachers alike interact with the virtual campus using all kinds of devices, either at the university, the library, the cafeteria, at home etc., it was thought that keystroke dynamics was the best-suited option from the available alternatives as it perfectly suits the purpose in terms of transparency, usability, and economic costs. Keystroke dynamics are security features that have the potential to help strengthen online assessment security by mitigating cheating and authenticating identity (Ahmed & Traore, 2007; Giot, Dorizzi, & Rosenberger, 2015; Miguel, Caball, Xhafa, & Prieto, 2015; Sewell, Frith, & Colvin, 2010).

User Identification and Authentication in OLE

Identity is a term that reflects uniqueness, sameness and distinctness. Hence, when an online assessment security system asks an answer to the "who you are?" question; it simply requires that the student provides a unique response which distinguishes him/her from every other student.

In e-assessment, it is insufficient to assume correctness of a student based only on an identity. The e-assessment security system requires prove that the identity claimed belongs to the owner who stored the information. Hence, when the security system asks an answer to the "is it really you?" question; it simply requests an evidence of the claimed identity.

Establishing students' identity and participation in their OLE is a challenging task in online courses where the instructors and students do not meet face to face. Identity verification involves matching the legitimate user's identity to a real-world identity, and Identity authentication confirms the person who completes the coursework is the same student who originally enrolled.

The currently existing four primary methods of user authentication are: 1) Knowledge factors, or something unique that the user knows; 2) Ownership factors, or something unique that the user has; 3) something unique that the user is, and 4) something unique that the user does (Anderson, 2008). However, when considering online examinations, each of these methods has several drawbacks.

In the keystroke dynamic process, there is thus a distinction between user identification and authentication. In the first case, the objective is to identify a user, typically with respect to an enrolment database of user keystroke templates (profiles). For example, in an access control situation where all users key-in the same access code, keystroke static authentication will be applied to identify the user as an extra security precaution. If the keyboard pattern is not in our database access can be denied. In the case of user authentication, we wish to check that the user is who they say they are, not to identify them.

User verification and authentication in the online environment is more challenging than what is required in many web-scale platforms such as online banking and shopping as the user never benefits when another person successfully uses their identity. In an OLE, the students may want others to complete their work for them, and thus may plan with unauthorized users to cheat the identity system.

Framework of Continuous Biometric Authentication Model

Continuous biometric authentication model shown in Figure 3 is a kind of intrusion detection system which monitors a series of events, with the aim of detecting unauthorized activities through the continuous monitoring of a user session.

There are six basic components are required to describe this model.

1. Initiators of activity on this model either user or administrator.
2. Sensor device collects keystroke data from the user and translates it into a signal that can be read by an observer or keyboard.
3. Feature extraction process will select optimum features and represent the user actions.
4. The detector compares profiles by analyzing keystroke data and then performs measurements for errors that may detect an unauthorized user. False acceptance rate and False rejection rate are used to determine the accuracy of the detector.

Figure 3. Continuous biometric authentication model

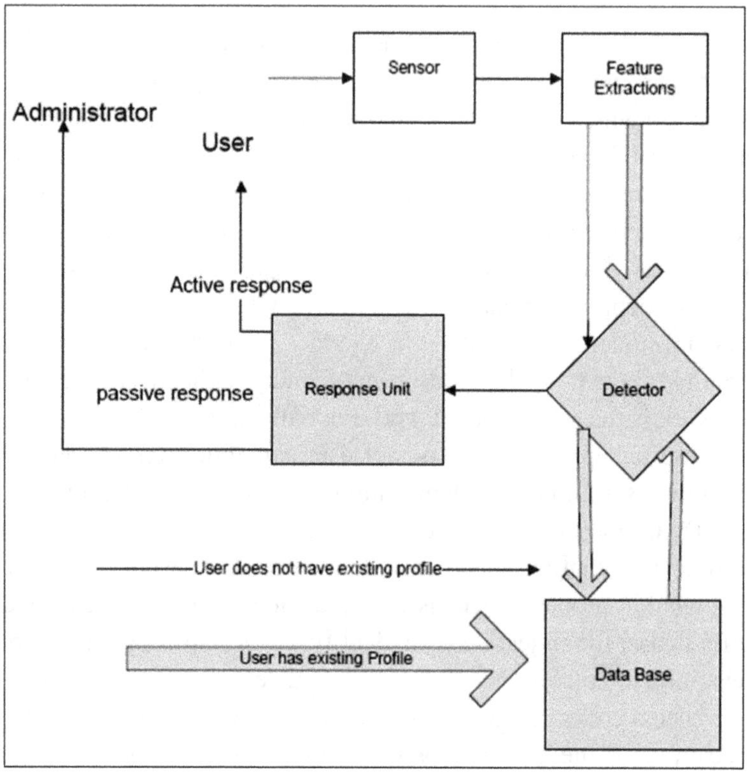

5. Biometric database stores user profiles during user registration phase. Then the continuous biometric system will use the database for comparison with the live data in the verification phase.
6. Response unit will take an appropriate response for detecting an unauthorized user.

Framework of Secured Biometric Authentication in Online Examination

To determine whether an e-assessment is secure, both from students' as evaluators' point of view, it is necessary to consider some system, account and test content security considerations in the planning process for introducing assessments into the online learning environment. Students should have an authorization ticket which contains their log-in credentials to access the assessment. During testing a proctor will use the online management system. Test proctors are provided own unique user ID and password to the online management system for administering tests. All screens data and test content a within this system should be treated as securely and confidentially. In the online management system, the proctors and test administrators can make groups and testing sessions in advance. The assessments should have seal codes which will restrict student's access to specific sections without obtaining the appropriate code.

The framework biometric authentication in online examination shown in Figure 4 involves three stages.

Step 1: Student Identification: The system will check the identity of the student by using keystroke dynamics biometrics before entering the exam. This will also check whether the student is eligible for that exam. It also involves the examination of the data quality; if it is insufficient, the data are rejected immediately, or appropriate user guidance is given to improve the quality.

Step 2: Exam Domain Login: The student will log into the exam domain with the credentials provided by the university domain login. If the username and password and profile information are correct, then the user will be able to log into the exam.

Step 3: Online Exam Proctor (Supervisor) Password: The supervisor password is given to the students who are successfully logged into the exam domain, for accessing their session and attending the specific exam.

Figure 4. Framework of secured biometric authentication in online examination

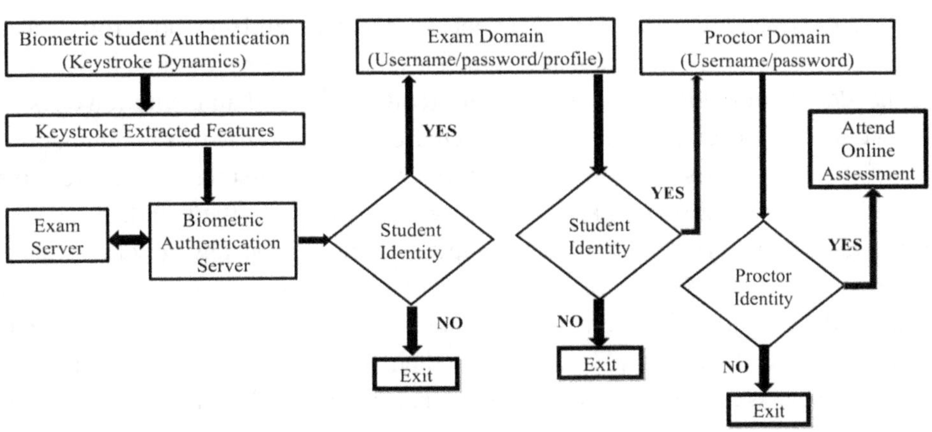

Online assessment security is aimed at decreasing the actual and probable occurrences of cheating (Caldarola & MacNeil, 2009) as well as securing an institution against cheating by students (Graf, 2002). Keystroke dynamics aids in preventing cheating by monitoring and analyzing a user's keyboard inputs based on his or her habitual rhythm typing patterns (Monrose & Rubin, 1997). Since its inception, keystroke dynamics has shown to be valuable in authenticating users (Ahmed & Traore, 2007; Flior & Kowalski, 2010; Graf, 2002; Gunetti & Picardi, 2005; Joyce & Gupta, 1990; Killourhy & Maxion, 2009), especially for the cost (Ali, Tappert, & Qiu, 2015; Gunetti & Picardi, 2005; Guven & Sogukpinar, 2003; Jenkins, Grimes, Proudfoot, & Lowry, 2014; Monrose & Rubin, 1997; Yu & Cho, 2004).

Applications of Keystroke Dynamics in the Online Learning Environment

Whilst cheating and plagiarism in education are not new phenomena, technology is widely seen as having facilitated its increase, however as part of a multi-faceted approach to addressing the issue, technology may also have an important role in preventing cheating and plagiarism. HEOA Act demand institutions ensure appropriate technologies are in place to guarantee a student's ID. In addition to the governmental regulations, accrediting bodies have made this a part of their accreditation practices.

TeSLA project (Trust-based e-assessment System for Learning) is a Horizon 2020 project, involving an association of technological and educational institutions, together with experts in data privacy and in quality assurance (Noguera et al. 2016). TeSLA aims to integrate a system for student authentication and authorship checking within institutional OLEs for addressing cheating in e-assessment in higher education. TeSLA uses a variety of biometrics including Face recognition, voice recognition, and keystroke dynamics to assure student authentication and uses Forensic Analysis (for writing style) and Plagiarism Detection to assure authorship. TeSLA does not use biometric techniques that require special devices such as digital fingerprint readers or high-definition cameras for iris recognition but rather relies on the use of commonly available devices such as webcam, keyboard, and microphone. TeSLA plug-in is made available in the online educational platform, Moodle, which is directly integrated into the most used assessment activities such as assignment, forum, and quiz. When assessment activities are being designed, the available TeSLA biometrics are presented as a list and the teacher can select the most appropriate TeSLA biometrics for that activity.

The Southern Association of Colleges and Schools standard 4.8.1 asserted that an institution offering distance or correspondence education needs to demonstrate that the student who enrolls in a course needs to be the same student who participates in and completes the course or program and receives the credit by verifying the identity of a student who participates (SACS, 2012). If an institution does not meet this standard, the likelihood of receiving accreditation decreases. In response to this standard, institutions like Coursera, a social entrepreneurship company that partners with universities to offer free courses online (Vrankuli, 2013), applies keystroke dynamics to verify the identities of their students by creating a biometric profile of their unique typing patterns by typing a short phrase. During the coursework submission, the student should enter the same short phrase which will be compared to their recorded typing samples for authenticating their identity. Coursera, for example, used an enterprise contract and an institutional strategy to track every key typed (KeyTrac, 2016) by their fifteen million-plus student users (EdSurge, 2015). But this approach is expensive for course providers and it caps the number of authentications and identifications that can be completed, limiting the effectiveness of potential continuous authentication.

Universities, e-learning providers and proctoring services often need a reliable solution to assure a student registering for a course is the same student who participates in the course or receives course credit. Currently, the students are verified with validating the ID cards through a webcam surveillance-session and personnel intensive techniques. Institutions can build their own biometric authentication solution which delegates the entire legitimation process to the student with KeyTrac facility. During this authentication process, the student is forced to type in a short text snippet into a textbox where all keystroke dynamics gets recorded. Then a student is redirected to the exam page. The recorded typing rhythm gets evaluated against the stored profile of a student. If the security check succeeds, the student can take the exam. Otherwise, the student is not allowed to appear for the exam.

Open Courseware, in many ways, was the starting point towards mainstream discussion and adoption of open learning, particularly in higher education. The MOOCs have emerged in recent years as a new way to educate students based on open, free, distributed and participatory courses, which provide certification to students. The online courses break with the barriers associated with traditional lessons and provide a new widely accessible education over the internet.

MIT and Harvard are also using keystroke biometric with their massively open online courses (MOOCs), by requiring students to type out a preselected phrase and then using that phrase to verify a student's identity (Talavera, 2014). Using keystroke dynamics to mitigate fraudulent academic activity and to ensure student identification is wise for an institution of higher learning to incorporate.

However, there is a discussion among all the academic sectors about the new educational scenario proposed by MOOCs. One of the controversial discussions comprises the challenges related to the "certificates" or "statements of completion" of courses without classroom attendance. The widely and accessible nature of the MOOCs increase the vulnerability of the systems and the authentication of the students emerges as a challenging and unsolved task. Researchers and MOOCs instructors have made efforts to propose and analyze convenient authentication approaches for the reliable authentication for the future of online education (Miguel, Caballe, & Prieto, 2013). The proposed methods show that biometric technologies seem to be the most attractive solutions as it allows authenticating users based on "something that users are" instead of "something that users know" such as PIN or passwords.

Keystroke dynamics are interesting for MOOCs authentication because of its transparency and it runs in the background without requiring explicit user interaction, continuous in which authentication can be performed over all the user activity, not only based on an initial access to the platform. Transparent authentication is strongly related to the user experience and usability of the platform. The authentication service must be simple and not affect the normal activity of the MOOC. As an authentication system based on PIN codes assumes that the user will not provide his/her code to any other user, which is critical for continuous authentication. That assumption is not valid in a scenario in which the user could be interested in providing his/her PIN code to another user to pass an exam.

EdX (5 million-plus users) and Udacity (four million-plus users), competitors of Coursera, do not employ keystroke dynamics—opting for face-to-face discussions and competency interviews. They do this partially because they believe the qualitative assessment gives them a better feel of authenticity than keystroke dynamics can, and because they have the money to do so. While not the choice of all institutions, keystroke dynamics is suggested as the most reasonable and cost-effective approach for higher-education institutions looking to improve their online assessment security (Monrose & Rubin, 1997).

Over the years there have been several different approaches on how to classify the rhythm a user has when typing on a keyboard. These include, among many others, machine learning techniques, statistical and distance measurements. Some of these methods, mainly in the machine learning area can have a feasibility problem in a way that they may need a lot of time and computer resources to be trained when models are big, something that it may not be always available in certain environments. Related to statistics is also the use of probabilities to classify keystrokes. The use of t-tests (Ahmed & Traore, 2005) and chi-squared (Jin, Andrew, Thian, &Tee, 2008) methods have also been explored. Clustering methods, like k-means and fuzzy c-means, have also been used with promising results (Eltahir, Salami, Ismail, & Lai, 2004; Miller,1994). According to (Chudá & Durfina, 2009; Gunetti and Picardi, 2005), especially when dealing with free text, the use of distance measurements tends to be the preferred technique.

Statistical techniques have all been used in attempts to increase the accuracy and effectiveness of keystroke authentication with data collection via a local network or server arrangement. (Bleha, Slivinsky, & Hussien, 1990; Joyce & Gupta, 1990; Monrose & Rubin, 1997; Obaidat & Sadoun, 1997). These studies results support the application of keystroke authentication in the online learning system.

(Thanganayagam & Thangadurai, 2015) proposed a method to strengthen the password authentication by introducing an additional layer of keystroke patterns used for authentication. Firstly, extract keystroke features from our database. Then calculate the mean and standard deviation of keystroke features to form the template. Hybrid model based on the combination of Gaussian probability density function (GPDF) and support vector machine (SVM) will convert test features into scores. Lastly, four fusion rules are applied to improve the result by fusing the GPDF and SVM scores. Best result with equal error rate of 1.612% is obtained with their database.

(Clarke & Furnell, 2007) investigated the use of keystroke dynamics in the application of mobile phones. They made use of the numerical keypad on many mobile phones before touch screens were introduced and tested several neural network-based analysis techniques: feed-forward MLP (FF MLP); radial basis function (RBF); and generalized regression neural networks, finding the FF MLP network to be the most stable and useful in this case.

The two literature investigates the acquisition of unimodal databases of keystroke dynamics and mouse dynamics. In (I. Traore et. al, 2012), 24 students were asked to perform a fixed task on their own computer. The classifier had an EER (Equal Error Rate) of 8.21%. The mouse dynamics isolated had EER of 22.41% and the keystroke dynamics isolated had EER of 24.78%. In (Mondal & Bours, 2016), 25 volunteers participated in the experiment, four different machine learning approach was used, Decision Tree (DT), Counter-Propagation Artificial Neural Network (CPANN), ANN and SVM. In this work, an identification accuracy was obtained of 62.2% in a closed-set experiment and a Detection and Identification Rate of 58.9% in an open-set experiment.

Keystroke dynamics research having been carried out on mobile phones due to high demands of advanced security in the modern technology. For example, numerical keypad systems (Mantyjarvi, Koivumaki, & Vuori, 2002; Grabham & White, 2008); button-based (Clarke & Furnell, 2007; Campisi, Maiorana, Lo Bosco, & Neri, 2009; Hwang, Cho, & Park, 2009a); and touch-based devices (Mantyjarvi, 2002; Saevanee & Bhatarakosol, 2008); and also web-based systems (Bleha, Slivinsky, & Hussien, 1990; Obaidat & Macchairolo, 1993; Curtin et al., 2006). It should be noted that research undertaken in this field tends to make use of different sets of data. The difference in methodologies provides a challenge for making direct comparisons among papers using the stated error rates alone.

Issues of Keystroke Dynamics Biometrics in the Online Learning Environment

Specifically, for a virtual learning environment, despite the anticipated benefits of online learning, security remains one of the major issues and a threat to the success of online learning (Alwi & Fan, 2010). Online learning offers more opportunities for academic dishonesty from remote locations than traditional face-to-face learning and assessment (Jung & Yeom, 2009). With an advent of technology and Internet, the unethical conducts (Alwi & Fan, 2010) have increased in online learning due to more opportunities for cheating in the online examination. Consequently, the threat level is higher for the online environment because it has increased the opportunities for deception due to non-rigorous authentication management.

The Higher Education Opportunity Act (HEOA 2008) demands that institutions offering online education should assure that students of record are those accessing the systems and taking online exams. The Act requires that institutions must follow any one of the methods such as use of secure login and password; conduct proctored examinations; and apply new technologies and practices that are effective in verifying student identification. To adhere this act, many institutions follows secure student login and password credentials through Learning Management Systems (LMS). While this meets HEOA regulation, basic LMS credentials are not an effective combatant against academic integrity violations as these credentials still do not document that the student who is authenticated is the one who is taking the exam. For following HEOA regulation to ensure a secure examination environment in online education, it is better to have a combination of live proctoring along with identity verification tools.

The major problem in this domain is lack of standardized protocol for keystroke system evaluation for providing accurate results and doing the comparisons. The experiments on keystroke dynamics may also face a risk of collecting bad samples. For instance, user's typing may vary substantially after a period, and their typing behavior may change not due to stress, but caused by other factors such as sleepiness, change of computer hardware and software, use of virtual keyboard, hardware defective, injury of hand or finger, or the user simply using a voice-to-text converter (instead of typing). Different keyboard designs may also affect the typing behavior too, and different keyboard types (qwerty or Dvorak) may also affect the classification algorithm designs (Kang & Cho, 2007). The variation of keyboard design parameters like the distance between keys, size of keys and requirement of pressure, are built to accommodate users with different physical abilities. Hence people with especially large or small hands may have difficulty in using standard keyboards. The system will have error rates because of the above variations, but keystroke dynamics will provide a valid solution to avoid the error rates.

Keyboard device only allows a small amount of information to be retrieved, so insufficient data collection could lead to less accurate data. For instance, a research by (Kang & Cho, 2007) shows that equal error rate would increase when text size or number of keystroke characteristics decreases. Therefore, most of the research choose to use fixed text analysis to ensure a minimum number of keystrokes per sample to produce good results in their building models.

For the text analysis, there are also some issues to consider, especially the length of the text samples. If it is too lengthy, then users will consider it a nuisance; if it is too short, then it will reduce the accuracy of the classifier. In addition, (Monrose & Rubin, 2000) argue the limitation of using free-text recognition, of which the input is unconstrained, i.e. the user may be uncooperative and environmental parameters are uncontrolled. (Revett et al., 2007) also observed that the experiment participants claim that periodic checking of their typing style is obtrusive and considered an unacceptable invasion of their privacy. Therefore, full consideration must be taken on the length of the text inputs, and participants must be briefed clearly on the expectations before they voluntarily take the roles in the OLEs.

SOLUTIONS AND RECOMMENDATIONS

A quick Google search for keystroke dynamics software yields multiple decently effective software packages available for free download. Of course, these free versions are basic packages limited to small quantities of users with more users available for a higher cost. For institutions of higher learning where users will most likely be in the thousands, an enterprise license will likely need to be negotiated, depending on specific institutional needs and goals.

Although we may not be able to solve the external problems that we couldn't control, such as the physical wellness and lifestyle of the users, we could take into considerations detecting a learner's states when he is doing e-learning using objective measurement. This will help the teacher or the adaptive system to analyze the effectiveness of the learning, to identify potential stress created by the design of the system or the learning materials. The objective measurement of stress could include keystroke dynamics-based analysis with other techniques. For instance, use of machine learning techniques allows the system to learn the user's typing behavior over time. We also propose a solution that combines both keystroke dynamics and text analysis, with mouse movement analysis since the mouse is also an essential part of computer hardware, which is also non-obtrusive and not costly. Most importantly, some researchers use this technique to model user behavior, and it is proven as good as keystroke dynamics.

There are some important implications for including keystroke dynamics with the online assessment that need to be addressed because it would force some important behavioral changes. For example, there would be greater awareness by students that they need to do their own authentic work. Knowing ahead of time that the system will be determining whether the student is answering the questions provides a deterrent effect, impressing on the students that the work must be their own (Flior & Kiwolski, 2010).

Additionally, administrators and instructors would have more concise evidence on where to focus their administrative resources. Keystroke dynamics reporting would flag problem areas (i.e., who is potentially cheating) allowing school administration to more precisely allocate scarce human resources. Using keystroke dynamics in online exams would alert administrators and instructors to areas of concern as well as give them a reason to suspect that the student who wrote the examination is not the student registered in the class. Armed with such evidence, administrators and instructors would potentially be able to mitigate the issue more meaningfully for all parties involved.

Another implication for including keystroke dynamics with online assessment is the saving of time. Students could theoretically take their exams in a non-proctored environment—freeing up both student and instructor time (Alexander, Bartlett, Truell, & Ouwenega, 2001). Online testing is convenient for students and instructors.

In knowing that a keystroke dynamics system is inexpensive to implement, practitioners can work with their local IT departments to develop and house a similar tool. If these institutions do not have the resources, they could partner with other institutions to work together to fund the project or apply for grant funds. Once implemented, keystroke dynamics would provide a clearer picture of possible cheaters and where it was occurring in a course. This would allow for a more focused allocation of resources, giving institutions a better understanding of potentially how much effort, time, and money is being spent on academic honesty issues.

Implementation of keystroke dynamics system also makes the need for policy development paramount. It would be an opportunity for the institution to develop the necessary administrative checks and balances that need to be in place so that management of the system is clear and meaningful.

Keystroke dynamics provide a greater sense of online assessment security as well as closer alignment with federal law. This does not mean that institutions of higher learning should abandon vigilance and oversight once keystroke dynamics are employed. It does mean, however, that there can be a greater sense of confidence that the students completing their courses online are the same ones who are supposed to be doing it. This sense of confidence should bring a stronger commitment to integrity and therefore provide assurance to an accrediting body, of the institution's efforts to uphold the quality of the learning, academic fidelity, and the authenticity of the experience by the individuals receiving credit for the work is complete.

FUTURE RESEARCH DIRECTIONS

Keystroke dynamics has received increasing attention in biometric user authentication research owing to its easy extensibility, and for other behavioral analysis tasks such as emotion prediction (Nahin, Alam, Mahmud, & Hasan, 2014; Kołakowska, 2013) and gender identification (Fairhurst, & Costa-Abreu, 2011).

Combining keystroke dynamics with online assessment security, while not without its problems, is potentially an inexpensive and impactful tool that higher education institutions could look more closely to help mitigate academic fraud. Implementing keystroke dynamics in online assessments is likely the next step for online education. Each institution needs to be aware of its assessment goals and strategies and how keystroke dynamics can help reach them. Administration and IT departments should work closely to develop clear and meaningful policies and execution strategies.

keystroke dynamics is a very secure authentication method when combined with other security methods. Keystroke dynamics could be paired with mouse dynamics to increase user authentication information data and improve the authentication process. Use of signature based on selected mouse movement characteristics is an emerging trend in mouse dynamics (Ahmed & Traore, 2007). Combining keystroke dynamics with mouse dynamics would increase the complexity of input data for analysis but surely limit the possibility of similar inputs for two different identities (Rybnik, Tabedzki, & Saeed, 2008). Having these two data points would potentially improve imposter detection accuracy percentages as well.

From the existing literature, there are two main gaps. First, there seems to be minimal research on authenticating an individual while they are participating in common assessment typing situations, such as free typing. Second, there seems to be minimal research on utilization of keystroke dynamics with typing authentication where injuries or impediments exist. Critically important to being able to address these gaps is knowing how these contexts affect keystroke dynamics verification in free typing generative list task and free typing explanatory activities as well as with minor and moderate injuries and impediments.

The existing researches suggest that keystroke dynamics can potentially lead to increased online assessment security. More exploration is needed, but what is clear is that while keystroke dynamics has been around for a while, little is published about its use in education journals. This means there is an opportunity for further research in this arena, in journals that reach the education audience, bringing more attention to the discipline and the topic. Also, more investigations should be explored in continuous authentication problems; typing rhythm of individuals and design of robust authentication systems.

Overall, keystroke dynamics has not been used extensively in online education and is still unproven in this arena. That will likely change as this important area of online assessment security is more thoroughly vetted and the value and significance of keystroke dynamics are realized (Giot, Dorizzi, & Rosenberger, 2015; Miguel, Santi, Fatos, & Josep, 2015).

With the number of touch-based systems and tablet computers increasing rapidly in the last few years, such research could help to create a universal signature that could be used across platforms without the need for multiple input data to each sensor. However, user authentication is also a critical issue for smart devices, such as smartphones or tablets. Thus, in future, the Keystroke dynamics-based user authentication should be verified using keystroke data collected from other input devices and keystroke data collected from various languages should also be tested. Future keystroke analysis authentication tools should take to the Internet as web-based security systems for aiding in the security of OLEs. For such systems to work effectively, they need to be able to complete keystroke analysis not only on traditional keyboards but also on touch-based devices. The investigation into the relationship between keystroke signatures obtained with traditional keyboards and those captured with touch-based systems could prove extremely useful. With the number of touch-based systems and tablet computers increasing rapidly in the last few years, such research could help to create a universal signature that could be used across platforms without the need for multiple input data to each sensor. This research could lead to the development of such web-based keystroke analyses tools being a great deal more flexible in their use and ability.

CONCLUSION

A keystroke dynamics method is convenient for users and security departures since keystroke dynamics monitoring is made in a hidden mode and does not requires any extra actions from users. One of the main advantages of this method is low cost of necessary hardware, because almost every computer has a keyboard. The concept of keystroke dynamics is not limited to the traditional keyboard but any interface in which keys must be pressed can benefit from similar techniques. Keystroke biometrics has also shown great potential as the features can be collected without the need for special hardware. That's why, the major contributions of this chapter come in the form of insight into the applications of keystroke dynamic in the OLE. The researchers believe that more advanced attacks can happen in the real world. The researchers, therefore, want to evaluate the security of keystroke dynamics against advanced attacks, by developing a program that helps attackers learn others typing characteristics. The domain of keystroke dynamics is still an emerging field, where most of the challenges need to be overcome for it to become an effective biometric. While the keystroke dynamics applications provide new features and benefits to the OLE, which will ensure that security issues in online learning can be solved by using keystroke dynamics with additional knowledge-based authentication. Keystroke dynamics policy development and implementation could use additional study, including the articulation of keystroke dynamics policy best practices in the OLE. This policy development will mature as authentication thresholds are tested, verified, and applied in real-life situations.

REFERENCES

Ahmed, A., & Traore, I. (2005). *Detecting Computer Intrusions Using Behavioral Biometrics. Technical Report*. University of Victoria.

Ahmed, A., & Traore, I. (2007). A new biometric technology based on mouse dynamics. *IEEE Transactions on Dependable and Secure Computing*, *4*(3), 165–179. doi:10.1109/TDSC.2007.70207

Ahmed, A., & Traore, I. (2014). Biometric recognition based on free text keystroke dynamics. *IEEE Transactions on Cybernetics*, *44*(4), 458–472. doi:10.1109/TCYB.2013.2257745 PMID:23757560

Alexander, M. W., Bartlett, J. E., Truell, A. D., & Ouwenga, K. (2001). Testing in a computer technology course: An investigation of equivalency in performance between online and paper and pencil methods. *Journal of Career and Technical Education*, *18*(1), 69–80. doi:10.21061/jcte.v18i1.600

Ali, L., Tappert, C., & Qiu, M. (2015). Authentication and identification methods used in keystroke biometric systems. *International Conference on High Performance Computing and Communications (HPCC)*, 1424-1429. 10.1109/HPCC-CSS-ICESS.2015.66

Alwi, N. H., & Fan, I. S. (2010). Threats analysis for e-learning. *International Journal of Technology Enhanced Learning*, *2*(4), 358–371. doi:10.1504/IJTEL.2010.035738

Anderson, R. (2008). *Security Engineering: A Guide to Building Dependable Systems*. Indianapolis, IN: Wiley Publishing, Inc.

Antal Margit & Nemes Lehel. (2016). The MOBIKEY Keystroke Dynamics Password Database: Benchmark Results. In *Software Engineering Perspectives and Application in Intelligent Systems* (pp. 35–46). Boston, MA: Springer.

Benjamin, M. (1994). Vital signs of identity. *IEEE Spectrum*, *31*(2), 22–30. doi:10.1109/6.259484

Bergadano, F., Gunetti, D., & Picardi, C. (2002). User Authentication through Keystroke Dynamics. *ACM Transactions on Information and System Security*, *5*(4), 367–397. doi:10.1145/581271.581272

Bleha, S., Slivinski, C., & Hussien, B. (1990). Computer-Access Security Systems Using Keystroke Dynamics. *IEEE Transactions on Pattern Analysis and Machine Intelligence*, *12*(12), 1217–1222. doi:10.1109/34.62613

Bours, P. (2012). Continuous keystroke dynamics: A different perspective towards biometric evaluation. *Journal of Information Security of Tech. Report*, *17*(1), 36–43.

Brown, M., & Rogers, J. (1993). User Identification via Keystroke Characteristics of Typed Names using Neural Networks. *International Journal of Man-Machine Studies*, *39*(6), 999–1014. doi:10.1006/imms.1993.1092

Caldarola, R., & MacNeil, T. (2009). Dishonesty deterrence and detection: How technology can ensure distance learning test security and validity. *Proceedings of the 8th European Conference on E-Learning*, 108-115.

Campisi, P., Maiorana, E., Lo Bosco, M., & Neri, A. (2009). User authentication using keystroke dynamics for cellular phones. *IET Signal Processing*, *3*(1), 333–341. doi:10.1049/iet-spr.2008.0171

Checco. (2006). Keystroke Dynamics and Corporate Security. *WSTA Ticker Magazine*.

Clarke, N. L., & Furnell, S. M. (2007). Authenticating mobile phone users using keystroke analysis. *International Journal of Information Security*, *6*(1), 1–14. doi:10.100710207-006-0006-6

Curtin, M., Tappert, C., Villani, M., Ngo, G., Simone, J., & Fort, H. S. (2006). Keystroke biometric recognition on long-text input: a feasibility study. *Proceedings of the International Workshop Scientific Computing and Computational Statistics (IWSCCS)*.

Daniela, C., & Michal, D. (2009). Multifactor authentication based on keystroke dynamics. In *Proceedings of the International Conference on Computer Systems and Technologies and Workshop for PhD Students in Computing*. ACM.

Dowland, P. S., Furnell, S. M., & Papadaki, M. (2002). Keystroke Analysis as a Method of Advanced User Authentication and Response. In A. El-Hadidi (Ed.), *Ghonaimy* (pp. 215–216). Security in the Information Society. doi:10.1007/978-0-387-35586-3_17

EdSurge. (2015). *Udacity, Coursera and edX now claim over 24 million students*. Retrieved September 17, 2018, from https://www.edsurge.com/news/2015-09-08-udacity-coursera-and-edx-now-claim-over-24-million-students

Elsadig, Salami, Faris, & Kin. (2004). Dynamic keystroke analysis using AR model. *IEEE International Conference Industrial Technology*, 3, 1555–1560.

Fairhurst, M., & Costa-Abreu, M. D. (2011). Using keystroke dynamics for gender identification in social network environment. *4th International Conference on Imaging for Crime Detection and Prevention*, 1-6. 10.1049/ic.2011.0124

Flior, E., & Kowalski, K. (2010). Continuous biometric user authentication in online examinations. *Proceedings of 7th International Conference on Information Technology: New Generations*, 488–492. 10.1109/ITNG.2010.250

Freidman, J. (2016). *Study: Enrollment in Online Learning Up, Except at For Profits*. Retrieved September 17, 2018, from https://www.usnews.com/education/online-education/articles/2016-02-09/study-enrollment-in-online-learning-up-except-at-for-profits

Gikandi, J., Morrow, D., & Davis, N. (2011). Online formative assessment in higher education: A review of the literature. *Computers & Education*, *57*(4), 2333–2351. doi:10.1016/j.compedu.2011.06.004

Giot, R., Dorizzi, B., & Rosenberger, C. (2015). A review on the public benchmark databases for static keystroke dynamics. *Journal of Computer Security*, *55*(C), 46–61. doi:10.1016/j.cose.2015.06.008

Giot, R., Hemery, B., & Rosenberger, C. (2010). Low cost and usable multimodal biometric system based on keystroke dynamics and 2D face recognition. *Proceedings of the International Conference on Pattern Recognition*, 1128–1131. 10.1109/ICPR.2010.282

Grabham, N. J., & White, N. M. (2008). Use of a novel keypad biometric for enhanced user identity verification. *IEEE International Conference on Instrumentation and Measurement Technology*, 12–16. 10.1109/IMTC.2008.4546995

Graf, F. (2002). Providing security for eLearning. *Journal of Computers & Graphics*, *26*(2), 355–365. doi:10.1016/S0097-8493(02)00062-6

Grijalva, T., Nowell, C., & Kerkvliet, J. (2006). Academic honesty and online courses. *College Student Journal*, *40*(1), 180–185.

Gunetti, D., & Picardi, C. (2005). Keystroke Analysis of Free Text. *ACM Transactions on Information and System Security*, *8*(3), 312–347. doi:10.1145/1085126.1085129

Guven, A., & Sogukpinar, I. (2003). Understanding users' keystroke patterns for computer access security. *Journal of Computer Security*, *22*(8), 695–706. doi:10.1016/S0167-4048(03)00010-5

Higher Education Opportunity Act (HEOA) of 2008. (n.d.). Retrieved September 17, 2018, from https://www2.ed.gov/policy/highered/leg/hea08/index.html

Hwang, S. S., Cho, S., & Park, S. (2009a). Keystroke dynamics-based authentication for mobile devices. *Computers & Security*, *28*(1), 85–93. doi:10.1016/j.cose.2008.10.002

Janakiraman, R., & Sim, T. (2007). Keystroke Dynamics in a General Setting. In S. W. Lee & S. Z. Li (Eds.), Lecture Notes in Computer Science: Vol. 4642. *Advances in Biometrics*. Berlin: Springer. doi:10.1007/978-3-540-74549-5_62

Jenkins, J. L., Grimes, M., Proudfoot, J. G., & Lowry, P. B. (2014). Improving password cybersecurity through inexpensive and minimally invasive means: Detecting and deterring password reuse through keystroke-dynamics monitoring and just-in-time fear appeals. *Journal of Information Technology for Development*, *20*(2), 196–213. doi:10.1080/02681102.2013.814040

Jin, L., Ke, X., Manuel, R., & Wilkerson, M. (2004). Keystroke Dynamics: A Software based Biometric Solution. *13th USENIX Security Symposium*.

Jorge, M. M., Caballé, S., Xhafa, F., & Prieto-Blazquez, J. (2015). Security in online web learning assessment. *World Wide Web (Bussum)*, *18*(6), 1–24.

Joyce, R., & Gupta, G. (1990). Identity authentication based on keystroke latencies. *Communications of the ACM*, *33*(2), 168–176. doi:10.1145/75577.75582

Jung, I. Y., & Yeom, H. Y. (2009). Enhanced security for online exams using group cryptography. *IEEE Transactions on Education*, *52*(3), 340–349. doi:10.1109/TE.2008.928909

Kang, P., Hwang, S., & Cho, S. (2007). Continual Retraining of Keystroke Dynamics Based Authenticator. In S. W. Lee & S. Z. Li (Eds.), *Advances in Biometrics* (Vol. 4642). Springer. doi:10.1007/978-3-540-74549-5_125

Karnan, M., Akila, M., & Krishnaraj, N. (2011). Biometric personal authentication using keystroke dynamics: A review. *Journal of Applied Soft Computing*, *11*(2), 1565–1573. doi:10.1016/j.asoc.2010.08.003

KeyTrac. (2016). *How KeyTrac records your typing behavior*. Retrieved September 17, 2018, from https://www.keytrac.net/en/technology

Khalaf, H. A., & Nouman, A.-H. M. (2013). Multifactor Strong Authentication Method Using Keystroke Dynamics. *International Journal of Science and Applied Information Technology*, *2*(2), 31–34.

Khan, B. H. (1998). Web-based Instruction (WBI): An Introduction. *Educational Media International*, *35*(2), 63–71. doi:10.1080/0952398980350202

Khan, B. H. (2000). Discussion of resources and attributes of the web for the creation of meaningful learning environments. *Cyberpsychology & Behavior*, *3*(1), 17–23. doi:10.1089/109493100316193

Killourhy, K., & Maxion, R. (2009). Comparing anomaly-detection algorithms for keystroke dynamics. *International conference on Dependable Systems & Networks*, 125-134.

King, G., Guyette, W., & Piotrowski, C. (2009). Online exams and cheating: An empirical analysis of business students' views. *Journal of Educators Online*, *6*(1), 1–11. doi:10.9743/JEO.2009.1.5

Kołakowska, A. (2013). A review of emotion recognition methods based on keystroke dynamics and mouse movements. *6th International Conference on Human System Interactions (HSI)*, 548-555. 10.1109/HSI.2013.6577879

Leggett, J., & Williams, G. (1988). Verifying identity via keystroke characteristics. *International Journal of Man-Machine Studies*, *28*(1), 67–76. doi:10.1016/S0020-7373(88)80053-1

Leggett, J., Williams, G., Usnick, M., & Longnecker, M. (1991). Dynamic identity verification via keystroke characteristics. *International Journal of Man-Machine Studies*, *35*(6), 859–870. doi:10.1016/S0020-7373(05)80165-8

Li, C., & Irby, B. (2008). An overview of online education: Attractiveness, benefits, challenges, concerns and recommendations. *College Student Journal*, *42*(2), 449–458.

Mantyjarvi, J., Koivumaki, J., & Vuori, P. (2002). Keystroke recognition for virtual keyboards. *Proceedings - IEEE International Conference on Multimedia and Expo*, *2*, 429–432. doi:10.1109/ICME.2002.1035630

Marsters, J. D. (2009). *Keystroke dynamics as a biometric* (Doctoral dissertation). Retrieved September 17, 2018, from https://eprints.soton.ac.uk/66795/

Miguel, J., Caballe, S., & Prieto, J. (2013). Providing information security to MOOC: Towards effective student authentication. In *Proceedings of the IEEE International Conference on Intelligent Networking and Collaborative Systems* (pp. 289–292). IEEE. 10.1109/INCoS.2013.52

Moini, A., & Madni, A. (2009). Leveraging biometrics for user authentication in online learning: A systems perspective. *IEEE Systems Journal*, *3*(4), 469–476. doi:10.1109/JSYST.2009.2038957

Monaco, J., Stewart, J., Cha, S., & Tappert, C. (2013). Behavioral biometric verification of student identity in online course assessment and authentication of authors in literary works. *Proceedings of IEEE Sixth International Conference on Biometrics: Theory, Applications and Systems*, 1-8. 10.1109/BTAS.2013.6712743

Mondal, S., & Bours, P. (2016). Combining keystroke and mouse dynamics for continuous user authentication and identification. *IEEE International Conference on Identity, Security and Behavior Analysis, ISBA*, 1–8. 10.1109/ISBA.2016.7477228

Monrose, F., & Rubin, A. (1997). Authentication via keystroke dynamics. *Proceedings of the 4th ACM Conference on Computer and Communications Security*, 48-56.

Monrose, F., & Rubin, A. (2000). Keystroke dynamics as a biometric for authentication. *Elsevier Future Generation Computer Systems*, *16*(4), 351–359. doi:10.1016/S0167-739X(99)00059-X

Nahin, A. F. M. N. H., Alam, J. M., Mahmud, H., & Hasan, K. (2014). Identifying emotion by keystroke dynamics and text pattern analysis. *Behaviour & Information Technology*, *33*(9), 987–996. doi:10.1080/0144929X.2014.907343

Newton, D. (2015). *Cheating in online classes is now big business*. Retrieved September 17, 2018, from https://www.theatlantic.com/education/archive/2015/11/cheating-through-onlinecourses/413770/

Nick, B. (2005). *Username and password verification through keystroke dynamics* (PhD Thesis). West Virginia University, ProQuest Dissertations Publishing.

Noguera, Guerrero-Roldán, & Rodríguez. (2016). Assuring authorship and authentication across the e-assessment process. *International Computer Assisted Assessment Conference*, 86–92.

Obaidat, M., & Macchairolo, D. (1993). An online neural network system for computer access security. *IEEE Transactions on Industrial Electronics*, 40235–40242.

Obaidat, M., & Sadoun, B. (1997). Verification of Computer User Using Keystroke Dynamics. *IEEE Transactions on Systems, Man, and Cybernetics*, *27*(2), 261–269. doi:10.1109/3477.558812 PMID:18255865

Panchumarthy, R., Subramanian, R., & Sarkar, S. (2012). Biometric evaluation on the cloud: A case study with human-id gait challenge. *Proceedings of the International Conference on Utility and Cloud Computing*, 219-222. 10.1109/UCC.2012.49

Revett, K., Gorunescu, F., Gorunescu, M., Magalhaes, S., & Santos, H. (2007). A machine learning approach to keystroke dynamics based user authentication. *International Journal of Electronic Security Digital Forensics*, *1*(1), 55–70. doi:10.1504/IJESDF.2007.013592

Romain, G., Mohamad, E.-A., Baptiste, H., & Christophe, R. (2011). Unconstrained keystroke dynamics authentication with shared secret. *Computers & Security, Elsevier*, *30*(6-7), 427–445. doi:10.1016/j.cose.2011.03.004

Ruiz, J. G., Mintzer, M. J., & Leipzig, R. M. (2006). The impact of e-learning in medical education. *Academic Medicine*, *81*(3), 207–212. doi:10.1097/00001888-200603000-00002 PMID:16501260

Rybnik, M., Tabedzki, M., & Saeed, K. (2008). A keystroke dynamics-based system for user identification. Proceedings of 7th Computer Information Systems and Industrial Management Applications, 225–230. doi:10.1109/CISIM.2008.8

SACS, Standards for Academic and Student Affairs. (2012). Retrieved September 17, 2018, from http://www.fau.edu/iea/pdf/accreditation/as_i2-1.pdf

Saevanee, H., & Bhatarakosol, P. (2008). User authentication using combination of behavioural biometrics over the touchpad acting like touch screen of mobile device. *IEEE International Conference on Computer and Electrical Engineering*, 82–86.

Sewell, J., Frith, K., & Colvin, M. (2010). Online assessment strategies: A primer. *Journal of Online Learning and Teaching / MERLOT*, *6*(1), 297–305.

Talavera Franco, R. (2014). *Facial biometrics: A solution or a problem for MOOCs assessments?* MOOCstream.

Thanganayagam, R., & Thangadurai, A. (2015). Fusion approach on keystroke dynamics to enhance the performance of password authentication. *International Conference on Electrical, Computer and Communication Technologies*, 1–6. 10.1109/ICECCT.2015.7226123

Traore, I., Woungang, I., Obaidat, M. S., Nakkabi, Y., & Lai, I. (2012). Combining Mouse and Keystroke Dynamics Biometrics for Risk-Based Authentication in Web Environments. *Fourth International Conference on Digital Home*, 138-145. 10.1109/ICDH.2012.59

U.S. Higher Education System. (2016). *In the National Science Foundation*. Retrieved September 17, 2018, from https://www.nsf.gov/statistics/2016/nsb20161/#/report/chapter-2/

Vrankuli, A. (2013). *Coursera looks to verify online student identity with photo, keystroke dynamics*. Retrieved September 17, 2018 from https://www.biometricupdate.com/201301/coursera-looks-to-verify-online-student-identity-with-photo-keystroke-dynamics

Yu, E., & Cho, S. (2004). Keystroke dynamics identity verification—its problems and practical Solutions. *Computers & Security*, *23*(5), 428–440. doi:10.1016/j.cose.2004.02.004

Zhe, J., Andrew, T. S., & Connie, T. (2008). Typing dynamics biometric authentication through fuzzy logic. *International Symposium on Information Technology*, 1–6.

ADDITIONAL READING

McMurtry, K. (2001). E-Cheating: Combating a 21st Century Challenge. *T.H.E. Journal*, *29*(4), 36–41.

Ramu, T., & Arivoli, T. (2013). A Framework of Secure Biometric Based Online Exam Authentication: An Alternative to Traditional Exam. *International Journal of Scientific & Engineering Research*, *4*(11), 52–60.

Ullah, A., Xiao, H., & Barker, T. (2016). *A classification of threats to remote online examinations*. IEEE 7th Annual Information Technology, Electronics and Mobile Communication Conference, (pp. 1-7). 10.1109/IEMCON.2016.7746085

Yassine, K., & El-Sabagh, H. (2017). A Novel Authentication Scheme for E-assessments Based on Student Behavior over E-learning Platform. *International Journal of Emerging Technologies in Learning*, *12*(4), 62–89. doi:10.3991/ijet.v12i04.6478

KEY TERMS AND DEFINITIONS

Authentication: The process of recognizing a user's identity.

Biometrics: The measurement and statistical analysis of people's unique physical and behavioral characteristics.

Digraphs: It is a combination of two letters representing one sound.

Distance Metrics: It is a generalization of the concept of physical distance used to measure the similarity or regularity among the points.

Dwell Time: The latency between the key pressing and releasing of a key.

E-Learning: Learning system based on formalized teaching but with the help of electronic media, typically on the internet.

Equal Error Rate (EER): An algorithm used to predetermine the threshold values for its false acceptance rate and its false rejection rate in a biometric security system.

False-Acceptance Rate (FAR): The percentage of imposters incorrectly matched to a valid user's biometric.

False-Rejection Rate (FRR): The percentage of incorrectly rejected valid users.

Fixed Text: Structured text to perform the analysis in a constrained way.

Flight Time: The period of silence between successive keystrokes events.

Free Text: Users are free to type whatever they want.

Hold Time: The mean time between key press and release for a particular key.

ICT: The infrastructure and components that enable modern computing with the help of information and communications technology.

Identification: The biometric system process in which a person can be identified by performing a one-to-many (1:n) search against the entire enrolled population.

Keyprint Profile: Baseline used for comparison, which includes only those data points somewhat outside the norm.

Keyprint Signature: Baseline used for comparison that includes all available data points.

Keystroke: A single press of one of the keys on a computer on a keyboard.

Keystroke Biometrics: The timing information about a key press and key release as a person is typing at a computer keyboard.

Keystroke Pattern: Keystroke measurements obtained yield information in the form of real numbers.

KeyTrac: Tool used to detect users based on their keystroke dynamics.

Latency: The elapsed time between the release of the first key and the depression of the second key.

Moodle: Modular object-oriented dynamic learning environment to provide an organized and central interface for e-learning in educational institutions around the world.

N-Graph: It means n consecutive key presses.

Online Assessment or E-Assessment: The systematic process of documenting and using practical data on the knowledge, skills in an e-learning environment.

Online Learning Environment: Learning platform that allows sharing of educational materials with the learners via the web. Examples include Moodle, WebCT, and Blackboard.

Proctor: An invigilator or supervisor at a university or college examination.

Signature Track: Build a signature profile that links the coursework to student's identity using keystroke dynamics.

Template: A mathematical representation of biometric data.

Trigraphs: It is a group of three letters that together form a single specific sound.

Verification: The biometric system authentication process which matches a captured biometric against the person's stored template (1:1).

Chapter 6
Biometric Authentication Techniques and Its Future

Shanthi Sivakumar
Malla Reddy College of Engineering and Technology, India

ABSTRACT

The number of users using the internet has drastically increased. Due to the large number of online users, demand has increased in various fields like social networks, knowledge sharing, commerce, etc. to protect the user's private data as well as control access. Unfortunately, the need for security and authentication for individual data also increased. In an attempt to confront the new risks unveiled by the networking revolution over the recent years, we need an efficient means for automatically recognizing the identity of individuals. Biometric authentication provides an improved level of security and paves the way to the future. Further, biometric authentication systems are classified as physiological biometric and behavioral biometric technologies. Further, the author provides ideas on research challenges and the future of authentication systems.

INTRODUCTION

With the fast increase in the growth of the Internet, the possibility of e-crimes such as unauthorized access, transaction fraud has been increased in recent years and the requirement for controlling the access of data and securing the user's private data has become more inevitable. The word 'biometric' relates to unique characteristics of human beings based on physiology or behavior which doesn't change over a period of time. Biometric Authentication has gained more popularity in recent years due to its personal identification. Personal identification is very crucial in many applications like computer forensics, computer security, financial transactions and it has become a major concern in the society. Biometrics or biometric authentication is used for uniquely verifying or determining the identity of humans depending upon their physiological or behavioural traits (Bolle, Connell, Pankanti, Ratha and senior, 2004). It is more reliable and capable of recognising a person by identifying the authenticity of physiological or behavioural traits possessed by the person. One of the main design issues to be considered here is to determine how the individual is to be identified. It provides confidentiality, security and data privacy when compared to that of the traditional systems.

DOI: 10.4018/978-1-5225-7724-9.ch006

LITERATURE REVIEW

Zviran and Erlich, 2006 discussed that the computer systems are shielded via user identification and authentication. He discussed about the widely used authentication technique known as knowledge based authentication and popularly called as KBA. It is used to identify the individual user and provide access to the protected material. Ullah et al., 2012; Zviran and Erlich, 2006 according to these authors, KBA is a simple and inexpensive technique and it uses user id, passwords and some challenging questions and it's easy to memorize. The author Zviran and Erlich, 2006,further discussed about the possession based authentication(PBA).It includes tokens and it is used to identify the user.But this mayn't be a valid method since there are chances of tokens being stolen by somebody. Flior and Kowalski, 2010 discussed some of the PBA examples such as dongles, key, tokens, memory cards etc.

Obaidat and Sadoun, 1998; Bours and Barghouthi, 2009, they discussed about the accessing of systems by using username and passwords. Davaanaym et al., 2009 provided the mostly used solution by means of username and password. Bours and Barghouthi, 2009, discussed about the fingerprints and iris recognition used in airports. Lamport, 1981 discussed about the password based authentication techniques.

Apampa et al., 2010 discussed that Biometrics is one the rapidly improving technology which is used to provide security in various applications.It depends on the physiological and behavioural characteristics. Gao, 2012; Zviran and Erlich, 2006 discussed various behavioural characteristics which includes mouse movement, voice,, signature, gait, keystroke, pulse etc., Marcialis et al., 2009 discussed about the soft and hard biometrics and characteristics such as voice, clothing colour, height, skin, hair colour, Signature and tattoos are the ones which change rapidly. Khit an et al., 2011 discussed that these soft features can be used to identify but we can't entirely depend on these features. Gao, 2012 discussed about the hard biometrics such as fingerprints, facial features, palm prints, retinas, vein patterns, iris patterns etc. The advantage is that these are not changeable easily and they can be firmly used for identification.

Authentication Techniques Used in Online Exam

According to the authors, Al-Saleem, S.M. and H. Ullah,they have proposed an integrated approach which combined the two such techniques such username, password and palm print technique. Webcam is being used here to monitor the examinee but there will be inconvenience to the examinee.

Ramu, T. and T. Arivoli they proposed again an integrated approach which combines the keystroke approach with the traditional approach username and password. This method is very inexpensive since no special hardware is required for this and this method doesn't cause inconvenience to the examinee. But the drawback of this method is that it not accurate.

Qinghai Gao, discussed about monitoring the students of distance learning exams by using the timestamp and IP address. Malpractices are checked by using the two factors ip address and timestamps of examinee. Here no special hardware is required is used for carrying out this technique. But the major drawback is that these IP addresses can be easily manipulated by VPNs.

A. Ullah, H. Xiao, M. Lilley and T. Barker, an integrated approach based on profile and username and password is used and it is very simple to implement and requires no special hardware. Here the strength of this method is based on the challenging question and if shared there is no use of it. Swathi Prathish, S.Athi Narayanan and Kamal Bijlani, proposed Yaw angle variation scheme and used webcam for capturing audio and video signals. Based on the analysis of these video and audio signal,an examinee has been checked whether he had indulged in malpractice or not.

Rudrapal, S. Debbarma and S. Das and N.Kar proposed for physically disabled people to help them in taking tests. It is based on voice and there is a chance of impersonation since only this factor is being considered.

C.C. Ko and C.D. Cheng, proposes exams for online and offline. Here encryption technique are used. symmetric and asymmetric cryptosystems and digital signature schemes are proposed by N Harini, Dr. T R Padmanaban and authentication is carried out using the trusted registration authority. E-monitoring services are used to prevent frauds.

EVOLUTION OF AUTHENTICATION SYSTEMS

User Authentication is normally verified by using these three categories shown in Figure 1, namely,

- Authentication based on knowledge - Something the user knows.
- Authentication based on Object -Something the user has.
- Authentication based on Biometric -Something you are.

Knowledge Based Authentication

Users adopted the PIN or password mechanism to secure their data. Single factor Authentication was used to authenticate the subject like PIN's. It is commonly used due to its simplicity and user friendliness. SFA is the weakest level of authentication. Passwords are one of the most commonly used methods by users, and it is more vulnerable to attacks like dictionary, brute force attack due to the simple and easy passwords. If the password is stolen or forgotten, the possibility of risk of unauthorized access becomes more. It finds its way in many applications and in many fields like financial transactions, health sectors, military services, banking sectors, law and social services, etc.

Object Based Authentication

Also known as token based authentication which the user has such as smart cards or keys. The problems faced by this method include like duplication of tokens or sharing or chances of getting lost or stolen and they must be carried always. The single factor authentication is not reliable to provide adequate security because of various threats. In order to overcome the drawback of SFA and to enhance identification techniques, two factor Authentication (2FA) has been introduced which uses the combination of username/password along with the personal factors like phones, smart cards, etc. Knowledge based and objects based authentication techniques have many limitations like managing too many passwords and PINs and remembering them is not an easy task. The limitations of authentication techniques based on knowledge, object is overcome by using the biometric- based Authentication.

Biometric Based Authentication

This method is based on the usage of the distinct characteristics of persons which varies from individual to individual. This method is one of the most widely used one due to the advantages like it is more convenient, the results are accurate, and it provides an audit trail. Multi factor Authentication MFA was

Figure 1. General classification of biometric systems

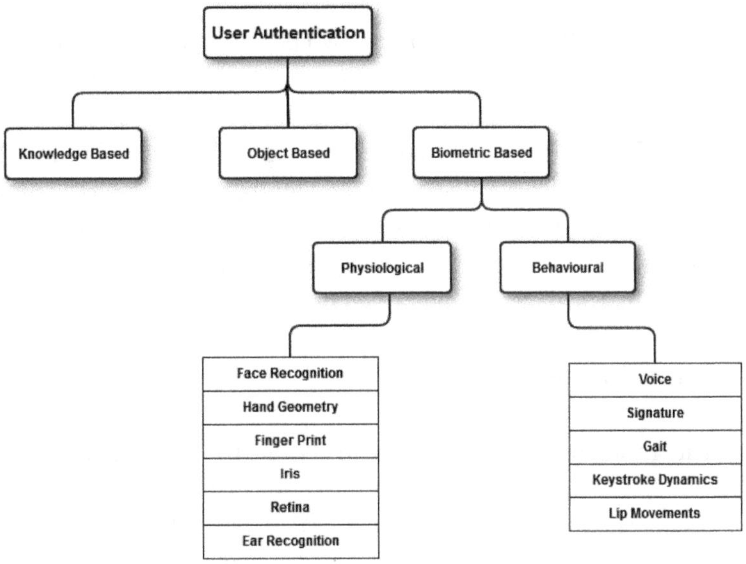

introduced to increase the security and protection which uses more than two factors and it is based on biometrics. Evolution of authenticated methods is discussed in Figure 2 .

Biometric Techniques are used to identify or verify the user based on the physiological or behavioural characteristics of a person (Jain, Flynn & Ross 2007).

1. Physiological characteristics have been correlated with the shape of parts of the body Ex: Fingerprints, iris recognition, hand geometry and Face recognition, recognition of veins, the retina, DNA, palm print.
2. Behavioral characteristics are associated with the behavior of a person. Ex: signature, gait, keystroke dynamics and voice.

Figure 2. Evolution of authentication methods

WORKING MECHANISM OF BIOMETRIC AUTHENTICATION SYSTEMS

The general mechanism is basically categorized into two phases mainly enrolment and authentication / verification phase and is shown in figure 3. During the enrolment phase, the user biometric features are collected from various individuals and the information is manipulated in order to identify the features of each and every individual. The attributes are captured during this phase and is analyzed mathematically by using various algorithms and then individual unique templates are being created and stored in the database for each and every individual. During the authentication/ verification phase the biometric data from the user is obtained and processed. The Matching process is carried out with the templates stored in the data base and the biometric data obtained from the user.

The Attributes of the Biometric Systems

Biometric system considers single or multiple biometric traits by considering many factors. In order to provide the authentication system, Several requirements are identified by the researchers that should be possessed by the biometric identifiers (Jain, Boelle, Pankanti, 1999).The theoretical and practical requirements proposed by (Jain, Boelle, Pankanti, 1999) is given as follows:

- **Universality:** Every person should have his own individual biometric trait
- **Collectability:** The trait should be collected, digitized and stored in the database.
- **Circumvention:** Biometric trait should not be replicated or reproduced by anyone.
- **Distinctiveness:** The biometric identifier of each person should be unique and it should be distinguishable from the other persons
- **Permanence: B**iometric identifiers need to stay constant over a long period of time. The trait should able to work even under conditions like age or related issues.
- **Reliability:** The biometric trait used by the person should be more reliable and not mask able.
- **Performance:** The biometric identifier used should have more accuracy, recognition accuracy, etc.

Figure 3. Block diagram of biometric methods

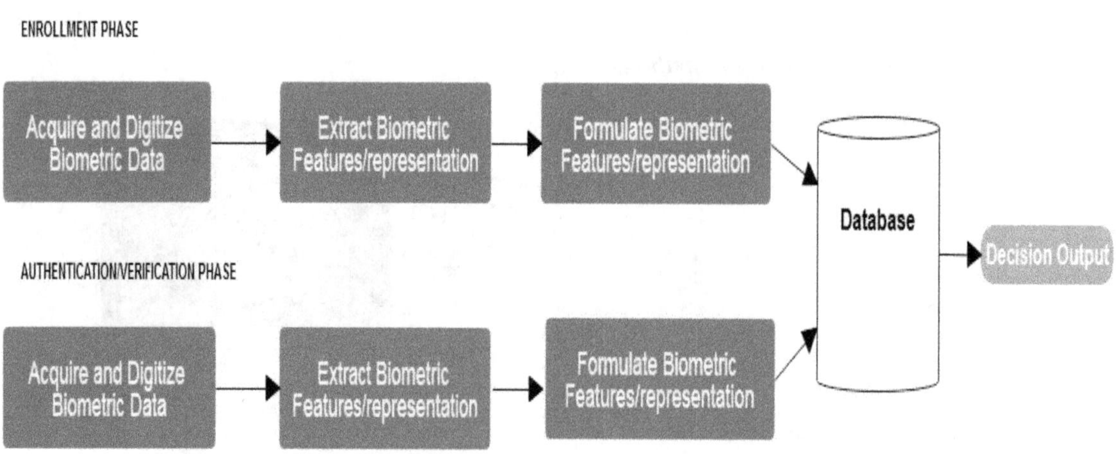

Evaluation Factors

There are many biometric techniques currently being in the society. The performance of the biometric system plays a very crucial role and it mainly depends on the accuracy. Though many biometric systems are used, none of them could be able to give the expected result in terms of the accuracy. Decision of the biometric system is generally classified either as genuine or imposter. So the decision outcomes can be true or false for the above listed genuine and imposter type.

The possible outcomes of biometric systems are given below:

- An individual(genuine) is accepted or rejected.
- An impostor is either rejected or accepted.

False Match Rate (FMR) or False Acceptance Rate (FAR)

The probability at which the input biometric pattern is mistakenly matched to a biometric pattern (non-matching) i.e., template stored in the database, erroneously.

FAR = %

No of Accepted Imposters * 100
Total no of imposter comparisons made

False Non-Match Rate (FNMR) or False Rejection Rate (FR)

The probability in which the Input biometric measurements fail to match to the corresponding biometric pattern already stored in the database.

FRR = %

No of Genuine persons rejected * 100
Total no of genuine comparisons made

FTE (Failure-to-Enroll Rate)

The percentage of an unsuccessful population who couldn't enrol their valid biometric template after a number of trials for a particular application.

FTA (Failure-to-Acquire Rate)

It is the probability rate that the biometric sensor in the biometric fails to work properly and couldn't extract the data presented by the user as an input.

Equal Error Rate (EER) or Crossover Error Rate (CER)

Rate at which the FAR (false accept) and FRR (false reject rates) are the same and they can be analyzed and represented using the ROC curve.

UNIMODAL BIOMETRIC TECHNIQUE

Unimodal biometric system uses the single source of biometric information to identify the persons. This technique is more reliable when compared to the traditional authentication systems. Biometrics is basically divided into two major types: Physiological and behavioural biometrics. Both metrics comes under the uni- modal biometric system, which uses a single biometric trait to identify the persons.

PHYSIOLOGICAL BIOMETRICS

Physiological characteristics are related to the shape of the body Ex: Fingerprints, iris recognition, hand geometry and Face recognition, recognition of veins, the retina, DNA, palm print and given in Figure 4.

Face Recognition

Face recognition is a popular upcoming technique and its performance is less when to the other techniques like iris recognition. The factors used for face recognition includes dimensions, lighting conditions, physiological attributes and size and shape of the nose, lips, eyes, etc. The steps involved in recognition of face include the detection of face, face extraction and face recognition. Two factors mainly affect the facial features, namely short and long term changes. Short term changes refer to the loss of weight or gain. Long term changes demote to the wrinkles and aging factors. Some of the techniques used to recognize the face includes the Feature based approach (specific features of face), holistic based approach (whole face), hybrid based (a combination of feature and holistic) and template based approach. Face recognition systems, technologies are dependent either on 2Dimensional or 3Dimensional representation of the face appearance. The advantages of face recognition include friendly, sensors commonly available and highly and most used modality. One of the popular biometric techniques is Face recognition due to its easy usage and more security [Chellappa, 1995]. Zhao et al. 2003 has provided the summary of the various existing techniques for recognizing the human face. Further, Abate et al. 2007 discussed challenges and the existing face recognition techniques. Research work on face recognition has been carried out using 2D and 3D. 3D face data research is first carried out by Cartoux et al. 1989 to identify humans. It's Research issues like face rotations or pose, facial expressions, occlusions and lightning conditions, etc greatly reduce recognition accuracy if not addressed properly (Singh 2008).

FingerPrint Recognition

Fingerprint recognition is one of the widely used biometric techniques applied in both computer systems and law enforcement, having a grown-up and widely accepted technology. It is a digitized model based on the ancient ink and paper system which is realized on numerous devices and platforms, like mobile phones laptops, PDA, etc. This method tries to recognize the individual characteristics of finger prints. Each and every individual is having unique fingerprint characteristics which are based on furrows and ridges (Jain, Flynn & Ross, 2007). These furrows and ridges determine the uniqueness of the user. Immutability and uniqueness are the two basic principles used in the identification of fingerprint. The techniques used for matching the fingerprint is basically divided into three types: Correlation based approach, Minutiae based approach and Image/pattern based approach. The main Challenge faced here is the recognition rate is greatly degraded at times due to wet finger or wrinkles. The performance issue need to be addressed here. Finger Print technique is widely adopted due to its accuracy rate, a user friendly, easy to use, requires less amount of storage space to store the templates. With the increased use of mobile applications, this biometric technique fits itself into mobile authentication techniques. Drawback of this method includes finger position placed on the device, finger scarf, skin conditions like dirt, sweat, moisture, etc.

Hand Geometry Based Recognition

Hand geometry recognition techniques consider the factors like thickness, length, breadth, area of the fingers, hand, etc. This method is dependent on the information that the shape of the hands need not be the same for other. Security provided by this biometric is very low and it is not scalable. When hand geometry combined with other biometric techniques to form multimodal biometric systems can be used as a secured and reliable system for verification (Biometric news portal, 2012). Issues related to hardware requirements and aging problems need to be addressed. This biometric is widely used in mobiles because of ease of use [Chen et al, 2007]. But the major drawback is the lack of distinctiveness and its applications are confined to limited applications only. Though, this trait can be combined with other traits.

Figure 4. Physiological biometric techniques

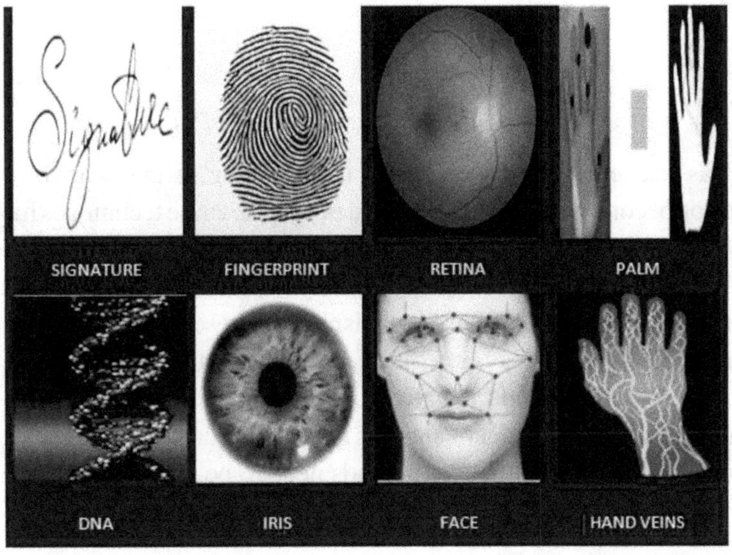

Iris Recognition

The authors Flom and Ara, 1987, proposed the concept of iris recognition. IRIS recognition is one of the popular biometric and gives the accurate result. Bowyer et al. 2008 has presented an extensive survey on the current iris recognition technologies. There are other methods also which have been extended for iris recognition. Quality enhancement is improved using Daugman's 2-D Gabor filter, proposed by Chenet al. 2006.1-D local texture patterns discussed by Du et al. and moment based iris blob matching proposed by sunet al. A survey of hand geometry verification is discussed briefly by Golfarelli et al. 1997; Jain et al. 1999; Sanchez-Reillo et al.2000; Pavesic et al.Iris recognition considers the colored portion of the eyes around the pupil. It is the most reliable type of biometric identification used for verifying the user. The iris recognition systems are believed to be unique and reliable. The biometric system tries to identify the furrows and ridges. The steps involved in identification include image acquisition (signal/noise ratio, resolution, illumination wavelength) iris localization and segmentation, template generation and biometric template matching (Vacca, 2007). Research issues related to watery eyes, contact lens etc need to be addressed (Iris Recognition, 2003).

Retina Recognition

Retina is made up of the network of neural cells at the back portion of the eye. Complicated capillaries in this part of the eye are having unique features which distinguish the persons from each other. Blood vessel patterns form a unique identification and even identical twins persons retina doesn't match other .Data acquisition is carried out by using a specific sensor and the user needs to peep the sensor and it performs the work of capturing the characteristics of the retina (Griaule Biometrics,2012). The main drawback associated with this approach is that the eyes should be in close proximity with the camera and dedicated hardware is used to implement this. This retina recognition is very difficult to spoof. The Main challenge is to identify the lens or wearing glasses. The drawback is that the user needs to focus his eye on the scanner. Due to this it may lead to some health issues and that's why this method is not widely accepted.

DNA Biometrics

The biometrics such as face recognition, fingerprint, iris scanning, retinal scanning, voice dynamics and handwritten recognition become popular and got rapid progress. These techniques have been incorporated into the system and these are based on the similarity of feature points measure, which gives inaccuracy results for a universal identification system. **DNA (Deoxyribonucleic acid)** becomes the best identifier. It is a genetic material presents in each living organisms and present in the cell's nucleus. The DNA of each individual human being consists of hereditary traits. It is intrinsically, digital and remains same throughout the individual's lifetime and even after the individual's death. DNA is a genetic code, which is unique to every individual and only identical twins have the same DNA. It needs a form of saliva, hair, blood sample, tissue sample, etc. Acquisition methods need to be addressed here.

Signature Based Recognition

Signature biometrics is a type of behavioural biometric and uses the signature to identify and authenticate the individual. Each and every individual's way of signing will be unique and it is considered as the distinct feature for identification. More emphasis is laid on how the signature is signed instead of the appearance of the signature. It is generally operated in two ways – namely Static and Dynamic signature recognition. Lack of accuracy, cost and reliability are the issues which need to be addressed.

Ear Based Recognition

Another technique based on the distinct shape and appearance of the ear. Usually each human have their unique visual appearance of ear. There won't be any change in the ear shape or appearance due to aging. So this biometric technique is one of the reliable ways of identifying the user. Ear recognition is carried out two steps similar to face recognition like ear feature extraction and matching process. This is user friendly and more comfortable approach when compared to other biometric systems. One drawback of this approach is less security. Pun and Moon, 2004; Yan and Bowyer, 2005 Discussed about the ear biometrics and Hurley at.al gave a recent survey on ear biometrics. Biel et al. and Israel at al. Discussed about the electrocardiogram biometric technique for recognizing humans. Anxiety of an individual is measured by considering the high and low stress of an individual based on its ECG features.

Palm Print

One more addition to the biometric technique is palm print recognition. Here, the palm of an individual is considered as a biometric trait and it is scanned using an optical reader an efficient sensing device. The template is stored in the database and the user traits are compared with the template for identifying the individual. One of the limiting factors is the storage space required by it to store the template. Some of the issues need to be addressed in this technique involves the identification of recognition rate in wet palm and the wrinkle palm. Shu and Zhang, 1998, discussed about the palm print recognition, and he considered some of the features like length and width of the fingers, ridges, wrinkles, texture, etc. Abuhaiba discussed about the Handwritten signature, and it is one of the mostly widely used biometric verification technique in forensic in our society. The features considered for signature verification includes pressure, speed and stroke of the signature, to verify the particular user. Research issues include rotating, stretching and out of positioning the palm. Contact based identifier device may lead to medical problems and hence contactless trait has been introduced in [Wu and Zhao, 2015].

BEHAVIOURAL BIOMETRICS

Behavioural characteristics are associated with the behaviour of a person. Ex: the signature, gait, keystroke dynamics and voice.

Gait Recognition

Gait recognition analyzes the way individual walks and can be captured easily at places where the resolution is less. This is a complex biometric trait where the video sequence needs to be collected in order to identify the person. The input is more intensive and it's expensive. Gait recognition may not be the same due to changes in body weight, injuries, etc. Camera viewpoint plays a major role in identification. If there is a change in the walking direction it influences the performance very badly. The user need not pay any attention during the identification by the gait. (Vera-Rodriguez et al, 2013) Discussed about the segmenting of humans into various components and put together a common distance metric is found leading to the improvement in gait biometric.

Keystroke Analysis Based Authentication

This authentication technique is the process of recognizing an individual depending on his typing characteristics. Two types of mode namely static or text dependent and dynamic or text independent mode can be used. The characteristics considered for analysis includes durations between the keystrokes, typing speed, use of number pad, frequency of errors, etc Using the static mode, the keystroke of pre-enrolled text is analyzed such as passwords which will be used during the login process. During the dynamic mode free text is allowed to authenticate a user.

Voice Recognition

Voice recognition is a technology which considers sounds, words and phrases of the human beings and these signals are converted into electrical signals and then converted into codes. This type comes under the behavioural biometric since it depends on the pronunciation and manner of speech. This recognition is carried out in the static or dynamic mode, i.e. text dependent or text independent. In the text dependent phase, the user will be allowed to talk a particular phrase whereas in the later he can speak freely. One of the major drawbacks is that the imposters can use the replayed voice to access the biometric system. Sometimes cold may affect the recognition of voice. Apple's Siri uses the voice recognition technology and modeled the question-answering system. Behavioural biometrics is shown in Figure 5.

Shortcomings of Unimodal Biometric Systems

Several problems faced by the unimodal biometric system is listed as follows:

- Due to the cloning of biometric characteristics the possibility of fraud has increased
- Error rates are too high for the unimodal system
- Lack of consistency and stability of the biometric characteristics
- Due to environmental conditions and wrong usage by the client's noisy signals are being captured by the sensors
- Safety of the Sensors is a major issue
- Lack of universality of some features in biometric systems
- Each system is having its own upper limit on recognition performances.

Figure 5. Behavioral biometric technologies

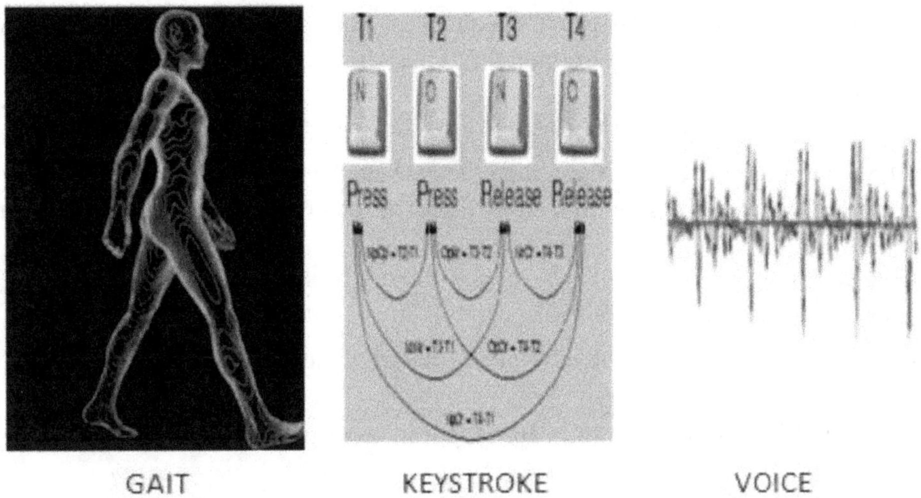

MULTIMODAL BIOMETRIC SYSTEM

Multimodal biometric system senses more number of features and followed by extraction of features and then matched with the database. Then all the scores of matching are added and the score is compared with the threshold value. If the score value is more or equal to the threshold value then the verified person is genuine or he is a fraud or imposter.The limitations of unimodal biometric systems can be overcome by using two or more biometric traits together, which is called as the multimodal biometric system. This technique is more reliable due to the usage of many traits for identification. Some of the drawbacks listed by the traditional biometric system like non universality, spoofing attacks are avoided. This leads to increase in accuracy, enhanced security and less enrolment issues. (Oloyede and Hancke, 2016) gave detailed review about the multimodal biometric system.

Some of the benefits of multimodal systems include privacy, recognition accuracy, and biometric data enrolment.

- **Recognition Accuracy:** The accuracy is always better than the unimodal biometric system. Due to multiple traits the accuracy and reliability of the multimodal biometric system is higher and it

Figure 6. Types of authentication

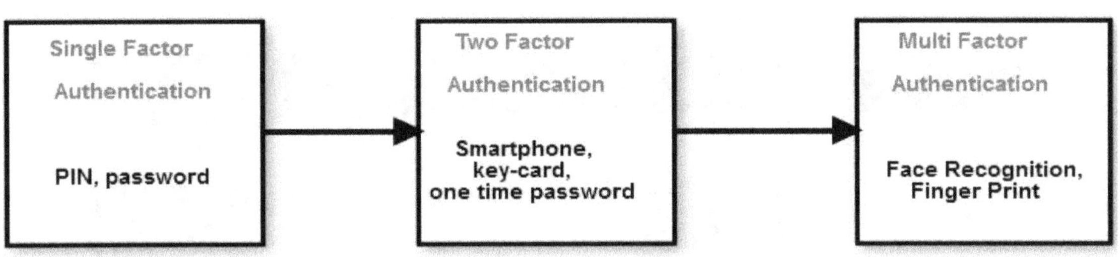

is difficult to forge. Experiments carried out have indicated that the accuracy of multimodality is nearing 100% in identification.

- **Privacy:** Vulnerabilities and attacks are reduced in Multimodal biometric systems and the template couldn't be stolen. Since more modalities have been the attacker faces difficulty in spoofing the modalities.
- **Biometric Data Enrolment:** If the quality of one modality is not good due to some poor environmental conditions or related issues then the data can be collected from the other modality. The problem of noisy data may be addressed by this .The accuracy is high and it covers universality.

Biometric Sources

Multimodal biometrics can work in one of the five scenarios shown in figure 7:

- **Multisensors:** Single biometric trait is used, but in order to capture the data multiple sensors are used.

Example, in facial recognition system multiple cameras are used to capture the face in different angles.

- **Multiple Algorithms:** Multiple algorithms are used to process the captured data. This saves the cost but adds up additional complexity computation.
- **Multiple Instances:** Multiple instances of the biometric modality is used here.
- **Multi Samples**: More samples of the biometric trait i.e.same trait is used here to capture the data.
- **Multimodal:** Here more than one biometric modality is used like face and fingerprint etc. This type of systems requires both sensors and algorithms to capture the process information.

FUSION WAYS

Three levels of fusion exist and are listed and shown as follows in figure 8:

Figure 7. Multiple biometric sources

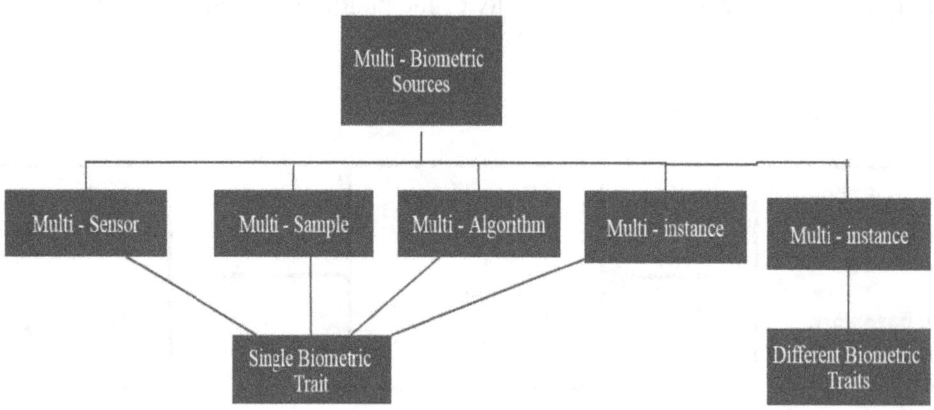

1. Fusion at the Feature Extraction Level
2. Fusion at the matching score
3. Fusion at the decision level

Fusion at the Feature Extraction Level

Here the information gained from the biometric modality is computed to produce a feature vector. Similarly, the other biometric data is also concatenated and a single feature vector is generated. The new feature vector represents the person's individuality in a diverse vector space. Feature reduction techniques can be utilized to collect the features.

Fusion at the Matching Score

Similarity Score is provided between the input feature vector and the template vector. This score can be joined with the other biometric modality in order to validate the authenticity of the user. Methods like weighted averaging can be employed to integrate the similarity scores of various matchers.

Fusion at the Decision Level

The biometric system makes its personal recognition technique based on its feature vector. Using the concept of vote scheme, we can decide the final recognition system.

Operation Modes

A multi-biometric system usually operates in three different modes:

Figure 8. Fusion levels

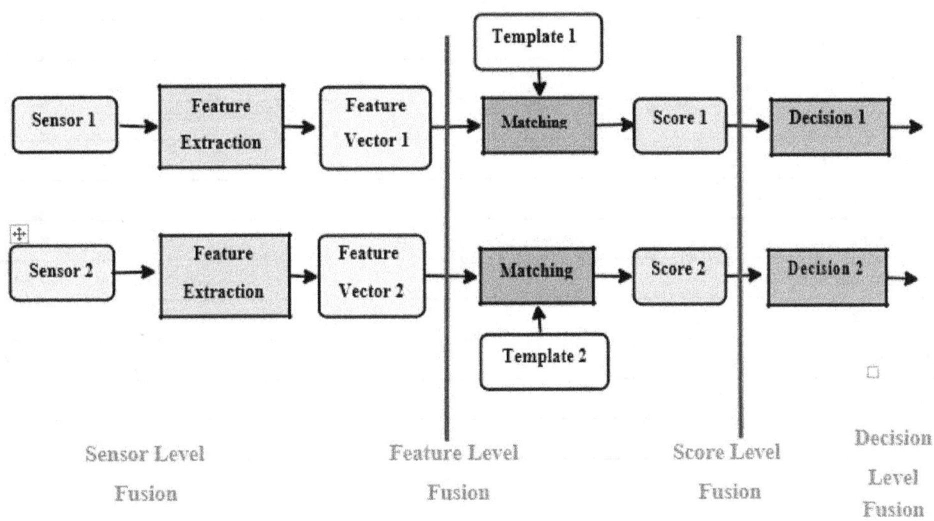

- **Serial Mode:** Each biometric trait is examined one after the other.
- **Parallel Mode:** Biometric traits are captured in parallel manner and then the results of all modalities are combined together to perform recognition.
- **Hierarchical Mode**: This mode uses a hierarchy –tree structure to represent the individual classifiers are then they are combined together. When additional number of classifiers is there, this mode is most preferable.Table 1 represents the various features for fusion.

SECURITY AND PRIVACY IN BIOMETRIC SYSTEMS

There exists quite a lot of security and privacy issues correlated with the usage of biometric. These issues are given below:

- Biometric is not secret. The Third party can capture the biometric data without much difficulty. Sometimes, the authenticated user might not be having an idea about the spying of her biometric information. For example, in the voice biometric conversations many be recorded or using a mobile app the voice biometric can be hacked. Similarly, users face or fingers or gait can be reproduced.
- Biometric data is permanent. We cannot able to make changes or it cannot be revoked. Unlike our traditional authentication systems like password, pin, it is impracticable to provide a fresh biometric trait, e.g., fingerprint. If the biometric trait is compromised once, then it becomes unusable. Further, the user can by no means be detached from the compromised data
- Biometric system can be used to identify an individual everlastingly.

The vulnerable attacks in the biometric system has been identified in 8 places and shown in the above figure 10 [Ratha et al, 2001].

Table 1. Biometric features for fusion

Voice, Face, Lips movement
Fingerprint and Face
Fingerprint, Face, Voice
Hand geometry, Fingerprint and Face
Fingerprint, Hand geometry and Voice
Voice, Hand geometry
Facial theromogram and Face
Iris and Face
Palm print, Hand geometry
Ear form, Voice
Voice, Lips movement

Figure 9. Comparison of biometric technologies

Biometric identifier	Universality	Distinctiveness	Permanence	Collectability	Performance	Acceptability	Circumvention
DNA	H	H	H	L	H	L	L
Ear	M	M	H	M	M	H	M
Face	H	L	M	H	L	H	H
Facial thermogram	H	H	L	H	M	H	L
Fingerprint	M	H	H	M	H	M	M
Gait	M	L	L	H	L	H	M
Hand geometry	M	M	M	H	M	M	M
Hand vein	M	M	M	M	M	M	L
Iris	H	H	H	M	H	L	L
Keystroke	L	L	L	M	L	M	M
Odor	H	H	H	L	L	M	L
Palmprint	M	H	H	M	H	M	M
Retina	H	H	M	L	H	L	L
Signature	L	L	L	H	L	H	H
Voice	M	L	L	M	L	H	H

Figure 10. Vulnerable points of attack

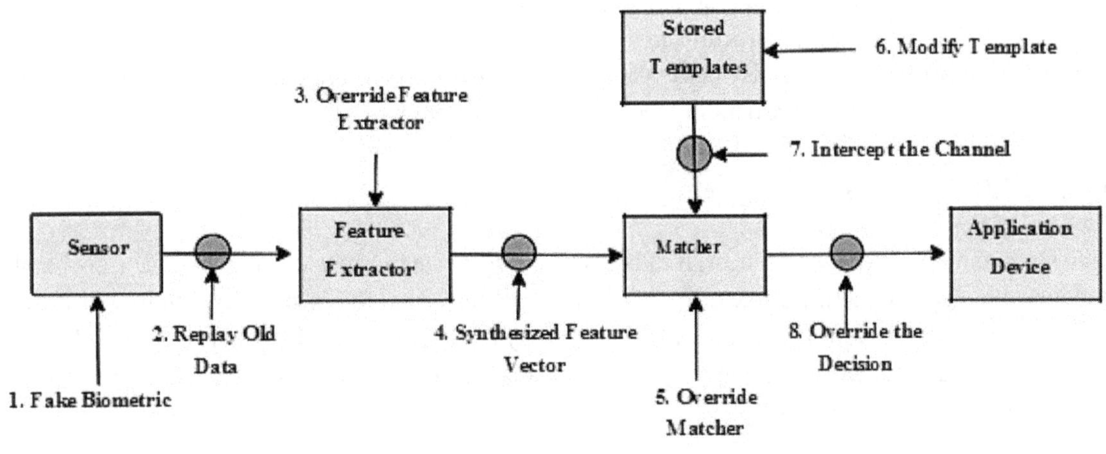

1. **False Biometric:** There is a possibility of fake biometric presented by the user with a purpose to deceive the system.There have been many such attacks of this kind reported so far.
2. **Resubmitting Stored Signals:** The biometric signal can be pre-recorded and The reproduced input is presented to the system such as face mask, fake finger, fake signature, etc at later times.
3. **Intervening Feature Extraction Process:** The attacker comprises the extraction module that retrieves various features. The attacker substitutes the original features with the preferred one of his own choice. Trojan horse is used here to attack the feature extractor.
4. **Tampering with Biometric Representation:** The templates of traits stored in the database are substituted by a longer one by the attacker

5. **Corrupting the Matcher:** Attacker may compromise the matching module and the attacker may produce a matching score by himself so that the matching scores are corrupted.
6. **Compromising Stored Templates:** An attacker may steal the template by gaining illicit right of entry to the stored templates.
7. **Communication Interception:** Attacker may intercept the communication happening between the two modules the matcher and the database. The information may be altered and passed between them.
8. **Overriding the Final Decision:** Sometimes the attacker may supersede the decision taken by the matcher module.

LIMITATIONS OF BIOMETRIC SYSTEMS

Biometric systems are having many limitations and are discussed below.Unimodal biometric couldn't demonstrate required performance due to various factors like lack of accuracy in the noisy environment, spoofing and non universality. The problems related to biometric systems are as follows:

Noisy in Sensed Data

If sensors are defective or not maintained properly, there is a possibility of noise data within the biometrics. Ex: accretion of dirt in fingerprint sensor leads to noisy data. Failure to generate the voice due to health issues may be a kind of noisy data. Variations in camera position, camera focus and the lighting environment may generate noisy data in face and iris recognition.

Lack of Individuality

Due to some fault or error in the sensor, it fails to collect accurate biometric data from the users resulting in inconsistent data. Sometimes like traits are being collected from the biometric system. For example, identical twins may have similar facial features or it may be the similarity between son and father. Similarly, in fingerprint recognition, many of the persons may have scars or cuts and similar biometric data are extracted leading to noisy data.

Non-Universality

In the biometric system, each and every individual should be competent to generate a biometric trait for identification. Though the biometric trait seems to be universal they are not. It is estimated that around 2% of the total population may not be able to generate a quality fingerprint. Such persons are susceptible to problems like improper registration, unsuccessful enrolments etc.

Spoof Attacks

Biometric traits are vulnerable to attacks. Sometimes the traits are spoofed by frauds or imposters. In fingerprint, imposters can use artificial finger prints to spoof the system and gain access to the biometric system.

ONLINE EXAM THREATS

Online exams are being carried out in almost all the fields instead of traditional paper and pen approach. This online exam is prone to many security threats [3].Some of the problems faced during the online exams includes collusion, inadmissible material, plagiarism, impersonation etc. Out of these problems, impersonation is one of the critical threats faced during the online exams. Impersonation implies the fake action and the intent is to duplicate the authenticated user and fraud the security system.

According to Miguel et al., [4] he proposed that online security threats in the examinations can be broadly considered in two different stages. In the first stage the threats are examined, and then solutions to the threats are identified and discussed to avoid the threats. One of the important risk involved in the online exams is because of the absence of physical communication or monitoring and leads to chances for cheating.

In order to qualify the exams or to increase their grades in the online exams, students involved in the activities of cheating. The security threats can be classified into intrusion and non-intrusion threats. Further, we can classify non-intrusion threats into collusion and non-collusion threats. Intrusion attackers are usually carried out by hackers, criminals and cyber attackers. Collusion threat happens when the authenticated person brings a third party person to write his exam called as impersonators or abettors for helping them in the exams. It is a preplanned cheating and involves the interaction between the student and the third part person. This is one of the major issues which need to be addressed soon.

In impersonation attacks the student involved shares his credentials or the third party impersonator is given access to perform his/her test. The main problem with this threat is that we couldn't able to find out the impersonator after the test is being completed. This threat also poses serious challenge to the online examinations.

Plagiarism is copying the material and ideas of some other person and claiming it as his work. Because of the improvement of internet, the chances of taking some body's idea without their permission have become very easy.

Techniques to Handle Online Threats

There are many authentication techniques currently used in the market to handle the online threats during examinations. Most of the authors have used the multimodal biometric approach to avoid the threats faced during the online exams. Most of them have discussed about the usage of username, password and along with that other authentication features such as iris, finger print, Keystroke, palm print etc. Security policies in online examinations should support authentication and authorization. Online learning and examinations is an upcoming research area which needs to be focused more. Online exams can be made more secure as well as user oriented by making use of new technologies. Some of them includes biometric authentication in personal learning environment, goals and security of online assessments.

Proposed Approach for Online Exams

One way of handling the online threats is to use the Keystroke dynamics based authentication technique to verify the user based on their typing pattern. In order to authenticate the user based on their typing speed samples need to be collected from the user in spite of the text being typed.

Usually two ways can be used to analyse the keystroke dynamics -static and dynamic. Static analysis uses predetermined sample data for authentication and dynamic involves continuous monitoring of the keystrokes. Some of the key features include Dwell time, latency of keystrokes, Flight time, typing error, force of keystrokes and we measure the standard in the form of FRR, FAR, ER etc.

In the proposed approach, three modalities are suggested to avoid the online threats. Using the static keystroke analysis, we can analyze the online attacker by comparing it against the samples. In addition to it, we can use a camera to recognize the iris of the person. We can attach the scanner to the mouse for finger print identification. So, this proposed method can be a good solution for avoiding the online exam threats.

APPLICATIONS OF BIOMETRIC SYSTEMS

Due to the increasing requirements in security, to improve the system performance and decrease the cost many biometric systems are currently available in the market. The usage of biometric techniques have been used around a broad sectors like commercial, health sector, government, education, forensic, military, financial applications etc. In health sector, biometrics are used for identification of patients, health care, storing the patients information etc. Forensic applications are used to identify the criminals, missing children, identify the victims. One of the biometric used is finger print and widely used (jain & kumar 2012).Others traits are also used in forensics such as gait, voice, DNA, face etc. Face trait can be used in many places like airport, malls, stadiums etc. Voice, DNA is used to recognize the criminals at a less price. In the commercial sectors, biometric systems finds its application in many areas like banking, financial sector, education, virtual reality,ATM's, e-commerce(woodward, Horn, gatune & Thomas 2003). Government sectors have broad applications in voting scheme, Registration,ID cards, Driving license, military applications etc(Grialue biometrics,2012).Some of the traits used are fingerprint, face etc.

Figure 11. Proposed biometrics solution for e-exam user's authentication

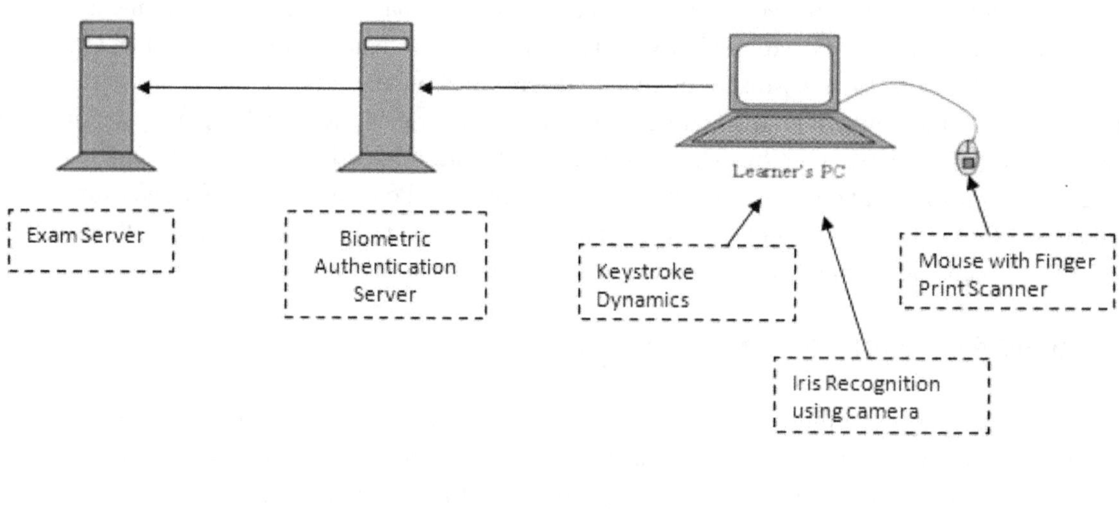

FUTURE TRENDS IN BIOMETRIC AUTHENTICATION

- **Heartbeat Biometrics:** One of the crucial upcoming biometric technologies is heart beat biometrics. In this technology, heart beat is considered as the key factor for identification of persons. Heart beat is also a unique to the human beings and this technology will replace the usage of many things like credit cards, car and house keys etc. This technology will be implemented in the car doors, house doors, Mobile and laptops in the future.
- **Sweat Recognition:** Another biometric authentication system introduced is sweat Recognition. It's a recognition technique uniquely used to identify the humans. It is very hard to be forged by anyone else. This technology doesn't require the people to have fingers, eyes or even could be people who couldn't memorize the things.
- **Odour Recognition:** This technique could also be a new addition in the biometrics in the upcoming future. This technique is similar to that of sweat recognition. It is a more reliable technique and it will be a widely used technique in the future.
- **Cognitive Biometrics:** Brain biometric comes under the cognitive biometrics. Brain waves have been used for biometric identification. The function of brain is to regularly monitor and helps in regulation of actions and reactions involved in the body. The activity of the brain can be recorded in two ways: by measuring the electrical activity of neurons, measuring the blood flow in the brain. Brain biometric system is carried out in 4 phases: signal acquisition, Feature Extraction, classification, Brain computer Interface.
- **Signal Acquisition:** Brain produces the electric fields by the electric currents and it is sensitive to Electroencephalogram (EEG).The recordings of EEG can be made portable with the use of less expensive devices.It has 5 different frequency components: Delta [0.5 - 4Hz], Theta [4 – 8 Hz], Alpha [8 – 14 Hz], Beta [14 – 30 Hz], Gamma [> 30 Hz].Slow waves are represented by Delta & Theta and Fast waves are represented by using Beta & Gamma components.
- **Feature Extraction:** The output of the first phase is that from the brain signals the various features are extracted. Numerous techniques are currently available in the society to extract the features like Time-Frequency Analysis, Analysis based on time, Time-Frequency-Space Analysis, Wavelet Transform., Analysis based on frequency.
- **Classification:** Artificial neural network along with the back propagation algorithm is used for classifying the extracted EEG features.
- **Brain Computer Interface:** BCI acts an interface between the machine and the brain. The main function is to BCI take out the electrical signals from brain and processes them in order to generate the signals to be fed to the computer. This technique is reliable and provides more security due to various characteristics like robust to attacks, unique, no duplication, continuous identification etc.
- **Robotics Biometrics:** This type of biometrics is used to classify and identify the robots. Methodologies used for the human beings have been used to authenticate the robots. Biometric systems may solely depend on single biometric trait to identify the humans or robots. But Some robots may have the same facial features but it may be having different gait, voice etc. So, the efficiency of single biometric trait may not sufficient enough to identify the robots. so we need to overcome the problems of single biometric system and introduce the multimodal artimetrics. This multimodal biometric will be useful when the patterns are more complex, abnormal behaviour,

abnormal data samples etc. An effective fusion technique can be used to combine the data from the multimodal biometric systems resulting in an efficient and reliable technique.

CHALLENGES AND RESEARCH FOCUS

Biometric authentication systems provided security and privacy in identifying the genuine and imposters and applied in many applications like transactions, border security etc. The expectations of the user community on biometrics are keeping on increasingly steadily. The four main challenges of the biometric authentication system are identified as follows:

1. Biometric sensors
2. Biometric Systems
3. Biometric systems Interoperatability
4. Communications and privacy

Biometric Sensors: The challenges involved in biometric sensors include the following:

The quality of data collected from the non-cooperative users, who are not aware that the biometric sample of his/her is being collected. Quality of the data collected at relaxed states. Problems associated with the rapid collection of biometric traits in harsh environment and in mobiles. The user biometric templates stored in the database should not be compromised and it can be revoked or replaced sometimes.

- **Biometric Systems:** The challenges involved in biometric systems include the following:

The recognition accuracy should be more under various operational environments. Determining the components required for a particular application. Problems faced in providing a clear explanation for the end users and the operators. Sometimes, the remote recognition is having problems like unattended enrolment and registration. Some of the research problems include the usage of improved matching algorithms, integrating multiple sensors, modalities and matching algorithms in a single system, Quality of data which is measured.

- **Biometric systems Interoperatability:** The challenges involved in biometric systems Interoperatabilty include the following: Need to integrate the components of the system and use it without modifying the functionality. To verify and validate the data collected from various resources. Need to safeguard and protect the data which is being collected for a long time and shouldn't be compromised. Some of the Research issues are conformance, data interchange formats, interoperatability testing etc.
- **Communications and Privacy:** Challenges include the privacy enabled solutions that need to be provided to the users in order to increase the confidence of the users and the various levels of security principles and privacy.

SUMMARY

Biometric Authentication techniques are gaining more popularity for identifying the persons through behavioral and Physiological identifiers. Different types of authentication systems like knowledge based, object based and biometric based are discussed in detail. This chapter briefly explained the different types of physiological and behavioral traits. Due to the different authentication techniques available in the society, it facilitates and reduces the identity threat in various fields like Government, public sectors, health sectors, financial transactions, Forensics etc. Strengths, weakness and challenges of various biometric authentication systems have been discussed comprehensively. Additionally, the unimodal and multimodal biometric systems along with advantages and drawbacks have been discussed elaborately. Further in multimodal systems, different types of fusion and operating modes have been discussed. This chapter also discusses the security, research issues and challenges which are correlated with the biometric systems. This chapter will help the readers get an overall idea about biometric authentication systems.

REFERENCES

A multi-matcher model for ear recognition. (n.d.). Available from: https://www.researchgate.net/publication/286142859_A_multi-matcher_model_for_ear_recognition

Abuhaiba, I. S. I. (2007). Offline Signature Verification Using Graph Matching. *Turk J Elec Engin, 15*(1), 89–104.

Ahuja, M. S., & Chabbra, S. (2011). A Survey of Multimodal Biometrics. *International Journal of Computer Science and its Applications, 1*, 157-160.

Biel, L., Pettersson, O., Philipson, L., & Wide, P. (2001). ECG analysis: A new approach in human identification. *IEEE Transactions on Instrumentation and Measurement, 50*(3), 808–812. doi:10.1109/19.930458

Biometric News Portal. (n.d.). *Fingerprint biometrics*. Retrieved from http://www.biometricnewsportal.com/fingerprint_biometrics.asp

Biometric recognition: challenges and opportunities. (2010). Retrieved from http://www.nap.edu/openbook.php?record_id=12720&page=2

Bolle, R. M., Connell, J. H., Pankanti, S., Ratha, N. K., & Senior, A. W. (2004). *Guide to Biometrics*. Springer Verlag. doi:10.1007/978-1-4757-4036-3

Bolle, R. M., Connell, J. H., & Ratha, N. K. (2002). Biometric perils and patches. *Pattern Recognition, 35*(12), 2727–2738. doi:10.1016/S0031-3203(01)00247-3

Cartoux, J. Y., Lapreste, J. T., & Richetin, M. (1989). Face authentication or recognition by profile extraction from range images. *Proceedings of the IEEE Workshop on Interpretation of 3D Scenes*, 194-199.

Chellappa, R., Wilson, C. L., & Sirohey, S. (1995). Human and Machine Recognition of Faces: A Survey. *Proceedings of the IEEE, 83*(5), 705–740. doi:10.1109/5.381842

Chen, W.-S., Chiang, Y.-S., & Chiu, Y.-H. (2007). Biometric Verification by Fusing Hand Geometry and Palmprint. *Proc. 3rd International Conference on Intelligent Information Hiding and Multimedia Signal Processing (IIHMSP)*, 403-406. 10.1109/IIHMSP.2007.4457734

Daugman, J. (2004). How Iris Recognition Works. *IEEE Transactions on Circuits and Systems for Video Technology*, *14*(1), 21–30. doi:10.1109/TCSVT.2003.818350

Daugman, J. (2004). How iris recognition works. *Circuits and Systems for Video Technology, IEEE Transactions*, *14*(1), 21–30. doi:10.1109/TCSVT.2003.818350

Daugman, J. (2004). *Iris recognition and anti-spoofing countermeasures*. 7th International Biometrics Conference, London, UK.

Du, Y., Ives, R. W., Euer, D. M., Welch, T. B., & Chang, C.-I. (2004). One Dimensional Approach to Iris Recognition. *Proceedings of the Society for Photo-Instrumentation Engineers*, *5404*, 237–247. doi:10.1117/12.542499

Feng, G., Dong, K., Hu, D., & Zhang, D. (2004). When Faces Are Combined with Palmprints: A Novel Biometric Fusion Strategy. *Proceedings of ICBA*. 10.1007/978-3-540-25948-0_95

Flom, L., & Safir, A. (1987). *Iris Recognition System*. U.S. Patent 4 641 349.

Frischholz, R. W., & Dieckmann, U. (2000). Bioid: A multimodal biometric identification system. *IEEE Comput.*, *33*(2), 64–68. doi:10.1109/2.820041

Gambhir,, A., Narke,, S., Borhade,, S., & Bokade,, G. (2014). Person Recognition Using Multimodal Biometrics. *International Journal of Emerging Technology and Advanced Engineering*.

Golfarelli, M., Maio, D., & Malton, D. (1997). On the error-reject trade-off in biometric verification systems. *Pattern Analysis and Machine Intelligence. IEEE Transactions on*, *19*(7), 786–796.

Hong, L., & Jain, A. K. (1998). Integrating faces and fingerprints for personal identification. *IEEE Transactions on Pattern Analysis and Machine Intelligence*, *20*(12), 1295–1307. doi:10.1109/34.735803

Hong, L., Jain, A. K., & Pankanti, S. (1999). Can multibiometrics improve performance? *Proc. AutoID'99*. Retrieved from http://www.griaulebiometrics.com/en-us/fingerprint_sdk http://www.computerscienceweb.com/

Hurley, D. J., Nixon, M. S., & Carter, J. N. (2009). Force field energy functionals for ear biometrics. *Computer Vision and Image Understanding*, *98*(3), 491–512. doi:10.1016/j.cviu.2004.11.001

Hurley, D. J., Nixon, M. S., & Carter, J. N. (2011). Force field energy functionals for image feature extraction. *Image and Vision Computing*, *20*(5–6), 311–318.

Israel, S. A., Irvine, J. M., Cheng, A., Wiederhold, M. D., & Wiederhold, B. K. (2005). ECG to identify individuals. *Pattern Recognition*, *38*(1), 133–142. doi:10.1016/j.patcog.2004.05.014

Jain, A., & Ross, A. (2002). Fingerprint Mosaicking. *Proc. IEEE International Conference on Acoustics, Speech and Signal Processing (ICASSP)*, 4, IV-4064-IV-4067.

Jain, A. K., Bolle, R., & Pankanti, S. (Eds.). (1999). *Biometrics: Personal Identification in Networked Society*. Dordrecht: Kluwer Academic. doi:10.1007/b117227

Jain, A. K., Ross, A., & Prabhakar, S. (2004). An Introduction to Biometric Recognition. *IEEE Transactions on Circuits and Systems for Video Technology*, *14*(1), 4–19. doi:10.1109/TCSVT.2003.818349

Jain, A. K., Ross, A., & Prabhakar, S. (2004). An introduction to biometric recognition. *IEEE Transactions on Circuits and Systems for Video Technology*, *14*(1), 4–20. doi:10.1109/TCSVT.2003.818349

Jain. (2003). Information fusion in biometrics. *Pattern Recognition Letters, 24*, 2115-2125.

Jitendra, J., Singh, B.K., & Ali, M.I. (2014). Voice Identification Secure System by Statistical Model of Speech Signal Using Normalization Technique. *International Journal of Engineering*.

Kumar, A., & Shekhar, S. (2011). Personal Identification Using Multibiometrics Rank-Level Fusion. IEEE Transactions on Systems, Man, and Cybernetics, Part C: Applications and Reviews.

Kumar, A., Wong, D. C., Shen, H. C., & Jain, A. K. (2003). *Personal verification using palmprint and hand geometry biometric*. Presented at the 4th Int. Conf. Audio- and Video-based Biometric Person Authentication, Guildford, UK. 10.1007/3-540-44887-X_78

Liu, X., Bowyer, K. W., & Flynn, P. J. (2005). Experimental Evaluation of Iris Recognition. *IEEE Computer Society Conference on Computer Vision and Pattern Recognition Workshops*, 158-165.

Maio, D., Maltoni, D., Cappelli, R., Wayman, J. L., & Jain, A. K. (2002). Fingerprint verification competition. *Proc. Int.Conf. Pattern Recognition (ICPR).*

Maltoni, D., Maio, D., Jain, A. K., & Prabhakar, S. (2003). *Handbook of Fingerprint Recognition*. New York: Springer-Verlag.

Miao, D., Sun, Z., & Huang, Y. (2014). *Fusion of Multibiometrics Based on A New Robust Linear Programming*. Presented at the Pattern Recognition (ICPR), 2014 22nd International Conference on, Stockholm, Sweden.

Obaidat, M. S., & Sadoun, B. (1998). *Keystroke Dynamics*. Keystroke Dynamics Based Authentication.

Oloyede, M. O., & Hancke, G. P. (2016). *Unimodal and Multimodal Biometric Sensing Systems: A Review* (Vol. 4). IEEE Access.

Oloyede, M. O., & Hancke, G. P. (n.d.). *Unimodal and Multimodal Biometric Sensing Systems-A Review*. doi:10.1109/ACCESS.2016.2614720

Phillips, P. J., Flynn, P. J., Scruggs, T., Bowyer, K. W., Chang, J., Hoffman, K., . . . Worek, W. (2005). Overview of the face recognition grand challenge. *IEEE Computer Society Conference on Computer Vision and Pattern Recognition, CVPR 2005*, 947–954.

Prabhakar, S., Pankanti, S., & Jain, A. K. (2003). Biometric recognition: Security and privacy concerns. *IEEE Journals on Security & Privacy*, *1*(2), 33–42. doi:10.1109/MSECP.2003.1193209

Pun, K., & Moon, Y. (2004). Recent advances in ear biometrics. *Proc. 6th Internat. Conf. on Automatic Face and Gesture Recognition*, 164–169. 10.1109/AFGR.2004.1301525

Ratha, N., Connell, J., & Bolle, R. (2001). Enhancing security and privacy in biometricsbased authentication systems. *IBM Systems Journal*, *40*(3), 614634. doi:10.1147j.403.0614

Sanchez-Reillo, R., Sanchez-Avila, C., Gonzalez-Marcos, A., Sanchez-Reillo, R., & Gonzalez-Marcos, A. (2000). Biometric identification through hand geometry measurements. *IEEE Transactions on Pattern Analysis and Machine Intelligence*, *22*(10), 1168–1171. doi:10.1109/34.879796

Sim, H. M., Asmuni, H., Hassan, R., & Othman, R. M. (2014). Multimodal Biometrics: Weighted Score Level Fusion Based on Non-Ideal Iris and Face Images. Expert Systems with Applications.

Singh, R., Vatsa, M., & Noore, A. (2008). Face recognition with disguise and single gallery images. *Image and Vision Computing*. doi:10.1016/j.imavis.2007.06.010

Srivastava, H.A. (2013). Comparison Based Study on Biometrics for Human Recognition. *IOSR Journal of Computer Engineering, 15*, 22-29.

Štruc, V., & Pavešić, N. (2009). Gabor-based kernel partial-least-squares discrimination features for face recognition. *Informatica*, *20*(1), 115–138.

Vera-Rodriguez, R., Mason, J., Fierrez, J., & Ortega-Garcia, J. (2012). Comparative analysis and fusion of spatio-temporal information for footstep recognition. *IEEE Transactions on Pattern Analysis and Machine Intelligence*, 99.

Wayman, J. L. (2001). Fundamentals of biometric authentication technologies. *International Journal of Image and Graphics*, *1*(1), 93–113. doi:10.1142/S0219467801000086

Yan, P., & Bowyer, K. W. (2005a). Empirical evaluation of advanced ear biometrics. Empirical Evaluation Methods in Computer Vision (EEMCV 2005), San Diego, CA.

Yan, P., & Bowyer, K. W. (2005b). Ear Biometrics Using 2D and 3D Images. *Proc. IEEE Conf. on Computer Vision and Pattern Recognition Workshop Advanced 3D Imaging for Safety and Security*.

Yan, P., & Bowyer, K.W. (2005c). Multi-Biometrics 2D and 3D Ear Recognition. *Proc. Audio and Video-Based Biometric Person Authentication*, 503–512.

Yang, X., Zhou, Y., Zhang, T., Shu, G., & Yang, J. (2008). Gait recognition based on dynamic region analysis. *Signal Processing*, *88*(9), 2350–2356. doi:10.1016/j.sigpro.2008.03.006

Zhang, D., & Shu, W. (1999). Two novel characteristic in palmprint verification:Datum point invariance and line feature matching. *Pattern Recognition*, *32*(4), 691–702. doi:10.1016/S0031-3203(98)00117-4

Zhang, H., & Shu, H. (1989). Effects of phonetic components on naming compound characters [in Chinese]. *Acta Psychologica Sinica*, *21*, 284–289.

Zhao, W., Chellappa, R., Phillips, P. J., & Rosenfeld, A. (2003). Face recognition: A literature survey. *ACM Computing Surveys*, *35*(4), 399–458. doi:10.1145/954339.954342

ADDITIONAL READING

Bailey, K. O., Okolica, J. S., & Peterson, G. L. (2014, June). User Identification and Authentication Using Multi-Modal Behavioral Biometrics. *Computers & Security*, *43*, 77–89. doi:10.1016/j.cose.2014.03.005

Bailey, K. O., Okolica, J. S., & Peterson, G. L. (2014, June). User Identification and Authentication Using Multi-Modal Behavioral Biometrics. *Computers & Security*, *43*, 77–89. doi:10.1016/j.cose.2014.03.005

Cappelli, R., Maio, D., Lumini, A., & Maltoni, D. (2007, September). Fingerprint Image Reconstruction from Standard Templates. *IEEE Transactions on Pattern Analysis and Machine Intelligence*, *29*(9), 1489–1503. doi:10.1109/TPAMI.2007.1087 PMID:17627039

Chang, K., Bowyer, K. W., Sarkar, K. W., & Victor, B. (2003, September). Comparison and Combination of Ear and Face Images in Appearance-Based Biometrics. *IEEE Transactions on Pattern Analysis and Machine Intelligence*, *25*(9), 1160–1165. doi:10.1109/TPAMI.2003.1227990

Chen, H., Valizadegan, H., Jackson, C., Soltysiak, S., & Jain, A. K. (2005). *Fake Hands: Spoofing Hand Geometry Systems*. Washington, DC, USA: Biometric Consortium.

Dinca, L. M., & Hancke, G. P. (2017, April). The Fall of One, the Rise of Many: A Survey on Multi-Biometric Fusion Methods. *IEEE Access: Practical Innovations, Open Solutions*, *5*, 6247–6289. doi:10.1109/ACCESS.2017.2694050

Dong, W., Sun, Z., & Tan, T. (2011, September). Iris Matching Based on Personalized Weight Map. *IEEE Transactions on Pattern Analysis and Machine Intelligence*, *33*(9), 1744–1757. doi:10.1109/TPAMI.2010.227 PMID:21173439

Espinoza, M., Champod, C., & Margot, P. (2011, January). Vulnerabilities of Fingerprint Reader to Fake Fingerprints Attacks. *Forensic Science International*, *204*(1-3), 41–49. doi:10.1016/j.forsciint.2010.05.002 PMID:21216360

Fatehpuria, A., Lau, D. L., & Hassebrook, L. G. (2006, April). Acquiring a 2D Rolled Equivalent Fingerprint Image from a Non-Contact 3D Finger Scan. *Proceedings of the Society for Photo-Instrumentation Engineers*, *6202*, 62020C. doi:10.1117/12.666127

Fox, N. A., Gross, R., Cohn, J. F., & Reilly, R. B. (2007, June). Robust Biometric Person Identification Using Automatic Classifier Fusion of Speech, Mouth, and Face Experts. *IEEE Transactions on Multimedia*, *9*(4), 701–714. doi:10.1109/TMM.2007.893339

Fridman, A., Stolerman, A., Acharya, S., Brennan, P., Juola, P., Greenstadt, R., & Kam, M. (2013). Decision Fusion for Multimodal Active Authentication. *IT Professional*, *15*(4), 29–33. doi:10.1109/MITP.2013.53

Liang, Y., Ding, X., Liu, C., & Xue, J.-H. (2016, January). Combining Multiple Biometric Traits with an Order-Preserving Score Fusion Algorithm. *Neurocomputing*, *171*, 252–261. doi:10.1016/j.neucom.2015.06.039

Maity, S., & Abdel-Mottaleb, M. (2015, February). 3D Ear Segmentation and Classification Through Indexing. *IEEE Transactions on Information Forensics and Security*, *10*(2), 423–435. doi:10.1109/TIFS.2014.2379437

Maltoni, D., Maio, D., Jain, A. K., & Prabhakar, S. (2003). *Handbook of Fingerprint Recognition*. NY: Springer Verlag.

Monwar, M. M., & Gavrilova, M. L. (2009, August). Multimodal Biometric System Using Rank-Level Fusion Approach. *IEEE Transactions on Systems, Man, and Cybernetics. Part B, Cybernetics*, *39*(4), 867–878. doi:10.1109/TSMCB.2008.2009071 PMID:19336340

Oloyede, M. O., & Hancke, G. P. (2016, September). Unimodal and Multimodal Biometric Sensing Systems: A Review. *IEEE Access: Practical Innovations, Open Solutions*, *4*, 7532–7555. doi:10.1109/ACCESS.2016.2614720

Paul, P. P., Gavrilova, M. L., & Alhajj, R. (2014, November). Decision Fusion for Multimodal Biometrics Using Social Network Analysis. *IEEE Transactions on Systems, Man, and Cybernetics. Systems*, *44*(11), 1522–1533. doi:10.1109/TSMC.2014.2331920

Paunwala, M., & Patnaik, S. (2013, July). Biometric Template Protection with DCT Based Watermarking. *Machine Vision and Applications*, *25*(1), 263–275. doi:10.100700138-013-0533-x

Poh, N., Kittler, J., Chan, C.-H., & Pandit, M. (2015). Algorithm to Estimate Biometric Performance Change Over Time. *IET Biometrics*, *4*(4), 236–245. doi:10.1049/iet-bmt.2014.0107

Raghavendra, R., Dorizzi, B., Rao, A., & Kumar, G. H. (2011, May). Designing Efficient Fusion Schemes for Multimodal Biometric Systems Using Face and Palmprint. *Pattern Recognition*, *44*(5), 1076–1088. doi:10.1016/j.patcog.2010.11.008

Rigas, I., & Komogortsev, O. V. (2015, December). Eye Movement-Driven Defense against Iris Print-Attacks. *Pattern Recognition Letters*, *68*, 316–326. doi:10.1016/j.patrec.2015.06.011

Roy, A., Memon, N., & Ross, A. (2017, September). Masterprint: Exploring The Vulnerability Of Partial Fingerprint-Based Authentication Systems. *IEEE Transactions on Information Forensics and Security*, *12*(9), 2013–2025. doi:10.1109/TIFS.2017.2691658

Sim, H. M., Asmuni, H., Hassan, R., & Othman, R. M. (2014, September). Multimodal Biometrics: Weighted Score Level Fusion Based On Non ideal Iris and Face Images. *Expert Systems with Applications*, *41*(11), 5390–5404. doi:10.1016/j.eswa.2014.02.051

Sultana, M., Paul, P. P., & Gavrilova, M. L. (2017). Social Behavioural Information Fusion in Multimodal Biometrics. *IEEE Transactions on Systems, Man, and Cybernetics. Systems*, 1–12. doi:10.1109/TSMC.2017.2690321

Tolosana, R., Vera-Rodriguez, R., Ortega-Garcia, J., & Fierrez, J. (2015). Preprocessing and Feature Selection for Improved Sensor Interoperability in Online Biometric Signature Verification. *IEEE Access: Practical Innovations, Open Solutions*, *3*, 478–489. doi:10.1109/ACCESS.2015.2431493

Wild, P., Radu, P., Chen, L., & Ferryman, J. (2016, September). Robust Multimodal Face and Fingerprint Fusion in the Presence of Spoofing Attacks. *Pattern Recognition*, *50*, 17–25. doi:10.1016/j.patcog.2015.08.007

Xu, Q., Chen, Y., Wang, B., & Liu, K. J. R. (2017, May). Radio Biometrics: Human Recognition Through a Wall. *IEEE Transactions on Information Forensics and Security*, *12*(5), 1141–1155. doi:10.1109/TIFS.2016.2647224

Xu, Y., Zhang, Z., Lu, G., & Yang, J. (2016, June). Approximately Symmetrical Face Images for Image Preprocessing in Face Recognition and Sparse Representation Based Classification. *Pattern Recognition*, *54*, 68–82. doi:10.1016/j.patcog.2015.12.017

Yan, X., Kang, W., Deng, F., & Wu, Q. (2015, March). Palm Vein Recognition Based On Multi-Sampling and Feature-Level Fusion. *Neurocomputing*, *151*, 798–807. doi:10.1016/j.neucom.2014.10.019

Chapter 7 Voice Biometric for Learner Authentication: Biometric Authentication

Kamaraj A.
Mepco Schlenk Engineering College, India

Selva Nidhyananthan S
Mepco Schlenk Engineering College, India

Kalyana Sundaram C.
Mepco Schlenk Engineering College, India

ABSTRACT

The objective of this chapter is to verify the identity of the claimed learner by extracting the prosodic features of the speech signal. TIMIT Acoustic-Phonetic Continuous Speech Corpus is used for learner verification using prosodic and articulation features such as energy, pitch, and formants. The prosodic feature includes pitch (F0), and articulation feature includes formants (F1-F7). From this database, for this project in the training phase, 200 learners were used and in the testing phase 160 learners were used. The pitch and formants were extracted using linear predictive analysis. The first seven formants were used for verification purpose. The feature set consists of eight features. The features are fed into the Guassian mixture model. In the Gaussian mixture model, parameters are estimated from the training and testing data using the iterative expectation-maximization. Log likelihood score is computed using these parameters, and then these scores are normalized to make decisions. The decision is made based on the threshold.

DOI: 10.4018/978-1-5225-7724-9.ch007

INTRODUCTION

Learner verification system must decide if a Learner is the person he/she claims to be. The person wishing to be verified first enters his claimed identity, and then on request from the verification system utters his verification phrase, and requests some decisions to be made in the event he is verified. The spoken utterance is processed to obtain a pattern which is compared to the stored reference pattern and then based on decision, accept or reject the claimed Learner. The major advantage of Learner verification system is voice mail, telephone banking, security purpose, authentication etc. The most important application is the forensics.

Speech Prosody

In linguistics, prosody is the rhythm, stress, and intonation of speech. Prosody may reflect various features of the learner or the utterance: the emotional state of the learner; the form of the utterance (statement, question, or command); the presence of irony or sarcasm; emphasis, contrast, and focus; or other elements of language that may not be encoded by grammar or choice of vocabulary.

In terms of acoustics, the prosodics of oral languages involve variation in syllable length, loudness and pitch. In sign languages, prosody involves the rhythm, length, and tension of gestures, along with mouthing and facial expressions. Prosody is typically absent in writing, which can occasionally result in reader misunderstanding. Orthographic conventions to mark or substitute for prosody include punctuation (commas, exclamation marks, question marks, scare quotes, and ellipses), and typographic styling for emphasis.

The details of a language's prosody depend upon its phonology. For instance, in a language with phonemic vowel length, this must be marked separately from prosodic syllable length. In similar manner, prosodic pitch must not obscure tone in a tone language if the result is to be intelligible. Although tone languages such as Mandarin have prosodic pitch variations in the course of a sentence, such variations are long and smooth contours, on which the short and sharp lexical tones are superimposed. If pitch can be compared to ocean waves, the swells are the prosody, and the wind-blown ripples in their surface are the lexical tones, as with stress in English. The word *dessert* has greater stress on the second syllable, compared to the noun *desert*, which has greater stress on the first (in its "arid land" meaning, but not in its "thing which is deserved" meaning); but this distinction is not obscured when the entire word is stressed by a child demanding "Give me dessert!" Vowels in many languages are likewise pronounced differently (typically less centrally) in a careful rhythm or when a word is emphasized, but not so much as to overlap with the formant structure of a different vowel. Both lexical and prosodic information are encoded in rhythm, loudness, pitch, and vowel formants.

Prosodic features are supra segmental. They are not confined to any one segment, but occur in some higher level of an utterance. These prosodic units are the actual phonetic "spurts", or chunks of speech. They need not correspond to grammatical units such as phrases and clauses, though they may; and these facts suggest insights into how the brain processes speech.

Prosodic units are marked by phonetic cues. Phonetic cues can include aspects of prosody such as pitch, pauses, and accents, all of which are cues that must be analyzed in context, or in comparison to other aspects of a sentence. Pitch, for example, can change over the course of a sentence. In English, falling intonation indicates a declarative statement while rising intonation indicates an interrogative statement. Pauses are important prosodic units because they can often indicate breaks in a thought and can also sometimes indicate the intended grouping of nouns in a list. Breathing, both inhalation and exhalation, seems to occur only at these pauses where the prosody resets. Prosodic units, along with function words and punctuation, help to mark clause boundaries in speech. Accents, meanwhile, help to distinguish certain aspects of a sentence that may require more attention. English often utilizes a pitch accent, or an emphasis on the final word of a sentence. Focus accents serve to emphasize a word in a sentence that requires more attention, such as if that word specifically is intended to be a response to a question.

"Prosodic structure" is important in language contact and lexical borrowing. For example, in Modern Hebrew, the XiXéX verb-template is much more productive than the XaXáX verb-template because in morphemic adaptations of non-Hebrew stems, the XiXéX verb-template is more likely to retain – in all conjugations throughout the tenses – the prosodic structure (e.g., the consonant clusters and the location of the vowels) of the stem.

Unique prosodic features have been noted in infant-directed speech (IDS) - also known as baby talk, child-directed speech (CDS). Adults, especially caregivers, speaking to young children tend to imitate childlike speech by using higher and more variable pitch, as well as an exaggerated stress. These prosodic characteristics are thought to assist children in acquiring phonemes, segmenting words, and recognizing phrasal boundaries. And though there is no evidence to indicate that infant-directed speech is necessary for language acquisition, these specific prosodic features have been observed in many different languages. Prosody is useful for listeners as they perform sentence parsing. Prosody helps resolve sentence ambiguity. For example, the sentence: "They invited Bob and Bill and Al got rejected," is ambiguous when written. But when the sentence is read aloud, prosodic cues like pauses and changes in intonation will make the meaning clear. The prosody of an ambiguous sentence biases a listener's interpretation of that sentence. Moving into national boundary in the above example will change the interpretation of the sentence. This result has been found in studies performed in both English and Bulgarian.

Prosody is also useful in expressing (for learners) and detecting (for listeners) sarcasm. The most useful prosodic feature in detecting sarcasm is a reduction in the mean fundamental frequency relative to other speech for humor, neutrality, or sincerity. While prosodic cues are important in indicating sarcasm, context clues and shared knowledge are also important.

LITERATURE SURVEY

Numerous measurements and signals have been propose and investigated for used in biometric recognition systems. Among the most popular measurements are fingerprint, face, and voice. While each has pros and cons relative to accuracy and deployment, there are two main factors that have made voice a compelling biometric. Learner verification is the process of verifying the claimed identity of a learner based on the speech signal from the learner. There are two types of learner verification systems: Text-Independent Learner Verification and Text-Dependent Learner Verification (Heck et. al, 1997). Text dependent learner verification system requires the learner saying exactly the enrolled or given password. This means that the training and testing word has to be same for learner verification process. Text independent Learner Verification is a process of verifying the identity without constraint on the speech content. Compared to Text dependent learner verification, it is more convenient because the user can speak freely to the system (Vuuren et. al, 1999). However, it requires longer training and testing utterances to achieve good performance. We have worked on text independent learner verification system (Newman et. al,). For the telephone banking and voice mail to identify the impostor text independent is most suitable.

The systems based on the analysis of physiological response as blood pressure, heart rate, respiration change present an initial phase where the signals are collected in configurations to be correlated with different emotional states and a subsequently recognition basing on the measure of indicators. The great importance and complexity of prosody in speech makes this subject an important area of research in speech synthesis applications. Researchers (Padmalaya Pattnaik et.al, 2012) investigated acoustic properties of speech prosody associated with five different emotions intentionally expressed in speech by male and female speakers. Results showed speech associated with love and sad emotions are characterized by longer utterance duration, and higher pitch. In this work the researchers failed to analyze universal prosody feature applicable for all five emotions considered. The profiles and effectiveness of these features are also investigated (Dogan Dalva et. al, 2014) for the sentence segmentation task on the Turkish BN data. In the work (Sarika et. al, 2015), the performances of various prosodic feature sets are compared on the sentence segmentation task. They implemented four feature extraction algorithms namely MFCC, LPC, PLP and GFCC and observed that GFCC features gave better result. The weakness of this approach is no speech rhythm feature is considered stand alone or combination feature. Prosody with fuzzy based hierarchical decision tree approach to improve the accuracy rate of speaker identification under noisy conditions is investigated (Manikandan et. al, 2016). But fuzzy cluster formation and levels of the tree are the additional complexities imposed in recognition task. Analysis of performance of Automatic Speech Recognition (ASR) in different emotional environments using prosody modification is done during testing (Vidyadhara Raju et. al, 2016). The limitations of this work are there is a need to explore and predict the prosody modification factor in real time. The database used is a limited vocabulary one and this study can be extended to large vocabulary datasets. Feature representation and model based adaptation techniques can also be explored to improve the ASR performance in Emotion conditions. A Phonetically Aware Sparse Representation of fundamental frequency and energy contour by using Continuous Wavelet Transform was examined (Berrak Si et. al, 2017).

Dialects represent the unique pronunciation patterns of a language, spoken among the community of speakers who belong to a particular geographical area. Dialects mainly exhibit grammatical, phonological, and prosodic differences among them. Pronunciation variations that occur in an individual's speaking styles are influenced by many surrounding factors related to the speaker, such as socioeconomic status, cultural background, geographical locations, education. Investigation of the significance of spectral and prosodic behaviors of speech signal has been carried out for dialect identification (Nagaratna et. al, 2018). Spectral features such as cepstral coefficients, spectral flux, and entropy are extracted from shorter frames. Prosodic attributes such as pitch, energy, and duration are derived from longer frames. IViE (Intona-tional Variations in English) speech corpus covering nine dialectal regions of British Isles has been considered, to evaluate the proposed approach. Since corpus is available in read and semi-spontaneous modes, the influence of spectral and prosodic behavior over these datasets is distinguishably articulated. Further, two distinct classification algorithms, namely support vector machine (SVM) and an ensemble of decision trees along with the SVM are used for identification of nine dialects. Dialect discriminating information captured from both features are used for constructing feature vectors.

Experiments have been conducted on individual and combinations of features. A better dialect recognition performance is observed with ensemble methods over a single independent SVM.

Parents, nursery staff, teachers or sports coaches will try to energize and direct behavior of children on a daily basis. They will use so-called motivational language to help children perform certain actions (e.g., tidy the room, learn the alphabet, try to write a sentence, swim) as well as try to encourage them to keep engaging in the proposed activity. In fact, it has been argued that parent-child interactions are characterized by the frequent use of motivational utterances. On numerous occasions will parents and children engage in motivational interactions, going from parental guidance in completing a new game, to involvement in children's school achievements, and influence on children's social behavior. Findings reported (Silke Paulmann et. al, 2018). support the hypothesis that distinct prosody patterns are used when motivating others in an autonomy-supportive or controlling way. They also extend past results to a new language and context, suggesting that motivational language is expressed similarly across languages, comparable to what has been found for emotional prosody. Finally, results suggest that the acoustic parameters intensity and speech rate are crucial when conveying these two distinct types of motivation. Specifically, controlling motivation is characterized by a louder voice and a faster speech rate than autonomy-supportive motivation. Pitch, on the other hand, does not seem to be a driving force.

Studies on the cognitive parallels between speech and music have primarily focused on perception. Research suggests that the two domains share 'basic processing mechanisms', including the ability to absorb and learn sound categories, to perceive regularities from rhythmic and melodic sequences, to integrate elements such as words and musical tones into syntactic structures, and to extract meaning from and emotional responses through sound. The preliminary investigation (Yundu Wang et. al, 2018) focused on two properties of production in both speech and music: timing variability and rate. Results of a comparison between speech rate and tempo were inconclusive for either language- based grouping or intra-speaker consistency. Overall tempo scores were higher for two Mandarin speakers than for two English speakers. Although English 1 had the fastest overall tempo in performance, this was not reflected in speech rate. English 2 had the fastest speech rate, but not a significantly fast tempo. It is interesting to note that English 1 was British, while English 2 and English 3 were from eastern and western United States, respectively. These results bring to attention the extent of prosodic variability between dialects of a language, including regional dialects.

Urdu/Hindi is an Indo-Aryan language that offers an interesting interplay of prosody, word order, and i(nformation) -structure. In Urdu/Hindi, all the major constituents in a sentence can scramble and this influences its i-structure. The sentence initial constituents are topicalized, the immediately preverbal position is reserved for focus, and the sentence final position is used to convey background information and given-ness. This is also the position for heavy NP shift. While this notion of i-structure is open to discussion, it shows that an investigation of prosody in Urdu/Hindi should take word order and i-structure into account. However, most of the existing re-searchers investigating the prosody of i-structure in Urdu/ Hindi do not consider the influence of structural positions. This has resulted in contradictory claims about the interplay of prosody, word order, and i-structure. The present research examines this interplay while controlling for variation in i-structure and word order by placing the target constituents only at the preverbal position in SOV sentences. It also investigates the production and perception of the prosody of narrow and corrective focus. The study (Farhat Jabeen et. al, 2018) investigates the production and perception of prosodic cues to realize narrow and corrective focus in Urdu/Hindi. We recorded SOV sentences with the target constituents at the preverbal position. Our results show that correctively focused nouns have longer syllable duration, wider F0 range, early alignment of F0 peaks, variant production of downtrend, and less steep post-focal compression as compared to narrowly focused nouns. We further set up a perception experiment to investigate if the difference in syllable duration of narrowly and correctively focused constituents is perceptible to Urdu/Hindi speakers. We manipulated syllable duration of the target constituents and presented them in contexts via a web-based interface. Twenty-nine respondents rated the naturalness of manipulated sentences in the given contexts. The analysis of respondents' ratings indicated that while they accepted both long and short durations in narrow focus, they rated long du-ration significantly better in corrective focus. Our results lend support to earlier claims about the prosodic cues of corrective focus in Urdu/Hindi and bring new evidence regarding the perceptual relevance of duration to cue corrective focus.

When one tries to imitate a given model utterance in context of language learning and actors' training, it is well-known that corrective feedback is effective to make efficient learning possible. Here, corrective feedback should include visualized comparison between a learner's utterance and a model one, which includes results of detecting the segments that exhibit segmental and/or prosodic gaps between them. When automating this process with speech technologies, an important principle is that technical errors should be avoided especially in the case where the automated process is introduced into educational context. In classes of language learning and actors' training, learners and trainees often compare their utterances prosodically with those from a model speaker. They want to know similarity of the pitch movement in their utterances to that in model utterances. To automate prosodic comparison, a classical but highly useful algorithm to compare F0 of a reference sound and that of an input sound is examined (Takuya OZURU et. al, 2018) for prosodic utterance comparison. The algorithm is Sub Harmonic Summation (SHS) and it is widely used for instrumental sounds and singing voices. In this paper, since both a reference stream and an input stream are vocal utterances, a modified algorithm of SHS is proposed and tested experimentally. It is interesting that the proposed method can calculate similarity in terms of pitch movement between the two utterances without extracting fundamental frequencies. Theoretical foundation and experimental verifications of the proposed method are presented. In experiments, it is shown that the method can detect speech segments produced with inadequate prosodic control better than the classical SHS, even without extracting fundamental frequencies.

Domains in linguistics are units to which rules apply, or to constituents of language structures, with no reference to 'clock time', though with implicit reference to 'rubber time' temporal relations. Domains in the 'clock time' of speech analysis are either the time domain or the frequency domain. The present approach (Takuya et. al, 2018) takes a syncretistic view of various time domain concepts, and addresses the multiple temporal domains which are constitutive for the field of prosody seen as a whole. The first major time domain (utterances, discourse) is just as dynamic as the others, in the sense that speech formulation and decoding strategies may change in real time, as with hesitation phenomena, and learning (for example of vocabulary) may occur even within very short sentential time domains. That the major time domains of language acquisition, and social, historical and evolutionary time, involve dynamic change is inherent in their definition. Much is known about short domain speech prosody, and very little is known about the longer major domains. First, a number of issues in prosodic phonology were discussed, pertaining to limitations on recursion types which underlie the oscillatory principles of 'rubber time' rhythm. Second, a number of 'rubber time' and 'clock time' approaches to speech timing hierarchies were discussed, from duration dispersion metrics through oscillator emulations of rhythm production in speech to an annotation-independent approach to rhythm analysis, the amplitude modulation spectrum analysis of speech, AEMS, and the Frequency Zone Hypothesis of rhythm, interpretable as emulation of perception. The incomplete exploratory analyses used to illustrate these concepts require much further research, for example of the development of the functionality of the AEMS in evolution.

Although it is fascinating to browse through the processing of multiple temporal domains, of course the everyday work of a scientist involves atomization of a given domain into manageable chunks, involving experimental, observational and hermeneutic analysis, and reconstruction of larger wholes. This keynote contribution is designed to encourage 'thinking outside the box' with unfamiliar strategies, looking at innovative methods which scholars have developed in neighboring disciplines as shown in this study, and, building on these, developing further innovative methods. The examples in this contribution are chosen more for illustration by exploratory visual inspection and illustration than for quantitative empirical proof, following the example of initial exploration in 'big data' projects, but here with 'tiny data'.There are many methodological boxes to be opened, re-thought and re-examined with care and new insight. And there are many wide-open domains for the study of prosody and its applications, especially in the longer major time domains. The future of prosody? It's about time.

Representation of Learner Verification

A learner verification system is composed of two distinct phases, a training phase and a test phase. Each of them can be seen as a succession of independent modules. First, speech parameters are extracted from the speech signal using exactly the same module as for the training phase. Then, the learner model corresponding to the claimed identity and a background model are extracted from the set of statistical models calculated during the training phase (Carey et. al, 1991). Finally, using the speech parameters extracted and the two statistical models, the last module computes some scores, normalizes them, and makes an acceptance or a rejection decision. The normalization step requires some score distributions to be estimated during the training phase or/and the test phase (Rosenberg et. al, 1992, Matsui et. al, 1994, Matsui et. al, 1995). Here the model we use is GMM-UBM model.

Application of Learner Verification

There are many applications to learner verification. The applications cover almost all the areas where it is desirable to secure actions, transactions, or any type of interactions by identifying or authenticating the person making the transaction. Currently, most applications are in the banking and telecommunication areas. Since the learner recognition technology is currently not absolutely reliable, such technology is often used in applications where it is interesting to diminish frauds but for which a certain level of fraud is acceptable. The main advantages of voice based authentication are its low implementation cost and its acceptability by the end users, especially when associated with other vocal technologies. Regardless of forensic applications, there are four areas where learner recognition can be used: access control to facilities, secured transactions, over a network (in particular, over the telephone), structuring audio information, and games. We briefly review those various families of applications. On-site applications regroup all the applications where the user needs to be in front of the system to be authenticated. Typical examples are access control to some facilities (car, home, warehouse), to some objects (locksmith), or to a computer terminal. Currently, ID verification in such context is done by mean of a key, a badge or a password, or personal identification number (PIN). Remote applications regroup all the applications where the access to the system is made through a remote terminal, typically a telephone or a computer. The aim is to secure the access to reserved services (telecom network, databases, web sites, etc.) or to authenticate the user making a particular transaction (e-trade, banking transaction, etc.). Organizing the information in audio documents is a third type of applications where learner recognition technology is involved. Typical examples of the applications are the automatic annotation of audio archives, learner indexing of sound tracks, and learner change detection for automatic subtitling. The need for such applications comes from the movie industry and from the media related industry. However, in a near future, the information structuring applications should expand to other areas, such as automatic meeting recording abstracting.

Data Collection

TIMIT is a corpus of phonemically and lexically transcribed speech of American English learners of different sexes and dialects. Each transcribed element has been delineated in time. TIMIT was designed to further acoustic-phonetic knowledge and automatic speech recognition systems. It was commissioned by DARPA and worked on by many sites, including Texas instruments Massachusetts institute of technology and, hence the corpus' name. There is also a telephone bandwidth version called NTIMIT (Network TIMIT).The TIMIT corpus of read speech is designed to provide speech data for acoustic-phonetic studies and for the development and evaluation of automatic speech recognition systems. TIMIT contain broadband recordings of 630 learners of eight major dialects of American English, each reading ten phonetically rich sentences. The TIMIT corpus includes time- aligned Orthographic, phonetic and word transcriptions as well as a 16-bit, 16 KHz speech waveform file for each utterance. It was recorded using a Sennheiser close-talking microphone at 16 kHz rate with 16 bit sample resolution. TIMIT contains a total of 6300 sentences (5.4 hours), consisting of 10 sentences spoken by each of 630 learners from 8 major dialect regions of the United States as in Table 1. All sentences were manually segmented at the phone level. The prompts for the 6300 utterances consist of 2 dialect sentences (SA), 450 phonetically compact sentences (SX) and 1890 phonetically-diverse sentences (SI). The training set contains 4620 utterances, but usually only SI and SX sentences are used, resulting in 3696 sentences from 462 learners. The test set contains 1344 utterances from 168 learners. The core test set, which is the abridged version of the complete testing set, consists of 192 utterances, 8 from each of 24 learners (2 males and 1 female from each dialect region). With the exception of SA sentences which are usually excluded from tests, the training and test sets do not overlap as shown in Figure 1. This speech corpus has been a standard database for the speech recognition community for several decades and is still widely used today, for both speech and learner recognition experiments. This is not only because each utterance is phonetically hand labeled and provided with codes for learner number, gender and dialect region, but also because it is considered small enough to guarantee a relatively fast turnaround time for complete experiments and large enough to demonstrate systems' capabilities. Here in this experiment 200 learners for training and 160 learners for testing impostors.

METHODOLOGY

Learner identification starts with speech signal pre-processing followed by Pre-Emphasis, Framing, Windowing, Prosody features extraction and GMM-UBM modeling. Pre-processing is the first and foremost process in any signal processing. It is the process of making the speech signal more suitable for further processing. Pre-processing of speech signals is segregating the voiced region from the silence or unvoiced portion of the captured signal. It is necessary in the development of a reliable speech or learner recognition system, to make the signal suitable for later stages. This is because most of the speech or learner specific attributes are present in the voiced part of the speech signals. Moreover, extraction of the voiced part of the speech signal by marking or removing the silence and unvoiced region leads to substantial reduction in computational complexity at later stages.

Figure 1.

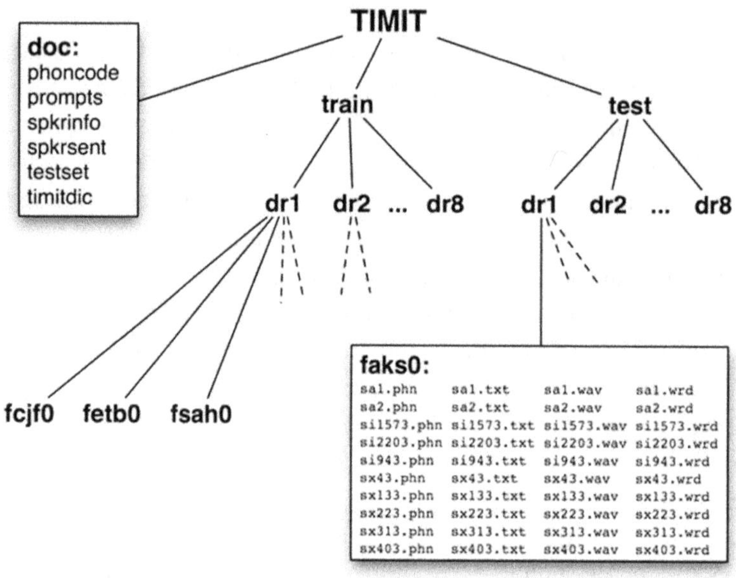

Other applications of classifying speech signals into silence or unvoiced region and voiced region are: Fundamental Frequency Estimation, Formant Extraction or Syllable Marking, Stop Consonant Identification and End Point Detection for isolated speech signals. One of the accepted ways of labelling a speech signal is the three state representation; (i) Silence region, where no speech is produced, (ii) Unvoiced region, where the resulting waveform is a periodic or random in nature as the vocal cords do not vibrate, and (iii) Voiced region, where the resulting waveform is quasi-periodic as the vocal cords are tensed and hence vibrate periodically. It should be made clear that the segmentation of the speech signal in the aforementioned regions is not very rigid; however, it has been noted that small errors in boundary locations seldom have any significant defect in most of the applications. The energy in the unvoiced portion of the speech signal is usually low, this region is usually clubbed together with the silence region and segregated from the voiced part. The removal of the unvoiced portion makes the processing easier. The extracted voiced part is then used for further processing.

Pre-Emphasis

Pre-emphasis is a form of spectral tilting that boosts the energy of high frequency components in the signals. Speech segments usually fall in the frequency range anywhere below 4 kHz. Therefore the uttered low frequencies are higher in amplitude than the high frequencies and higher frequencies of speech signal are generally weak. It thereby results in the fact that high frequency energy may not be used to extract features for recognition. But for effective functioning of the learner recognition systems, higher frequency components are necessary.

Pre emphasis does the job of taking care of these discrepancies. It equalizes the spectral tilt in speech and the signal is spectrally flattened. In other words, speech signals have a spectrum that falls off at high frequencies. In some applications it is desirable that this high-frequency falloff be compensated by 'pre-emphasis'. It is performed by a FIR filter as in equation1.

Table 1. TIMIT acoustic-phonetic continuous speech corpus

Item Name:	TIMIT Acoustic-Phonetic Continuous Speech Corpus
Authors:	John S. Garofolo, Lori F. Lamel, William M. Fisher, Jonathan G. Fiscus, David S. Pallett, Nancy L. Dahlgren, and Victor Zue
LDC Catalog No.:	LDC93S1
NIST Catalog No.:	1-1.1
ISBN:	1-58563-019-5
Data Type:	Speech
Sample Rate:	16000 Hz
Sampling Format:	1-channel pcm
Data Source(s):	microphone speech
Application(s):	speech recognition
Language(s):	English
Language ID(s):	Eng
Distribution:	1 CD, Web Download
Member fee:	$0 for 1993 members
Non-member Fee:	US $250.00
Reduced-License Fee:	US $150.00
Extra-Copy Fee:	US $150.00
Non-member License:	Yes
Online documentation:	Yes
Licensing Instructions:	Subscription & Standard Members, and Non-Members
Citation:	JohnS.Garofolo,etal.1993 TIMIT Acoustic-Phonetic Continuous Speech Corpus Linguistic Data Consortium, Philadelphia

$$y(n) = x(n) - \alpha\, x(n-1) \tag{1}$$

where

$x(n)$ is the speech signal
$y(n)$ is the output pre-emphasized speech
α is the adjustable parameter and it is chosen as 0.97.

Framing

Speech signal appears to be stationary when looked at a short time point of view. But normally, it is not stationary. A speech signal is divided into frames os short duration so that stationary property is satisfied for the Fourier transformation.

Framing Considerations

The speech signal is segmented into frames of appropriate samples. While framing; overlapping between frames is essential. If the frames have no overlap, there may be loss of information, due to presence of small gap between adjacent frames. The number of samples in a frame can be obtained as in equation 2,

$$n = t_{st} f_s \tag{2}$$

where t_{st} is the time period of a frame and f_s is the sampling frequency of the signal.

The first frame consists of the first *N* samples. The second frame begins M samples after the first frame and overlaps it by *N-M* samples. In this experiment, our input signal is cut into frames as shown in Figure 2. Each frame is of 256 samples. The entire speech signal is cut into 16ms.And framing is done by overlapping method. Here, 50% overlapping is done for framing the signal.

Windowing

Windowing of a simple waveform like cos ωt causes its Fourier transform to develop non-zero values (commonly called spectral leakage) at frequencies other than ω. The leakage tends to be worst (highest) near ω and least at frequencies farthest from ω. If the waveform under analysis comprises two sinusoids of different frequencies, leakage can interfere with the ability to distinguish them spectrally. If their frequencies are dissimilar and one component is weaker, then leakage from the larger component can obscure the weaker one's presence. But if the frequencies are similar, leakage can render them irresolvable even when the sinusoids are of equal strength. The rectangular window has excellent resolution characteristics for sinusoids of comparable strength, but it is a poor choice for sinusoids of disparate amplitudes. This characteristic is sometimes described as low-dynamic-range.

At the other extreme of dynamic range are the windows with the poorest resolution. These high-dynamic-range low-resolution windows are also poorest in terms of sensitivity; this is, if the input waveform contains random noise close to the frequency of a sinusoid, the response to noise, compared to the sinusoid, will be higher than with a higher-resolution window. In other words, the ability to find weak sinusoids amidst the noise is diminished by a high-dynamic-range window. High-dynamic-range

Figure 2.

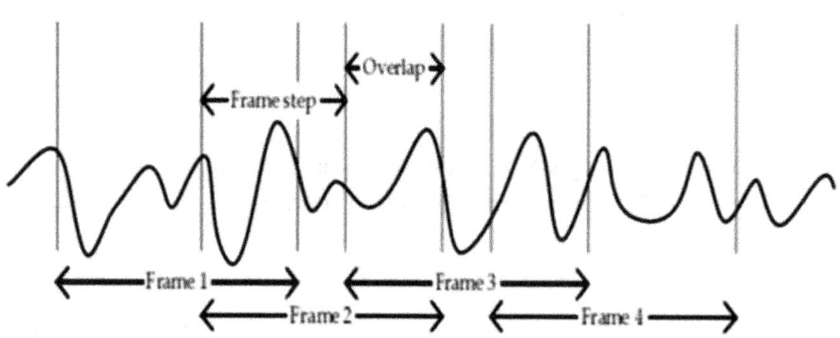

windows are probably most often justified in wideband applications, where the spectrum being analyzed is expected to contain many different components of various amplitudes.

In between the extremes are moderate windows, such as Hamming and Hanning. They are commonly used in narrowband applications, such as the spectrum of a telephone channel. In summary, spectral analysis involves a tradeoff between resolving comparable strength components with similar frequencies and resolving disparate strength components with dissimilar frequencies. That tradeoff occurs when the window function is chosen.

Hamming Window

In this experiment, Hamming window is used. Hamming Window has the form

$$w(n) = 0.54 - 0.46\cos(\frac{2\pi n}{N-1})\ 0 \leq n \leq N-1 \quad (3)$$

It is a combination of a rectangular window and a Hamming window. This window has similar properties to the Hamming window with the added feature that it suppresses the first side lobe. However, the fall-off rate of the side lobes is only 20dB per decade compared to 60dB per decade for the Hamming window. This hamming window function in time domain is shown in the Figure 3.

The windowed signal is given by,

Figure 3.

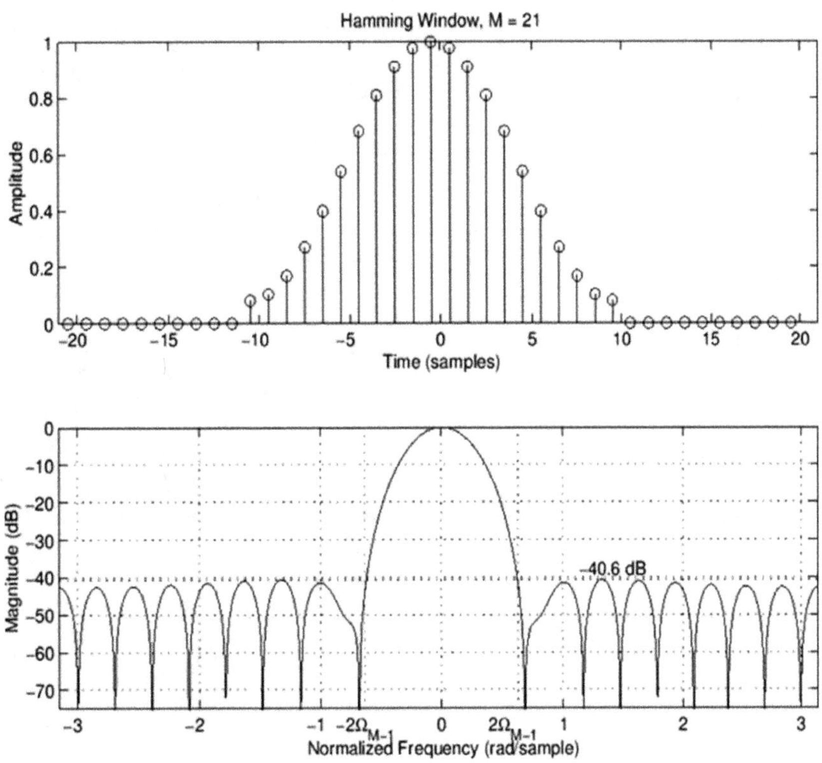

$$S_w(n) = y(n) * w(n) \tag{4}$$

where

$y(n)$ is the pre-emphasized signal
$w(n)$ is the window used.

Comparison of Hamming Window With Rectangular Window

The side lobes of hamming window are lower than rectangular window and so leakage effect is decreased. In hamming window, high resolution is not required. The main lobe of rectangular window is narrow; it is half as wide as hamming window. As the main lobe narrows, the ability to distinguish two closely spaced frequency components increases.

Moreover as the spectral resolution improves, the window energy spreads into its side lobes, increasing spectral leakage and decreasing amplitude accuracy. A trade off occurs between amplitude accuracy and spectral resolution. This problem is overcome by hamming window shown in Figure 3 as its main lobe are wider. Figure 4 shows; the bandwidth of the Hamming window is about twice the bandwidth of a rectangular window. The side lobe attenuation for the hamming however is 20 dB better outside the pass band.

Figure 4.

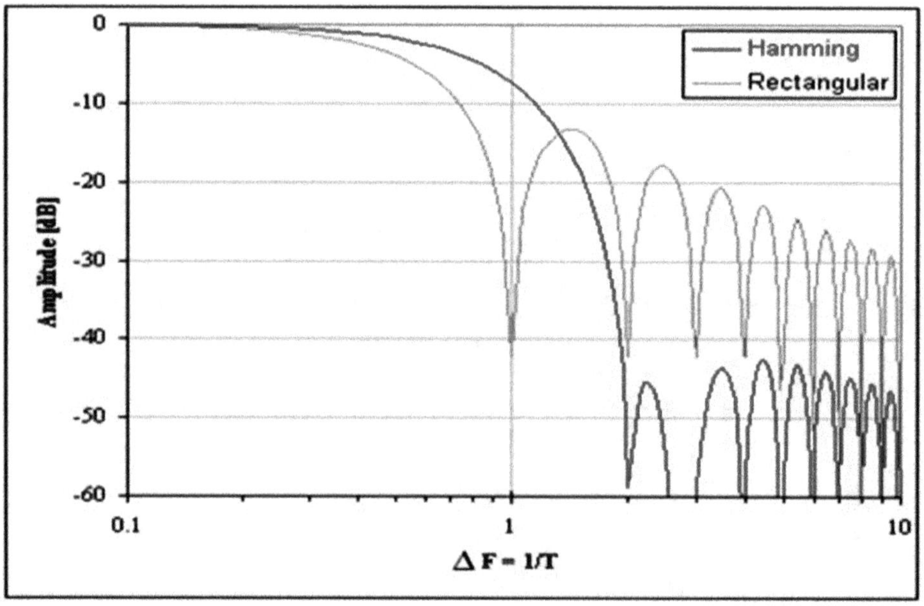

Feature Extraction

When the input data to an algorithm is too large to be processed and it is suspected to be notoriously redundant (much data, but not much information) then the input data will be transformed into a reduced representation set of features. Transforming the input data into the set of features is called feature extraction.

Prosody and Articulation Features

Prosody is conveyed through three different elements: intonation, rhythm and stress which are perceived by listeners as changes in fundamental frequency, sound duration and loudness, respectively. Humans use different methods to produce these linguistic phenomena that determine prosody Changes in the respiratory system and laryngeal muscles, for example, have an important role in controlling the fundamental frequency. Fundamental frequency is determined physiologically by the number of cycles that the vocal folds make in a second, and they are the natural result of the length of these structures. On the other hand, vocal intensity is directly related to the subglottic pressure of the air column. This subglottic pressure, in turn, depends on several factors such as amplitude of vibration and tension of vocal folds. It is well-known that prosody plays an important role in the speech act and communication in everyday speech.

Articulation is the act or manner of producing a speech sound, especially a consonant. Articulators change the structure of the formants in the speech, which is crucial for learner identification.

Figure 5 shows all feature groups used in this paper and corresponding number of features in each group. The four feature groups Energy, Pitch, Formants and Zero Crossing Rate (ZCR) are most common in learner verification. Here we have used feature groups namely energy, ZCR, Pitch (*F0*) and Formants. In Formants the first seven frequencies (*F1-F7*) are used.

In the feature extraction step the input sample is first pre-processed and the four features pitch, energy, ZCR, format frequencies are extracted by using suitable algorithms for each feature. The overall feature extraction process is shows in Figure 6.

Figure 5.

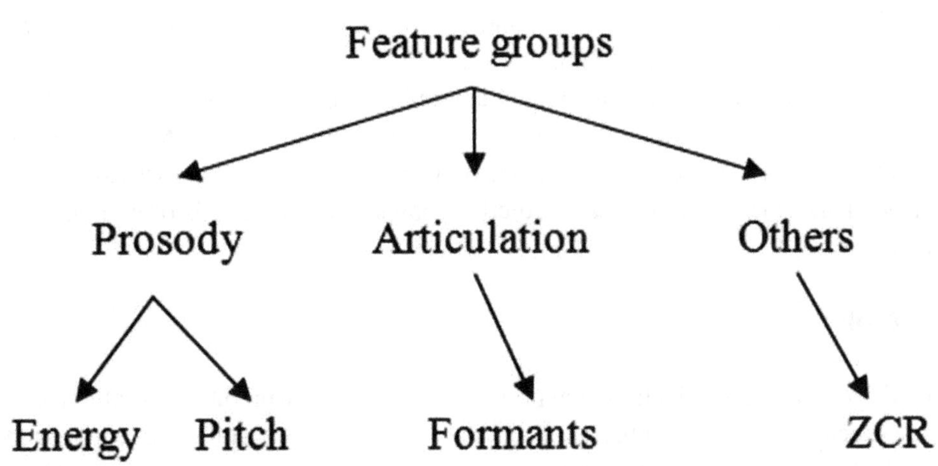

Figure 6.

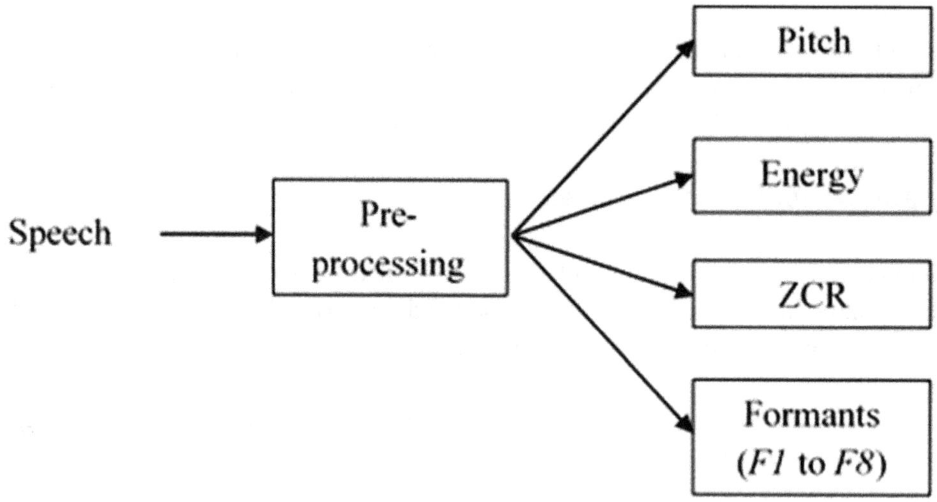

Linear Predictive Coefficient Analysis

One of the most powerful speech analysis techniques is the method of linear predictive analysis. The method has become the predominant technique for estimating the basic speech parameters, e.g., pitch, formants, spectra, vocal tract area functions, and for representing speech for low bit rate transmission or storage. The important of this method lies both in its ability to provide extremely accurate estimates of the speech parameters, and in its relative speed of computation.

The basic idea behind linear predictive analysis is that speech sample can be approximated as linear combination of past speech samples. By minimising the sum of squared differences (over a finite interval) between the actual speech samples and linearly predicted ones, a unique set of predictor co efficient can be determined. (The predictor co efficient are the weighting coefficients used in the linear combination.

The philosophy of linear prediction is intimately related to the basic speech synthesis modelling which it was shown that speech can be modelled as the output of a linear, time varying system excited by either by quasi- periodic pulses (during voiced speech), or random noise (during unvoiced speech). The linear prediction method provides a robust, reliable, and accurate method for estimating the parameters that characterise the linear, time varying system.

It was suggested that a linear predictor could be applied in a differential quantisation scheme to reduce the bit of the digital representation of the speech waveform. In adaptive DPCM coding the emphasis is on finding a predictor that will reduce the variance of the difference signal so that quantisation error can also be reduced. This general set of linear predictive analysis is technique is often referred to as linear predictive coding or LPC.

Basic Principles

The basic model for speech production is appropriate for the linear predictive analysis. In this model, the composite spectrum effects of radiation, vocal tract and glottal excitation are represented by a time-varying digital filter whose steady state system function is of the form

$$H(z) = \frac{S(z)}{U(Z)} = \frac{G}{1 - \sum_{K=1}^{p} a_k z^{-k}} \tag{5}$$

The system can be excited by an impulse train for voiced speech or a random sequence of unvoiced speech. The pitch period and voiced/unvoiced parameters can be estimated using linear predictive analysis. The speech samples *s* (*n*) can be given by using simple difference equation

$$s(n) = \sum_{k=1}^{p} a_k s(n-k) + Gu(n) \tag{6}$$

The linear predictor with predictor coefficients α_k , and order p is defined as a system whose output is

$$\bar{s}(n) = \sum_{k=1}^{p} \alpha_k s(n-k) \tag{7}$$

The system function of this linear predictor is

$$P(z) = \sum_{k=1}^{p} \alpha_k z^{-k} \tag{8}$$

The prediction error is defined as

$$e(n) = s(n) - \bar{s}(n) = s(n) - \sum_{k=1}^{p} \alpha_k s(n-k) \tag{9}$$

Thus the prediction error filter is the output of the system whose transfer function is

$$A(z) = 1 - \sum_{k=1}^{p} \alpha_k z^{-k} \tag{10}$$

Comparison of equations suggests that if, then and in such condition, prediction error filter A(z) will be an inverse filter for the system H(z)

$$H(z) = \frac{G}{A(z)} \tag{11}$$

The basic problem of linear prediction here is to determine a set of predictor coefficients directly from speech signal in such a manner as to obtain a good estimate of spectral properties of speech signal through the use of above equations.

Pitch Estimation

Pitch frequency is extracted using linear predictive coefficient analysis. Simple Inverse Filtering Tracking Algorithm is employed to extract this feature. The input signal is low pass filtered and then the sampling rate is reduced by decimation process as shown in Figure 7. The decimated output is then analyzed using the autocorrelation method with a value of $p = 4$ for the filter order. The signal is then inverse filtered to get an approximately a flat spectrum. The purpose of the linear predictive analysis is to spectrally flatten the input signal. The short-time autocorrelation of the inverse filtered signal is computed and the largest peak in the appropriate range is chosen as the pitch period. To obtain additional resolution the signal is interpolated.

Formants

Formants are the distinguishing or meaningful frequency components of human speech and of singing. By definition, the information that humans require to distinguish between vowels can be represented purely quantitatively by the frequency content of the vowel sounds. In speech, these are the characteristic partials that identify vowels to the listener. Most of these formants are produced by tube and chamber resonance, but a few whistle tones derive from periodic collapse of Venturi effect low-pressure zones. The formant with the lowest frequency is called F_1, the second F_2, and the third F_3. Most often the two first formants, F_1 and F_2, are enough to disambiguate the vowel. These two formants determine the quality of vowels in terms of the open/close and front/back dimensions (which have traditionally, though not entirely accurately, been associated with the position of the tongue). Thus the first formant F_1 has a higher frequency for an open vowel (such as [a]) and a lower frequency for a close vowel (such as [i] or [u]); and the second formant F_2 has a higher frequency for a front vowel (such as [i]) and a lower frequency for a back vowel (such as [u]). Vowels will almost always have four or more distinguishable formants; sometimes there are more than six. However, the first two formants are most important in determining

Figure 7.

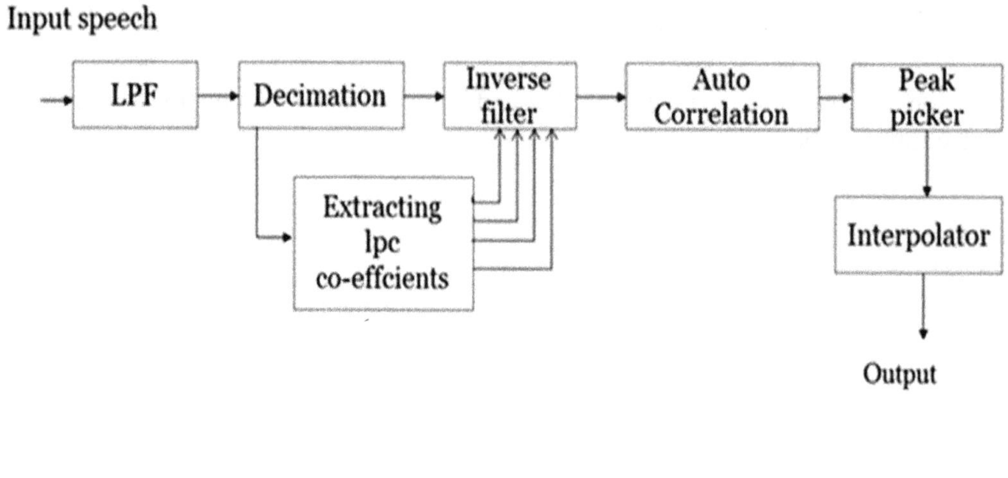

vowel quality, and this is often displayed in terms of a plot of the first formant against the second formant, though this is not sufficient to capture some aspects of vowel quality, such as rounding. Nasals usually have an additional formant around 2500 Hz. The liquid usually has an extra formant at 1500 Hz, while the English "r" sound ([ɹ]) is distinguished by virtue of a very low third formant (well below 2000 Hz).

Plosives (and, to some degree, fricatives) modify the placement of formants in the surrounding vowels. Bilabial sounds (such as /b/ and /p/ in "ball" or "sap") cause a lowering of the formants; velar sounds (/k/ and /g/ in English) almost always show F_2 and F_3 coming together in a 'velar pinch' before the velar and separating from the same 'pinch' as the velar is released; alveolar sounds (English /t/ and /d/) cause less systematic changes in neighboring vowel formants, depending partially on exactly which vowel is present. The time-course of these changes in vowel formant frequencies are referred to as 'formant transitions'. If the fundamental frequency of the underlying vibration is higher than a resonance frequency of the system, then the formant usually imparted by that resonance will be mostly lost. This is most apparent in the example of soprano opera singers, who sing high enough that their vowels become very hard to distinguish.

Control of resonances is an essential component of the vocal technique known as overtone singing, in which the performer sings a low fundamental tone, and creates sharp resonances to select upper harmonics, giving the impression of several tones being sung at once. Spectrograms are used to visualize formants.

Vocal Tract Filtering and Formant Frequencies

The vocal tract is made up of the oral cavity from the larynx to the lips and the nasal passage coupled with the oral passage (through the vellum). The oral tract can take on various different configurations depending upon the shape and movement of the tongue, mouth, teeth, lips, jaw, etc. The vocal tract provides frequency shaping to the output from the larynx and also generates new sources for sound production (impulsive source).

The vocal tract can be modeled as a linear filter with resonances. The resonances frequencies of the vocal tract are called formant frequencies. Graphically, the peaks of the vocal tract response correspond roughly to its formant frequencies. Therefore, if the vocal tract is modeled as a time-invariant, all-pole linear system, then each of the conjugate pole pairs corresponds to a formant frequency (resonance frequency). The formant frequencies vary with different vocal tract configurations and therefore, formant frequencies vary in speech with time as the vocal tract changes its shape. The peaks of the vocal tract response in each configuration correspond roughly to its formant frequencies The first resonance of the vocal tract is called the first formant frequency (or $F1$), the second resonance of the vocal tract is called the second formant frequency (or $F2$), and so on. For perfectly voiced and periodic speech (as in sustained vowels) the vocal tract can be accurately modeled by the stable all-pole model. To solve the transfer function of the vocal tract, parametric modeling, and so linear prediction, has been invented. This class of technique attempt to optimally model the spectrum as an autoregressive process.

Formant Extraction

Formant features can be interpreted as adaptive non-uniform samples of the signal spectrum that are located in the resonance frequencies of the vocal tract and normally happen to have higher signal-to-noise ratios than the other parts. The number and the position of these frequencies along the frequency axis might differ depending on the phonemes and the position of the window along the phoneme (i.e. beginning or ending part of a phoneme). Along with the formants (the resonance frequency), we might use the bandwidth and/or magnitude of the spectrum in that particular frequency to encode the properties of the speech and use them in different applications such as speech recognition, enhancement, noise reduction, hearing aid adaptive filters, etc. There are several methods of formant extraction such as peak picking, HMM and LP model pole extraction. The main method used in this work is the peak picking method.

Formants are estimated by the predictor parameters. The way of estimating formants is to obtain the spectrum, and choose the formants by peak picking method. Formants *F1-F7* is extracted in this method.

Energy

The amplitude of the signal varies appreciably with time. In particular the amplitude of the unvoiced segments is generally much lower than amplitude of voiced segments. The short-time energy of the speech signal provides a convenient representation that reflects these amplitude variations. In general, we can define the short-time energy as

$$E_n = \sum_{n=1}^{N} x(n).x^*(n) \tag{12}$$

Zero Crossing Rate

Zero Crossing is said to occur if successive samples have different algebraic signs. The rate at which zero crossing occurs is a simple measure of frequency content of a signal. This is particularly true of narrow band signals. For example a sinusoidal signal of frequency F_0, sampled at a rate F_s, has F_s / F_0 samples per cycle of the sine wave.

Each cycle has two zero crossings so the long-time average rate of zero-crossings is

$$Z = 2F_s / F_0 \tag{13}$$

An appropriate definition is

$$Z_n = \sum_{m=-\infty}^{\infty} |\text{sgn}[x(m)] - \text{sgn}[x(m-1)] \tag{14}$$

where

$$\text{sgn}[x(n)] = \begin{cases} 1, x(n) \geq 0 \\ -1, x(n) \leq 0 \end{cases}$$

GMM-UBM Model

Gaussian Mixture Models (GMMs) are among the most statistically mature methods for clustering (though they are also used intensively for density estimation). In this model, introduce the concept of clustering, and see how one form of clustering in which assume that individual data points are generated by first choosing one of a set of multivariate Gaussians and then sampling from them can be a well-defined computational operation (Reynolds et. al, 1992). Then see how to learn such a thing from data, and discover that an optimization approach not used in any of the previous Andrew Tutorials can help considerably here. This optimization method is called Expectation Maximization (EM) (Reynolds et. al, 1995). Here spend some time giving a few high level explanations and demonstrations of EM, which turns out to be valuable for many other algorithms beyond Gaussian Mixture Models.

The following section aims at describing the Gaussian mixture model and emphasizes its use in the field of learner verification system. The previous section aimed at extracting features from an input audio speech using the prosodic and articulation features. The GMM algorithm takes as an input a sequence of vectors provided by the extracted feature and uses it to create one model per learner, which is called the Gaussian mixture model. In this section, we will describe the Gaussian mixture model and its parameterization. First, the Gaussian mixture model is a "mixture density", characterized as a sum of M component densities. Each component density is a product of a "component Gaussian" with a "mixture weight" (Reynolds et. al, 2000) . Each individual component Gaussians represents *acoustic classe*. These classes reflect specific vocal tract configuration proper to a learner and are therefore, useful for modeling learner identity. Second, a Gaussian mixture density provides a good estimation independently speaking of the time differences between recording sessions. In other words, the GMM is not susceptible to natural vocal changes provoked by several factors such as aging (Sarkar et. al, 2010, Rosenberg et. al, 1996, Gauvain et. al, 1994) or when a given learner gets a cold. The entire GMM-UBM model structure is shown in Figure 8.

Figure 8.

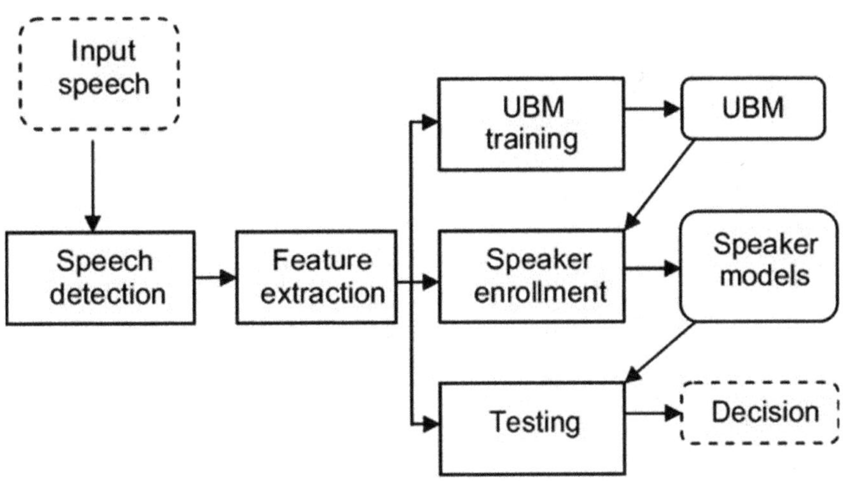

Maximum Likelihood Parameter Estimation

For a given training vectors and a GMM configuration, we have to estimate the parameters of the GMM, λ, for the best matches for the distribution of the training feature vectors. The most popular and well-known method is maximum likelihood (ML) estimation (Leggetter et. al, 1997).

The main purpose of ML estimation is to find the model parameters which maximize the likelihood of the GMM given the training data (Dempster et. al, 1977). For a sequence of T training vectors *X= x1, x2, x3,..., xT*, the GMM likelihood can be defined as

$$p(\boldsymbol{x}|\lambda) = \prod_{t=1}^{T}(X_t \mid \lambda). \tag{15}$$

The learner-specific GMM parameters are estimated by the Expectation-Maximization (EM) algorithm using training data spoken by the corresponding learner. The basic idea of the EM algorithm is, beginning with an initial model λ, to estimate a new model λ' such that $p(\boldsymbol{x}|\lambda') \geq p(\boldsymbol{x}|\lambda)$. The new model then becomes the initial model for the next iteration and the process is repeated until some convergence threshold is reached (Newman et. al, 1996). On each EM iteration, the following re-estimation formulas are used which guarantee a monotonic increase in the model's likelihood value,

Mixture Weights:

$$w_i \cdot \frac{1}{T}\sum_{i=1}^{T} pr\ (i \mid X_t, \lambda). \tag{16}$$

Means:

$$\mu_i .= \frac{\sum_{i=1}^{T} pr(i \mid X_t, \lambda) X_t}{\sum_{i=1}^{T} pr(i \mid X_t, \lambda)}. \tag{17}$$

Variance (diagonal covariance):

$$\sigma_i^2 .= \frac{\sum_{i=1}^{T} pr(i \mid X_t, \lambda) X_i^2}{\sum_{i=1}^{T} pr(i \mid X_t, \lambda)} .- \mu_i^2 . \tag{18}$$

The a posteriori probability for component *i* is given by

$$pr\left(i \mid X_t, \lambda\right) = .\frac{w_i b_i\left(x\right)}{\sum_{k=1}^{M} w_k b_k\left(x\right)}. \tag{19}$$

IMPLEMENTATION

In this experiment, for the learner verification using prosodic features, a TIMIT database was used which contains 630 learners and each learner contains 10 different sessions. The architecture contains of two phase, a training phase and testing phase as available in Figure 9. In the training phase features are extracted from the samples of database and the extracted values are compared in the GMM model with the values which are extracted from the testing phase. From the GMM model the log likelihood score values are obtained. After obtaining the score values, in order to obtain the values in range score normalization are done. Based on the threshold values the decision is made.

$\theta > 0, accept$

$\theta < 0, reject$

Implementation Results

In this experiment learner verification is done using prosodic features and good results are obtained. The following graphs are our intermediate results:

Figure 10 shows the session of the speech signal obtained from TIMIT database. This is subjected to pre-emphasis.

Figure 11 shows the pre-emphasized signal which boosts the high frequency components.

Figure 9.

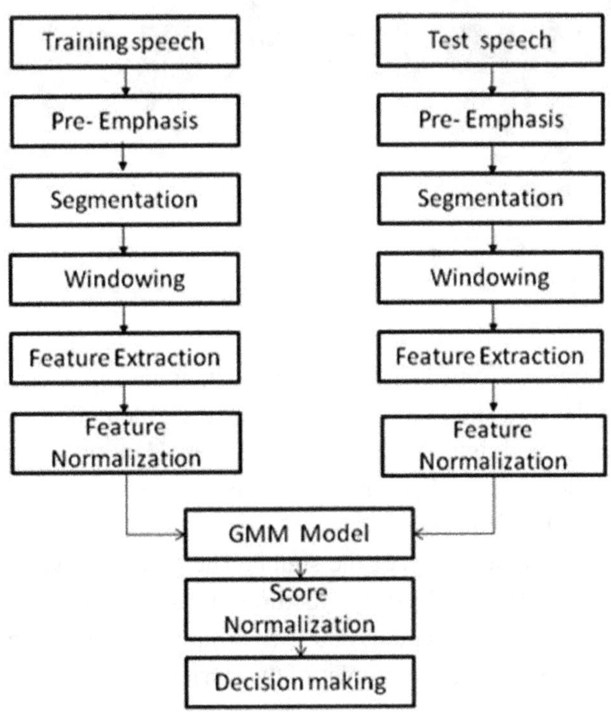

Figure 10.

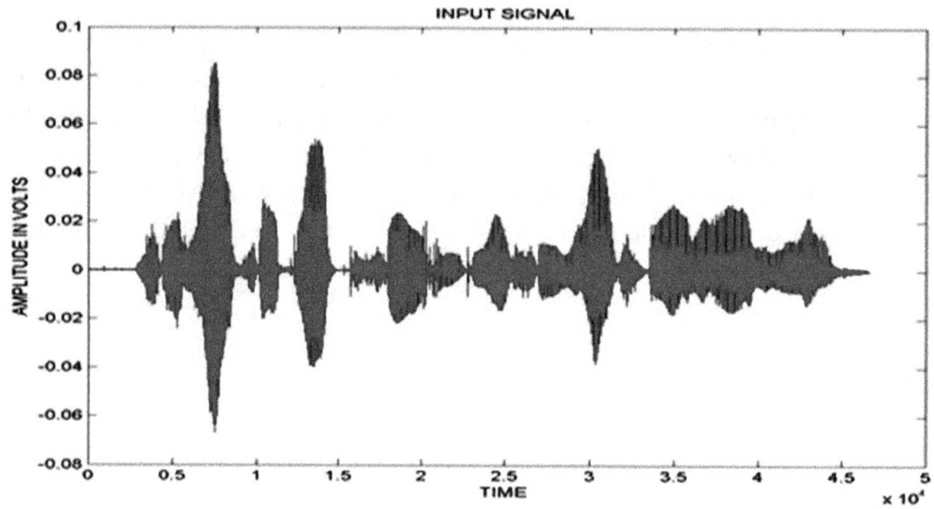

Figure 11.

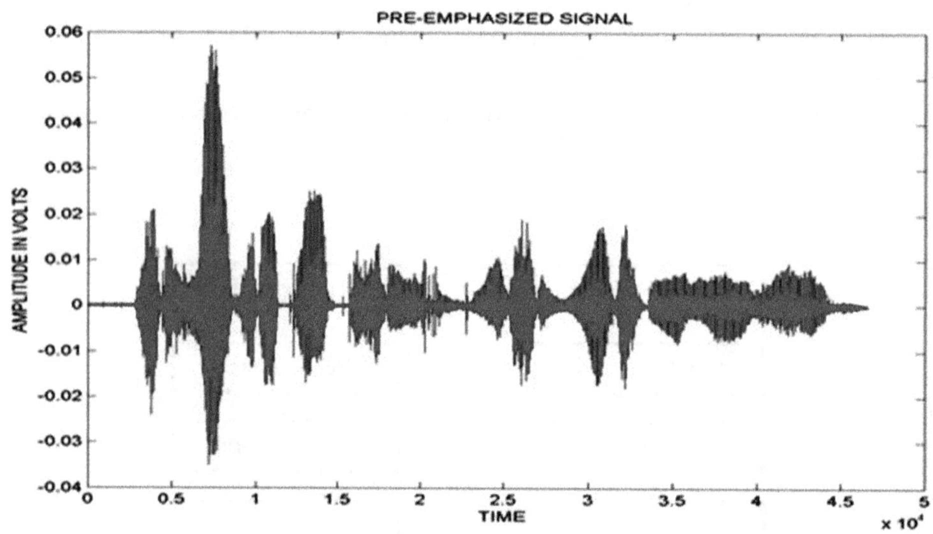

Figure 12 shows the segmented signal of input signal. The input signal contains 45000 samples. The signal is cut into 256 samples as a single frame. Here during segmentation 50% overlapping is done.

Figure 13 shows the segmented signal after windowing. Here Hamming window technique is applied. Thus our pre processing is done. After that this signal is fed as input to the feature extraction process.

Figure 14 shows that energy of the single frame.

Figure 15 shows the pitch and formants of the signal. The highest peak is the pitch which is picked by using peak picking method. The results of the methods are shown in Figure 16 & 17.

The experiment was carried out by training 200 speech samples with three sessions for each learner from the TIMIT database. Then 360 speech samples of the learners are tested individually with the learner

Figure 12.

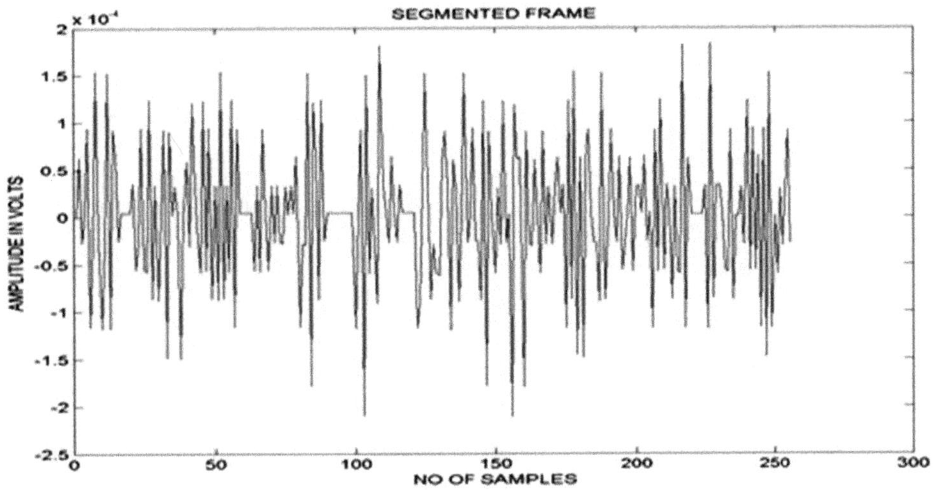

Figure 13.

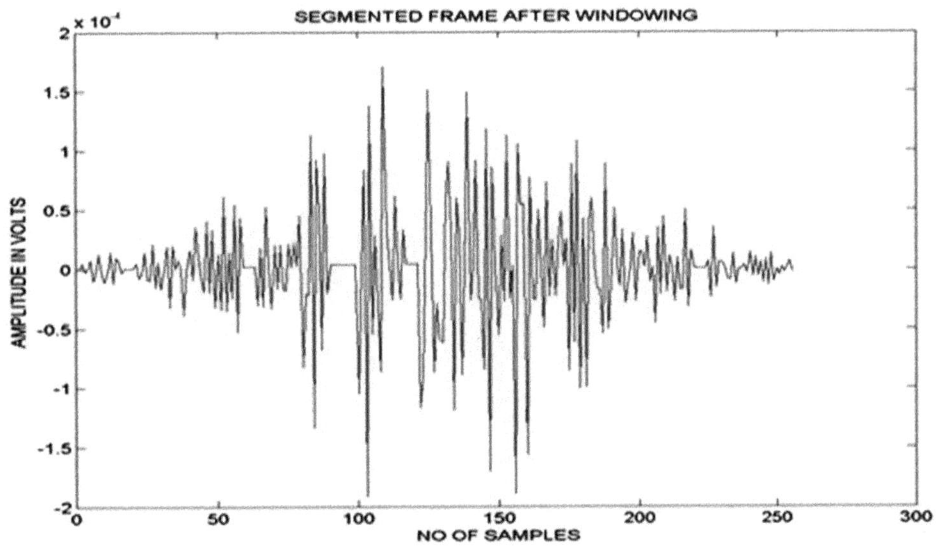

model, based on the matching of features log likelihood scores were computed. We have also compared the speech samples from the background model with the learner model. These log likelihood scores provided a range of values that helped in calculating an average imposter score, with which the scores were then normalized. From the normalized score it was inferred that the speech samples belonging to the learner arrived at a positive score whereas the imposter learner's score was negative.

For this analysis we have extracted 8 features from the speech samples pitch and formants (*F1-F7*) using linear predictive coefficients method. Initially Learner Verification was performed only with the pitch feature and whose results are shown in Figure 18 & 19. Later all the features were fused together to yield a better result.

Figure 14.

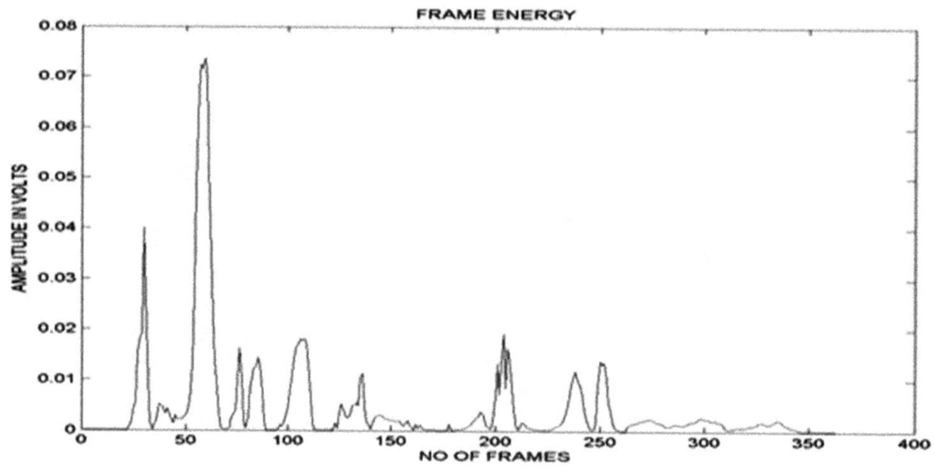

Figure 15.

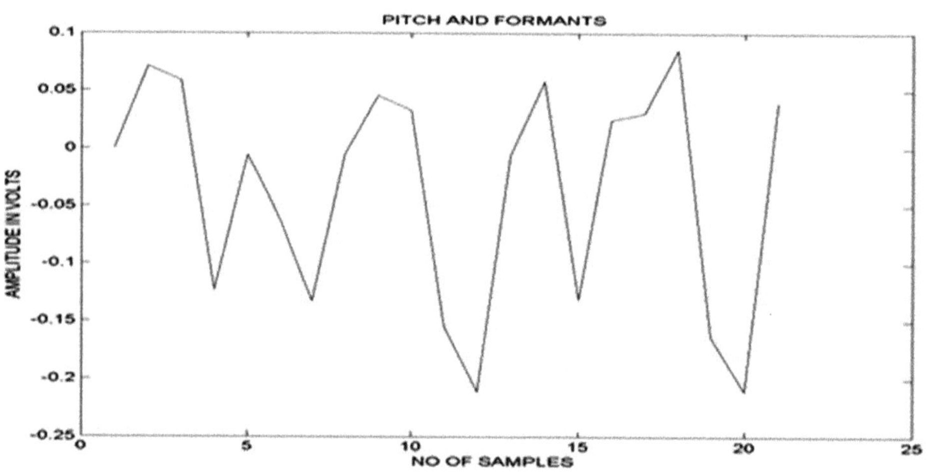

DET Plot

A popular method to find performance of the learner verification system is the equal error rate (EER) which corresponds to the operating point where probability of False Acceptance Rate (FAR %) or P(fa) is equal to False Rejection Rate (FRR%) or P(fr). Graphically, it corresponds to the intersection of the DET curve with the first bisector curve. The EER performance measure rarely corresponds to a realistic operating point. However, it is a quite popular measure of the ability of a system to separate impostors from true learners. Another popular measure is the half total error rate (HTER) which is the average of the two error rates FAR and FRR (Isobe et. al, 1999). Here we used the first method (EER) to show the performance in our experimental results. Furthermore, the two errors rates are functions of the decision threshold. So, it is possible to represent the performance of a system by plotting FAR as a function of FRR. This curve is known as the system operating characteristics which is monotonous and decreasing.

Figure 16.

```
100 iterations, log-likelihood = 421.804
100 iterations, log-likelihood = 450.178
100 iterations, log-likelihood = 404.359
100 iterations, log-likelihood = 369.729
100 iterations, log-likelihood = 403.482
100 iterations, log-likelihood = 468.535
100 iterations, log-likelihood = 440.562
100 iterations, log-likelihood = 422.823
100 iterations, log-likelihood = 366.321
100 iterations, log-likelihood = 455.891
100 iterations, log-likelihood = 403.118
100 iterations, log-likelihood = 401.165
NORMALISED SCORE
    2.3333

REGISTERED SPEAKER
```

Figure 17.

```
100 iterations, log-likelihood = 411.314
100 iterations, log-likelihood = 450.218
100 iterations, log-likelihood = 394.326
100 iterations, log-likelihood = 357.79
100 iterations, log-likelihood = 403.836
100 iterations, log-likelihood = 484.485
100 iterations, log-likelihood = 439.907
100 iterations, log-likelihood = 414.64
100 iterations, log-likelihood = 367.736
100 iterations, log-likelihood = 455.51
100 iterations, log-likelihood = 403.539
100 iterations, log-likelihood = 399.941
NORMALISED SCORE
   -2.1379

IMPOSTER
```

Again, it has become a standard to plot the error curve on a normal deviate scale in which case the curve is known as the detection error trade-offs (DETs) curve. The DET curve representation is therefore more easily readable and allows for a comparison of the system's performance on a large range of operating conditions. Here we also showed the DET curves of our experiments.

Detection tasks can be viewed as involving a tradeoff between two error types: missed detections and false alarms. An example of a speech processing task is to recognize the person who is speaking, or to recognize the language being spoken. A recognition system may fail to detect a target learner or language known to the system or it may declare such detection when the target is not present (Martin et. al, 1997).

Figure 18.

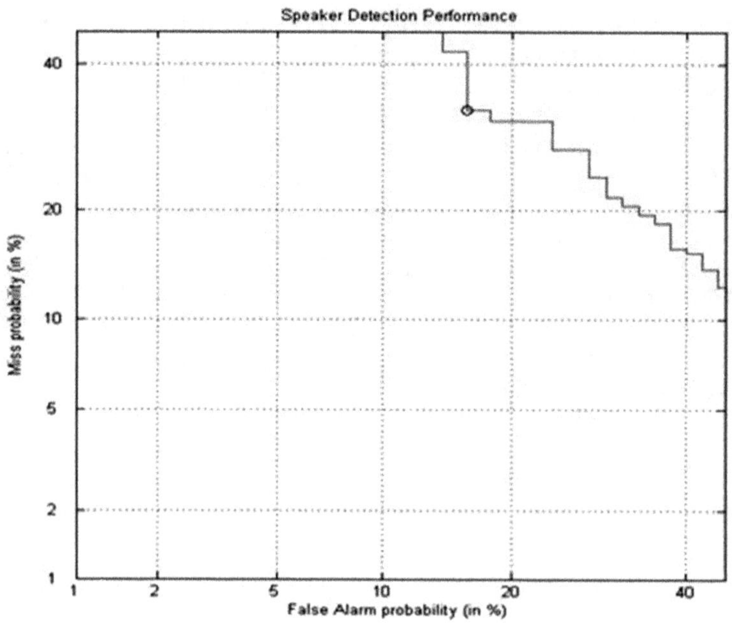

Figure 19.

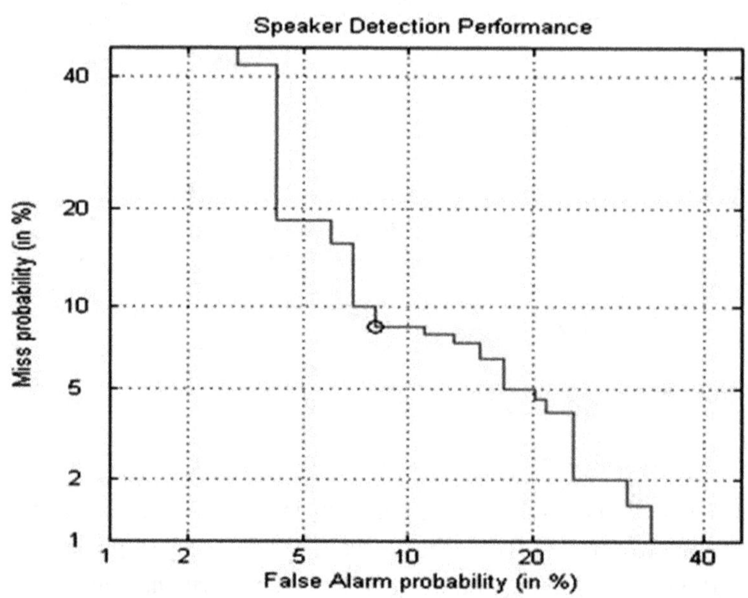

When there is a tradeoff of error types, a single performance number is inadequate to represent the capabilities of a system. Such a system has many operating points, and is best represented by a performance curve. The ROC Curve traditionally has been used for this purpose. Here ROC has been taken to denote either the Receiver Operating Characteristic or alternatively, the Relative Operating Characteristic. Generally, false alarm rate is plotted on the horizontal axis, while correct detection rate is plotted on the vertical. We have found it useful in speech applications to use a variant of this which we call the DET (Detection Error Tradeoff) Curve, described below. In the DET curve we plot error rates on both axes, giving uniform treatment to both types of error, and use a scale for both axes which spreads out the plot and better distinguishes different well performing systems and usually produces plots that are close to linear.

General Evaluation Protocol

Our evaluations of speech processing systems are comparable to fundamental detection tasks. Participants are given a set of known targets (learners or languages) for which their systems have trained models and a set of unknown speech segments. During the evaluation the speech processing system must determine whether or not the unknown segment is one of the known targets. The system output is likelihood that the segment is an instance of the target. The scale of the likelihood is arbitrary, but should be consistent across all decisions, with larger values indicating greater likelihood of being a target. These likelihoods are used to generate the performance curve displaying the range of possible operating characteristics.

Special Points

A number of special points may be included on the DET curve. These points are not limited to speech processing tasks and can be applied to the fundamental detection task. For example, it may be of interest to designate points corresponding to a fixed false alarm rate or fixed missed detection rate, perhaps a performance objective for an evaluation. The grid lines on the example curves may be viewed this way. Confidence intervals, or a confidence box, around such points may also be included. A weighted average of the missed detection and false alarm rates may be used as a kind of figure of merit or cost function. The point on the DET Curve where such an average is minimized may be indicated as like in Figure 20. In our evaluations, the speech processing systems must also provide a hard yes or no decision as well as a likelihood score for each decision. The operating points of the hard decisions may be indicated on the curves. The proximity of these points to the weighted average points described above is an indication of how appropriately the system implementers chose the hard decision operating points to optimize the chosen cost function.

Averaging Across Targets

The DET curves presented all involved multiple targets, and required systems to provide likelihood scores on the same scale for all targets. For some applications, requiring a common scale may be considered undesirable. Furthermore, if all targets do not occur with about equal frequency, it is arguable that combining data from multiple targets may present a misleading indication of performance. The alternative is to generate separate curves for each target, and then generate an average curve across targets from these. If the same non-targets are used with each target, then the ordinate values may be averaged for each

Figure 20.

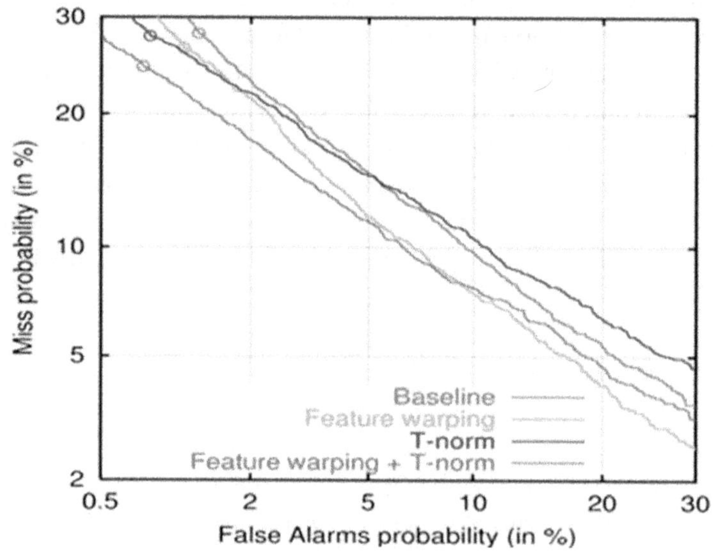

abscissa value. This situation will not hold, however, if each target example also serves as a non-target example for each of the other targets. In this case, interpolation may be used to obtain a common set of abscissa values for the individual target curves which may then be averaged. We prefer, however, to combine data from multiple targets directly. This requires systems to develop a common likelihood scale for all targets, which we believe desirable for many applications. We believe that with a large number of targets and roughly equal occurrences of all targets overall performance is effectively represented.

Performance Evaluation

DET curve represents the performance of the learner verification system. It involves tradeoff between two error types: miss detection and false alarm. Equal error rate (EER) corresponds to the point where the false alarm probability equals the miss probability. It is the measure of ability of the system to distinguish the imposters from the true learners. Totally 360 learners were tested and their performance was tabulated. It is inferred that when we fuse several features the performance of the system gets increased. The performance evaluation is shown in Table 2.

Table 2. Performance analysis of the learner verification system

Sl.No	Feature	No of learners	EER (%)
1	Baseline[3]	100	19
2	Pitch (proposed)	360	16.67
3	Prosody (proposed)	360	8.1

CONCLUSION AND FUTURE SCOPE

TIMIT database is used here for the learner verification. Features for an input utterance Are obtained from statistics of prosodic and articulation information. The Articulation Information includes formants and Zero Crossing Rate(ZCR). The prosodic information includes Pitch and Energy of the signal. The Pitch and Formants are extracted using Linear Predictive coefficient method. GMM-UBM model is used to produce the log-likelihood ratio. The score normalization is done to bring the score values in the range. The threshold is set to make accept/reject decision.

The obtained result shows that the, different prosodic and articulation features can produce good results in learner verification. Table 2 shows that EER for 360 sample and it is better than the baseline paper (Ruchi Chaudhary et. al, 2012) which uses MFCC feature.

It is found that through experimental results the performance rate of the learner verification system can be improved by combining various prosodic features. When using the pitch feature alone the equal error rate is about 16.67%. But when combining the formant features with pitch the error rate is reduced to 8.1%.From the performance evaluation it is clear that the efficiency increases with the combination of pitch based features. It is expected that the performance can be further improved by employing the jitter and shimmer measurements along with the prosodic features.

There are many methodological boxes to be opened, re-thought and re-examined with care and new insight. And there are many wide-open domains for the study of prosody and its applications, especially in approaches to speech timing hierarchies, from duration dispersion metrics through oscillator emulations of rhythm production in speech to an annotation-independent approach to rhythm analysis, the amplitude modulation spectrum analysis of speech, AEMS, and the Frequency Zone Hypothesis of rhythm, interpretable as emulation of perception. Using these analyses the voice biometric authentication in online learning can further be enhanced.

REFERENCES

Carey, M., Parris, E., & Bridle, J. (1991). A speaker verification system using alphanets. *Proceedings of the International Conference on Acoustics, Speech, and Signal Processing*, 397–400.

Chaudhary. (2012). Performance study of Text-independent Speaker identification system using MFCC & IMFCC for Telephone and Microphone Speeches. *International Journal of Computing and Business Research, 3*(2).

Dalva, D., Revidi, I. D., & Guz, U. (2014). DQG Haka Gurkan, Extraction and Comparison of Various Prosodic Feature Sets on Sentence Segmentation Task for Turkish Broadcast News Data. *11th International Joint Conference on Computer Science and Software Engineering (JCSSE)*, 70-73.

Dempster, A., Laird, N., & Rubin, D. (1977). Maximum likelihood from incomplete data via the EM algorithm. *Journal of the Royal Statistical Society. Series A (General), 39*, 1–38.

Gauvain, J. L., & Lee, C.-H. (1994). Maximum a posteriori estimation for multivariate Gaussian mixture observations of Markov chains. *IEEE Transactions on Speech and Audio Processing*, 2(2), 291–298. doi:10.1109/89.279278

Heck, L. P., & Weintraub, M. (1997). Handset-dependent background models for robust text independent speaker recognition. *Proceedings of the International Conference on Acoustics, Speech, and Signal Processing*, 1071–1073. 10.1109/ICASSP.1997.596126

Isobe, T., & Takahashi, J. (1999). Text-independent speaker verification using virtual speaker based cohort normalization. *Proceedings of the European Conference on Speech Communication and Technology*, 987–990.

Jabeen, F., & Braun, B. (2018). *Production and Perception of Prosodic Cues in Narrow & Corrective Focus in Urdu/Hindi*. 9th International Conference on Speech Prosody 2018, Poznań, Poland.

Leggetter, C., & Woodland, P. (1995). Maximum Likelihood Linear Regression for Speaker Adaptation of Hmms. *Computer Speech & Language*, *9*, 171–186. doi:10.1006/csla.1995.0010

Manikandan & Chandra. (2016). Speaker Identification using a Novel Prosody with Fuzzy based Hierarchical Decision Tree Approach. *Indian Journal of Science and Technology*, *9*(44).

Matsui, T., & Furui, S. (1995). Likelihood normalization for speaker verification using a phoneme and speaker-independent model. *Speech Communication*, *17*(1-2), 109–116. doi:10.1016/0167-6393(95)00011-C

Matsui, T., & Furui, S. (1994). Similarity normalization methods for speaker verification based on a posteriori probability. *Proceedings of the ESCA Workshop on Automatic Speaker Recognition, Identification and Verification*, 59–62.

Nagaratna, B. (2018). Dialect Identification U sing Spectral a nd Prosodic Features on Single and Ensemble Classifiers, Research Article - Special Issue - Computer Engineering And Computer Science. *Arabian Journal for Science and Engineering*, *43*, 4289–4302.

Newman, M., Gillick, L., Ito, Y., McAllaster, D., & Peskin, B. (1996). Speaker verification through large vocabulary continuous speech recognition. *Proceedings of the International Conference on Spoken Language Processing*, 2419–2422. 10.1109/ICSLP.1996.607297

Pattnaik & Dash. (2012). A Study on Prosody Analysis. *International Journal of Computational Engineering Research, 2*(5), 1594-1599.

Paulmann, S., Vrijders, B., Weinstein, N., & Vansteenkiste, M. (2018). How parents motivate their children through prosody. *9th International Conference on Speech Prosody*, 13-16. 10.21437/SpeechProsody.2018-51

Reynolds, D. A. (1995). Speaker Identification and Verification using Gaussian Mixture Speaker Models. *Speech Communication, 17*(1-2), 91–108. doi:10.1016/0167-6393(95)00009-D

Reynolds, D. A., Quateri, T. F., & Dunn, R. B. (2000). Verification using Adapted Gaussian Mixture Models. *Digital Signal Processing*, *10*(1-3), 19–41. doi:10.1006/dspr.1999.0361

Reynolds, D. A., Rose, R. C., & Smith, M. J. T. (1992). PC-based TMS320C30 implementation of the Gaussian mixture model text-independent speaker recognition system. *Proceedings of the International Conference on Signal Processing Applications and Technology*, 967–973.

Rosenberg, A. E., DeLong, J., Lee, C. H., Juang, B. H., & Soong, F. K. (1992). The use of cohort normalized scores for speaker verification. *International Conference on Speech and Language Processing*, 599–602.

Rosenberg, A. E., & Parthasarathy, S. (1996). Speaker background models for connected digit password speaker verification. *Proceedings of the International Conference on Acoustics, Speech, and Signal Processing*, 81–84. 10.1109/ICASSP.1996.540295

Sarika, S. (2015). Survey Paper on Automatic Speaker Recognition Systems, International Journal Of Engineering And. *Computer Science*, *4*(3), 10895–10898.

Sarkar, A. K., & Umesh, S. (2010). Investigation of Speaker- Clustered UBMs based on Vocal Tract Lengths and MLLR matrices for Speaker Verification. *The Speaker and Language Recognition Workshop.*

Si, Li, & Tan. (2017). Transformation of Prosody in Voice Conversion. *Proceedings of APSIPA Annual Summit and Conference.*

Takuya, O., Nobuaki, M., & Daisuke, S. (2018). Prosodic Comparison of Utterances without Extracting Fundamental Frequencies based on Vocalized Subharmonic Summation. *9th International Conference on Speech Prosody 2018*, 172-176.

Vidyadhara Raju, V. V., Gangamohan, P., Suryakanth, V. G., & Vuppala, A. K. (2016). Application of prosody modification for Speech Recognition in different Emotion conditions. *IEEE Region 10 Conference (TENCON) — Proceedings of the International Conference*, 951-954.

Vuuren, S. (1999). *Speaker Verification in a Time-Feature Space* (Ph.D. thesis). Oregon Graduate Institute.

Wang, Y., & Payne, E. (2018). Investigating prosody in music and speech. *9th International Conference on Speech Prosody 2018*, 47-551.

ADDITIONAL READING

Nolan, F., & Jeon, H.-S. (2014). Speech rhythm: A metaphor? *Philosophical Transactions of the Royal Society of London. Series B, Biological Sciences*, *369*(1658), 20130396. doi:10.1098/rstb.2013.0396 PMID:25385774

Patel, A. D. (2008). *Music, Language, and the Brain*. New York: Oxford University Press.

Rabiner, L. R., & Schafer, R. W. (2011). *"Digital Processing of speech signals" (sixth impression).* Pearson Education and Dorling Kindersley Pvt. Ltd.

Section 3
Innovations and Technologies

Chapter 8
Keystroke Dynamics:
A Behavioral Biometric Model for User Authentication in Online Exams

Senthil Kumar A. V.
Hindusthan College of Arts and Science, India

Rathi M.
Dr. N. G. P. Arts and Science College, India

ABSTRACT

Online learning has entirely transformed the way of learning by the students. Online tests and quizzes play an important role in online learning, which provides accurate results to the instructor. But, the learners use different methods to cheat during online exams such as opening a browser to search for the answer or a document in the local drive, etc. They are not authenticated once they login and progress to attend the online exams. Different techniques are used in authenticating the students taking up the online exams such as audio or video surveillance systems, fingerprint, or iris recognition, etc. Keystroke dynamics-based authentication (KDA) method, a behavioral biometric-based authentication model has gained focus in authenticating the users. This chapter proposes the usage of KDA as a solution to user authentication in online exams and presents a detailed review on the processes of KDA, the factors that affect the performance of KDA, their applications in different domains, and a few keystroke dynamics-based datasets to authenticate the users during online exams.

INTRODUCTION

Authentication is the process to ensure an individual's identity. This provides access control for systems by verifying the user's credentials with the credentials stored in a database. Only authenticated users are allowed to access the protected resources of the organizations which are called as authorization process. The terms authentication and authorization are often used interchangeably; Authentication confirms the identity of a registered user before permitting access to the protected resource, whereas authorization confirms that the authenticated user has the permission to access the requested resources or not. Security

DOI: 10.4018/978-1-5225-7724-9.ch008

of the computers is compromised through unauthorized access of highly confidential information. User authentication is the primary method in preventing unauthorized access of information. It is generally classified into three categories. i) Knowledge Based Authentication (Something the user knows): User name and passwords is the best example for this method. This is very popular as it authenticates the user directly and it also does not require any special hardware for authentication. ii) Object Based Authentication (Something the user possesses): The users are identified by presenting or applying physical objects i.e. electronic chip cards, RFID tags etc. iii) Biometric Based Authentication (Something the user is): Identifying an individual based on his/her unique physiological and/or behavioral characteristics.

Now-a-days, the usage of hard copy of the book materials is getting reduced as purchasing the book is expensive and also it becomes outdated in the modern learning environment. The book materials reproduced and stored in another storage media may not help the users, as the device may become inaccessible at the time of requirement. But, online learning is an environment where these drawbacks could be overcome. The course contents can be uploaded online and the user can access them at any time. At the end of the course, the users are expected to undertake the exam online. Few users attend the exam genuinely and few others involve in forgery by asking some of their friends (an impostor) to take up the exam instead of them, so that they can score high. This acts as a great challenge to the one who is conducting the exam. Hence, it becomes important to authenticate the users during the entire session of the online exam. The most effective method of user authentication is based on biometrics. Keystroke Dynamics is a behavioural biometric authentication method which is used to authenticate the user during the online exams. This chapter presents the significance of online learning environment, literature review on the various biometric techniques that are applied to prevent cheating in online examination systems, and it particularly focuses on keystroke dynamics based authentication model, which has been used as a strong authentication model by the researchers. Applications of keystroke dynamics in different domains, the factors that affect the performance of the keystroke dynamics model have also been discussed.

BIOMETRIC TECHNOLOGIES: AN OVERVIEW

The following section discusses on the various biometric technologies available for authentication which are broadly categorized based on their physical characteristics or behavioural characteristics.

Physiological Biometrics

Physiological biometrics are those features that describe 'who the user is' depending on their physical attributes such as fingerprints, face, iris, retina scanning, hand geometry etc.

Facial Recognition

The most widespread biometric characteristic used by humans to make a personal identification is the facial recognition. It uses the spatial geometry of distinctive features of the face. It is a type of computer vision that uses the face to identify or to authenticate a person. It does not require any additional hardware, as it can be used with existing image capture devices like webcams, security cameras, etc. The features of a face include the distance between the eyes, width of the nose, position of cheekbones, jaw line, chin etc. The face recognition software uses deep learning algorithms to compare live captured image to the stored face print in order to verify an individual's identity. To improve the accuracy and to resist spoofing by photos or mask, 3-D modelling can be used in face recognition software. Difficulties that arise with this technique are image lighting and background makes it difficult to locate the face, user not looking into the camera directly, using different types of cameras to read the face, age factor etc. This method can also be applied in situations when it is necessary to monitor the person without his knowledge of the fact that he is being observed. This makes face recognition suitable for many applications using surveillance cameras.

Fingerprints

Individuals always carry their fingerprints with them and they cannot be lost or forgotten like magnetic strip cards or passwords. A fingerprint is the pattern of ridges and furrows on the surface of a fingertip that is distinct even for identical twins. Ridges are the upper skin layer segment and furrows are the lower skin layer segments. Fingerprint recognition devices are available at a low cost, with which users need not provide or type passwords; only a touch using finger is sufficient for authentication. Fingerprint sensor detects the ridges and furrows and identifies fingerprint characteristics, identified as the "minutiae". Minutiae points are native ridge characteristics that seem as moreover a ridge ending or a ridge junction. A complete fingerprint comprises about 100 minutiae points on average. These are stored in a template and a subset of these is used for verification. Fingerprint authentication systems are broadly used with applications such as smart phones, laptops, USB flash drives as well as smart locks and other home automation devices.

Hand Geometry

Hand geometry depends on the way that each individual's hand is formed uniquely in contrast to others and it doesn't change after a time frame. The physical geometric attributes of a person's hand are estimated. Biometric characteristics such as the shape of human hand, lengths, widths, thickness of the fingers, creases and ridges of the skin surface etc. can be extracted and used for identification and verification. Hand geometry scanning devices can be either mechanical or image-edge detection. The user needs to put his hand on the reader's surface and lines up with the correct area of the fingers utilizing which the user can be verified. Dissimilar to fingerprint imaging frameworks, hand geometry readers don't consider the ecological factors, for example, messy hands, dry climate, or dry skin. The uses of hand geometry incorporate money vault, time and participation administration, parking garage, PoS applications and so on.

Retinal Pattern

In the posterior portion of the eye, a thin tissue composing of neural cells called the retina is located.The complex structure of the blood vessels that supplies blood to retina is unique and stable for everyone. A retinal scan uses low intensity beam of visual or infrared light. Retinal blood vessels are more absorbent of this light than the rest of the eye. The amount of reflection varies during the scan and the pattern of variation is stored in a database as a template. These unique patterns on the retina are used for comparison when the users try to access the resources and hence can be used for authentication. This authentication method is applied in government sectors, commercial sectors, various intelligence agencies to identify people at airports, ATM centres, etc.

Iris Recognition

Iris recognition is another method in strategy for affirming the identity of a person. The iris is the annular region of the eye bounded by the pupil and the sclera (white of the eye) on either side. Iris of each person is unique and even to genetically identical twins. Iris recognition systems take high resolution images of the iris of a person's eye. The boundaries of the centre of the pupil are identified, the features are extracted and these patterns are stored in the database.In this method of authentication, the biometric template for the scanned image is matched with biometric templates stored in the database. Specific hardware is required to capture the iris image. Biometric iris acknowledgment can work at long distance, thus they are utilized oftentimes in border control arrangements where it can distinguish explorers as they enter and exit via land, ocean and air.This method of authentication has also been implemented in mobile devices.

BEHAVIORAL BIOMETRICS

Behavioral biometrics is related to the behaviour pattern of a person, such as signature, keystroke dynamics, voice, stylometry, mouse dynamics etc.

Signature

Each individual has an exclusive style of handwriting. Signature recognition is a behavioural biometric method which is one of the most accepted methods of identification used in many business transactions. The signature can be collected either in the static mode or dynamic mode. In the Static signature identification mode, the users write the signature in a paper and this signature is digitized and used for authentication. This technique is referred to as an "off-line" mode of recognition. This mode of signature verification is used to validate credit cards, bank cheques, passport verification, historical documents etc. Features like center of gravity, edges, curves etc., are extracted and used for authentication. In the dynamic signature identification mode, users write the signature in the digitized tablet, or by means of stylus-operated PDAs. Features like shape, direction, timing, acceleration, pressure etc., are measured. Now-a-days, to authenticate the mobile phone users, signature recognition is used where the user can sign using a finger or an appropriate pen in the smart phones.

Keystroke Dynamics

Keystroke dynamics is a behavioural biometric method used together with username and password. It confirms the identity of an individual based on the manner and the rhythm of typing on a keyboard. If another individual accesses the keyboard, the system is capable of recognizing the impostor. The measurements used for keystroke dynamics are dwell time, flight time, latency etc. These are discussed in detail later in this chapter.

Voice Recognition

Voice recognition is a unique biological factor that is used as biometric authentication method by measuring the sounds a user makes while speaking. The characteristics such as natural accent, flow of their voice, frequency are measured when the user speaks in a conversation or when the user is made to converse a passphrase. The software records the voice data at the time of enrolment and extracts unique features to create a template. This template is used for verification purpose. This method of authentication is normally used to operate a device, or write without using a keyboard, mouse, or press any buttons, perform commands etc. This technology is applied in transliterating voice to text, to set up reminders, to respond to simple questions, to play music, to share weather or traffic information, to search the internet etc.

Stylometry

Stylometry is defined as the statistical analysis of writing style of an individual. Stylometry analysis, checks whether a target document was written by a specific individual or not. It is one of the most effective forms of verification of authorship. Features are extracted from sample texts and they are categorized based on the owners of the documents. Some of the stylistic features include, statistical measures based on word or character (lexical), grammatical functors, punctuation, part-of-speech (syntactic), text organization and layout (structural), important key words and phrases pertaining to certain topics (content-specific) etc. These features can be measured qualitative and quantitatively and can be used for authentication of users.

Mouse Dynamics

Mouse movement of an individual is captured and the characteristics are extracted which is used as a behavioural biometric method. These features are stored as a template in the database and verified against the user's mouse movements. If the user's mouse movement data does not match the authenticated user then the user will be unauthenticated and rejected by the system.

BACKGROUND

Advancement in Information and Communication Technology (ICT) has enabled the educational institutions to provide the educational resources to the students through online. Examinations are conducted to examine the knowledge of the student in a particular subject. Students are expected to do their best in the exams to be competent enough in today's world. Online examination is an essential and vital element of online learning. Security concerns associated with online learning is on raise as there will be no physical invigilators to monitor the students during the exams. There is a chance that the students can easily deceive during the online exams. Some of the reasons why students' cheat during online exams include aspiration for a better rank, parents' expectation that their children have to score well in the exams, engaging themselves in other activities without learning for the exams, history of failure, accessibility of online information easily, possibility of being caught is less etc.

Some of the techniques followed by the students for cheating during the exams include creating programs which could fetch the right answers for the questions, trying to access google by opening a separate window, hiding the answers in water bottles etc. Cluskey et al. (2011), had discussed the various methods that are handled by the student during online exams. Few of those are: the questions may be attempted by one set of the students, who record the answers and distributes the answers to the remaining students, some students get the test banks and solution manuals through underground sources. Moten et al. (2013), had claimed that students' copy and paste entire passage from the internet without citations on essay or short answer assessments and the author had also claimed that few websites allows student to submit exam questions and purchase the answers. Learners may obtain assistance from others to improve their examination results or have another individual take the examination for them (Fayyoumi & Zarrad, 2014).

Even though different types of biometric authentication methods exist, this chapter focuses on Keystroke Dynamics based authentication method for authenticating the users in online exams. It is one of the behavioural biometric methods which use the typing pattern of an individual to authenticate the user. This method can be applied in different dimensions like online banking transactions, in smart phones, in detecting the emotions of the users while typing, in Online learning etc.

RELATED WORK

E-exam management systems was implemented by Bawarith et al. (2017), to detect and avoid cheating in online exams. To keep track of students in the exam session, the authors had used fingerprint reader authenticator for continuous authentication and eye-tribe tracker to check whether the student is in front of the screen and looking at it or looking elsewhere. They have also proposed that other authentication techniques like keystroke authentication, face recognition can also be used for assessment of the students. The parameters like total time on out screen and the number of times on out screen were computed to detect the examinee status as cheating or non-cheating. The authors conducted experiments and they have achieved the result as sensitivity of 100%, specificity of 95.56%, precision of 95.74%, accuracy of 97.78%, and f-measure of 97.83%.

Khlifi & El-Sabagh (2017), had introduced an efficient secure model for monitoring online evaluation including e-assessment or e-exam. The authors have integrated the knowledge method with the behavioural aspects for identity verification and authentication of the students during the e-exams. They have collected the students' personal information during the Online learning hours like DOB, age, their interests etc. and stored these in the database. These information were posted as challenging questions during the exam session and used for continuous random authentication. The proposed scheme has been evaluated using the MATLAB simulator and the results have shown that proposed approach can effectively improve the authentication process.

In this paper, the author Rodchua (2017), had discussed the existing tools to authenticate students' identification and proctoring in online exams like Secureexam Remote Proctor, ProctorU, ProctorCam, WebAssessor etc. The author had also discussed about the plagiarism detection softwares like SafeAssign, TurnItIn to compare students' assignments, the lockdown browsers like Respondus to control the students from copying, printing, accessing other applications etc., KioWare Kiosk system that restricts the user from accessing the operating system and desktop. Lee-Post & Hapke (2017) have discussed on the techniques for online integrity solutions like biometric authentication, surveillance systems and predictive analytics.

Atoum et al. (2017) had presented a multimedia analytics system that does online exam proctoring automatically. The author had used different hardware components like webcam, microphone etc.

and other components like the user's voice detection, phone detection, active window detection etc. to continuously monitor the behavioural aspects of the students during the online exams. Support Vector Machines (SVM) is used as the classifier model to detect cheating. The experimental results show that the author had achieved better results.

The authors Karim & Shukur (2015), have conducted an extensive review on the authentication methods used to improve online learning security and online exams. In the authentication process of online exams, the authors have identified that different authentication techniques were used such as palm print, IP address with timestamp, Keystroke Dynamics along with username and password, challenge questions, biometric properties like fingerprint, face, iris, voice recognition methods, Keystroke Dynamics and stylometry etc., The authors had also discussed the threats that can occur during the online exams like the original student taking up the exam and assistant answering the questions, some other person taking up the exam etc.

Bhagat & Katankar (2014) had proposed the novel framework for online examination. The author had visual cryptography to make system more secure and reliable. A 3-layer security is provided in the student authentication system for online examination. The first layer being user id and password authentication, second layer generates the shares of stego image where the student has to upload one share during the exam, and the third layer is to protect the share. RSA algorithm is used for encryption and decryption of shares.

Gao (2012), proposed a randomized fingerprint minutiae subtemplate to authenticate user in online education and therefore to protect the privacy of the original fingerprint stored in the database. Since the subtemplate is a random selection of its original template and it only contains partial information about a user's fingerprint, it protects the user's private biometric data.

Rudrapal et al. (2012) had proposed a system using voice recognition technique to authenticate the user in online examination. To authenticate the user through voice, Fast Fourier Transform method is used. The author had claimed that this system will be helpful for blind people and the people who don't use keyboard for interaction with the system.

Dehnavi & Fard (2011) had presented a multimodal biometric model to authenticate the users during the virtual classes. This was done by using mouse, keystrokes and face biometrics. Bal & Acharya (2011) had used iris recognition as the biometric model for authenticating the user in online exams. Jung & Yeom (2009) had used Security Control system in the Online Exam (SeCOnE) which was based on e-monitoring system and group cryptography.

BIOMETRIC AUTHENTICATION IN ONLINE LEARNING

Focus on applications like online learning and online exams are gradually increasing. Authentication in online exams is ineffective using knowledge based or object based authentication methods. Few drawbacks in the knowledge based method are poor selection of password characters, sharing the passwords with others, forgetting the password, reusing the same password in many applications, usage of devices that can capture the passwords, possibility to guess the password using brute force attack or dictionary attack etc. It is clear that this method of authentication is insufficient in critical situations. The threats with the object based method are the physical objects which are used for authentication may be transferred to others or they may be lost which the hackers can use for their benefits. Biometric authentication method would be more effective for authenticating the users in online exams as this type of authentication cannot be stolen, guessed or lost.

Based on the number of factors used in the authentication methods, authentication methods can be classified as one-factor, two-factor or multi-factor authentication. One-factor authentication methods count on one factor such as fingerprint, face, voice, password etc., while those based on two factors such as voice and facial recognition together or face and keystroke dynamics together etc., is two-factor authentication. Similarly, combining multiple factors like usernames and passwords, PIN numbers and one or more of the biometric characteristics is called as multi-factor authentication. For Example, see Table 1.

This section discusses on online learning environment, the advantages of learning through online, issues related with online learning, the tools that are available for online proctoring and the authentication techniques that are proposed or implemented in online exams.

Online Learning

Online learning is a learning system with the help of electronic resources. It can also be called as 'virtual learning' or 'E-learning'. It is an environment which uses Information and Communication technologies (ICT's) as a platform for teaching and learning activities. It can be defined as education through technology. The major component of online learning is the computers and the Internet. Teaching can be based on inside or outside of the classrooms. Physical presence of a teacher in a classroom for imparting education has slowly been replaced by the advancement of Information and communication technology (ICT). Online learning has its root in distance learning and is a piece of the transformation brought by the new media: Web. Teachers and coaches found the possibilities to propel learning at the approach of the new Web innovations. Online educational courses can be taken by house-wives, students, faculties, office goers etc. Through online learning, education is delivered to a large number of people at the same or different times. Earlier days, it was not accepted by most of the people as it replaced the human. But, with the rapid growth and advancement in technology, it is now widely accepted. Information can be shared via the Internet, which is accessible anywhere, anytime. Some of the advantages and issues of online learning are discussed below.

Advantages of Online Learning

- Online education can be taken up by anyone at any time. Students, faculties, housewives, employees and others can use their convenient time in schools, colleges, home, office etc for learning additional content during the working days, evening hours, or weekends. This method of learning is best suitable for everybody.

Table 1.

Fingerprint / Face / Voice / Password :	One-factor authentication
Password + Fingerprint / Voice + Face / Face + Keystroke Dynamics :	Two-factor authentication
Password + PINs + Keystroke Dynamics / Password + OTPs + Face :	Multi-factor authentication

- In classroom teaching, the lecture can be listened only once and if a lecture class is missed, then the topic has to be prepared by the students themselves. It is not in the case of online education, the contents can be accessed any number of times and it also helps the students to revise during the exams.
- Teaching materials in online learning are planned based on the interest and needs of the learner. The online educators cover a large group of audience through their training with less effort. The learners are able to receive the same type of training with this learning mode.
- The cost incurred for training the trainers or the learners attending the classes is reduced as there is no cost for travel, accommodation etc.
- Lessons are wrapped in a single session which enables the completion of course faster.
- Learners can learn at their own speed according to their understanding capability.
- Learners can focus on some specific area instead of learning all areas.
- Corporates trained their employees using online learning environment where they need not travel to the location and waste their time. The employees were able to earn online degrees and enrich their skill sets.
- Time is another important factor for both the learners and teachers. Online learning provides a flexible environment where everyone can learn the course at their convenient time. Online learning tools such as video tutorials, PDF documents, slideshows etc. are available to the users 24 x 7.
- Attending webinars or chatting with the professors online in the discussion forum enables students to increase their talents.
- Online learning is an economical and more flexible environment which has increased the ratio of higher education.

Issues in Online Learning

Eventhough there are lot of benefits in online learning, the issues are also on raise. Few of them are discussed below.

- Users with traditional mindset of attending classroom teaching find it difficult to adapt in virtual classroom through online mode
- The technical issue like slow bandwidth of the internet causes delay in loading the content online which may frustrate the users.
- Due to their flexibility in time, users don't fix up a time for online learning and also they are unable to discuss on the topic with their peers which will lead to incompletion of the course, late submission of assignments etc.
- The quality of the course content may not be in par with the user's expectations.
- The integrity of the course content or any private data of the user need to be protected from the unauthorised users.
- The invigilator of the online exam has to be very keen in monitoring the users or students behaviour during the exam in the exam hall.

ONLINE EXAM PROCTORING TOOLS

Online proctoring, sometimes called remote proctoring, refers to proctors supervising an exam over the Internet through a webcam for which the software is installed on the computer. It verifies the authenticity of the exam taker and ensures that they do not use any unauthorized resource during the exams and prevent them from any form of cheating. To begin the exam, the exam taker confirms his identity through his ID to the camera, facial recognition, and sometimes by keystroke analysis too. During the exam, the camera records all the actions. Different types of proctoring methods are i) Live Online Proctoring: Candidates are monitored by a qualified proctor from a remote location through audio-video and screen share feeds at real time. ii) Recorded Proctoring: The audio-video and screen share feeds of the test takers are recorded during the test and later reviewed. iii) Advanced Automated Proctoring: Audio-video and screen share feeds of the exam takers are recorded as well as monitored for any suspicious activity through latest technologies. It observes the exam takers eye movements or any suspicious behaviour that exhibit cheating. The proctor issues yellow flags when minor deviation is observed and red flag when clear cheating is observed. The few tools that are available in the market are described below.

Securexam Remote Proctor (SRP)

The student is authenticated using a video camera, microphone and a fingerprint scanner. The exam taker is verified using his fingerprint and also this system records 360° real-time audio and video of the environment during the complete duration of the exam. All suspicious activities like an unknown voice or motion and sound are monitored through this audio/video surveillance system.

ProctorU

ProctorU was developed by Andrew Jackson University and employs username and password, photo ID card, with the human proctor matching the student with his ID card in real-time, audio-video surveillance system using TokBox (Rodchua et al., 2011). The keystroke pattern of the exam-taker is also monitored to make sure it is the same person. Exam takers personal data are gathered and during the exam, they are posted with challenging questions such as their DOB, address etc, to establish their identities.

Webassessor™

Webassessor™ from Kryterion Inc provides authentication based on facial image of the exam takers which are captured using webcams, their keystroke patterns to authenticate the exam taker, secure browser control to ensure the inaccessibility of internet and other resources on the hard disk, etc. If there is any disparity, the video surveillance system alerts the proctors. Ability for faculty to review recorded test sessions

AUTHENTICATION TECHNIQUES

User authentication is one of the real worries for E-evaluation on the grounds that unapproved people can get to and oversee data. For this reason, there need to be some mechanism to authenticate these users whether they are valid user or not. Students learn the courses online and take up their exams also online.

If the students clear the exams, they are provided with a certificate which will increase their credit scores. Hence, there are possibilities that an impostor may take up the exam and get good score. To avoid this, strong authentication methods together with username and password are required. Table 2 presents the summary of the authentication techniques proposed or implemented in online exams.

KEYSTROKE DYNAMICS

Although there are many authentication methods employed or proposed by the researchers as shown in Table 1, the authors are interested in working with Keystroke Dynamics, as this does not require any special hardware to authenticate the user and it is an effective method of authentication. Keystroke dynamic based authentication method is very powerful which is based on the typing pattern of the individual on the devices like keyboard, mouse or smart phones. Biometric authentication can be performed based on the keystroke of the user. The pattern of errors that two users make are distinct i.e typing adjacent letters together, typing the characters in reverse, holding down the shift key etc. Hence Keystroke Dynamics is a strong authentication method. The general methodology of this authentication method is described below.

1. Involve individuals or subjects for data collection
2. Record the keystroke timing information.
3. Preprocess the extracted data to reduce noise and outliers.
4. Extracting the necessary features from the recorded keystroke.

Table 2. Authentication techniques proposed (P) / implemented (I) in Online Exams

References	Authentication Techniques
(Bawarith et al., 2017)	Fingerprint and Eye tribe tracker (I)
(Khlifi & El-Sabagh, 2017)	Challenging questions (P)
(Lee-Post & Hapke, 2017)	Biometric Authentications(P)
(Traoré, Awad, & Woungang, 2017)	Mouse dynamic biometric, Keystroke dynamic biometric, and Face biometric (P)
(Al-Saleem & Ullah, 2014)	Palm-based biometrics (P)
(Bhagat & Katankar, 2014)	Visual Cryptography (P)
(Fayyoumi & Zarrad, 2014)	Face recognition (P)
(Krause, 2014)	Stylometry (I)
(Ramu & Arivoli, 2013)	Keystroke Biometrics (P)
(Gao, 2012)	Fingerprint (P)
(Rudrapal et al., 2012)	Voice Recognition (P)
(Bal & Acharya, 2011)	Iris Recognition (I)
(Canales et al., 2011)	Stylometry (I)
(Flior & Kowalski, 2010)	Keystroke Signatures(I)
(Asha & Chellappan, 2008)	Fingerprint and mouse dynamics (P)
(Kumlander, 2008)	Keystroke Dynamics (P)

5. Extracted data can be used for training the classifier.
6. Test the classifier performance for accuracy.

Enrolment and verification are two important phases in KDA. Figure 1 describes the Keystroke dynamic authentication model. In the enrolment phase, data is collected from the user, such as username, password, user's typing behaviour etc. For each user's typing behaviour, a template is created by extracting the keystroke timing features. This template is stored in the database as user's profile. In the verification phase, whenever the user accesses the system, the system gathers the user's keystroke timing features and a classifier is used to compare with the template in the database. If the data collected during the enrolment phase and the verification phase matches, the user is a valid user otherwise the user is an impostor. The individual processes in KDA are discussed in detail below.

CAPTURING KEYSTROKE DATA USING STATIC OR DYNAMIC TEXT

Keystroke Dynamic based authentication can be applied in two different ways by capturing keystroke data using static text or dynamic text. Static keystroke authentication methods provide static text as passwords. It acts as an additional layer of security together with username and password where the individual's keystroke behaviour is also monitored using a passphrase. The user is asked to type the passphrase several times (Syed Idrus et al., 2014) and the characteristics of this keystroke behaviour are stored in the database. Later, this behavioural pattern is used for verification. But, this is done only during the initial login sequence and there is a possibility that once after login session, the user may change.

In continuous keystroke authentication method, the user is not restricted to type static text or password. The user can be monitored continuously with a long text as input or his entire keystroke behaviour is monitored continuously during the entire logon session. This can be used in detecting any kind of unusual activities during the session.

Figure 1. Keystroke dynamic authentication model

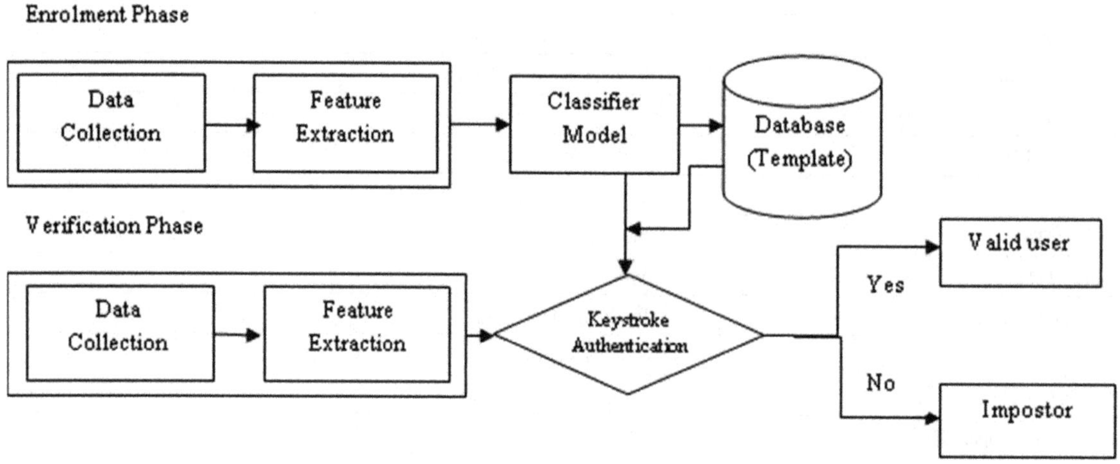

Data was collected from 40 students by the authors (Canales et al., 2011) where stylometry was used in the authentication process. Gao (2012) collected data from 30 female graduate students of age ranging from 25 to 27 years. Data was collected from 30 participants of different age groups, different backgrounds by Bawarith et al., (2017). Krause (2014) collected data from 45 authors and used them for training data and test data. Video streams were recorded for 11 individuals where face recognition is used as the authentication mechanism by Traoré et al. (2017).

RECORDING KEYSTROKE TIMING AND FEATURE EXTRACTION

Keystroke Dynamics extracts many features which can be detected when the user presses keyboard keys. The figure 2 shows the pictorial representation of the features used for Keystroke Dynamics. Few of them are duration of the keystroke i.e hold time or dwell time, latency between consecutive keystrokes i.e, flight time, typing speed, frequency of errors, pressure on the keys etc. Mostly Keystroke Dynamics is concerned with the timing features of the user's typing data. These features are described below.

Dwell Time: (Key Press- Key Release)

This is also known as hold time. It is the span of time for which the key is held down i.e. the measure of time amongst pressing and releasing a solitary key. In figure 2, time interval between key press and key release of the character 'A' is the dwell time.

Down-Down Time: (Key Press - Key Press)

It is the key down-key down time i.e the time interval between key press of A and key press of B. In figure 2, 'flight time' represents the down-down time.

Figure 2. Features extracted from keystrokes

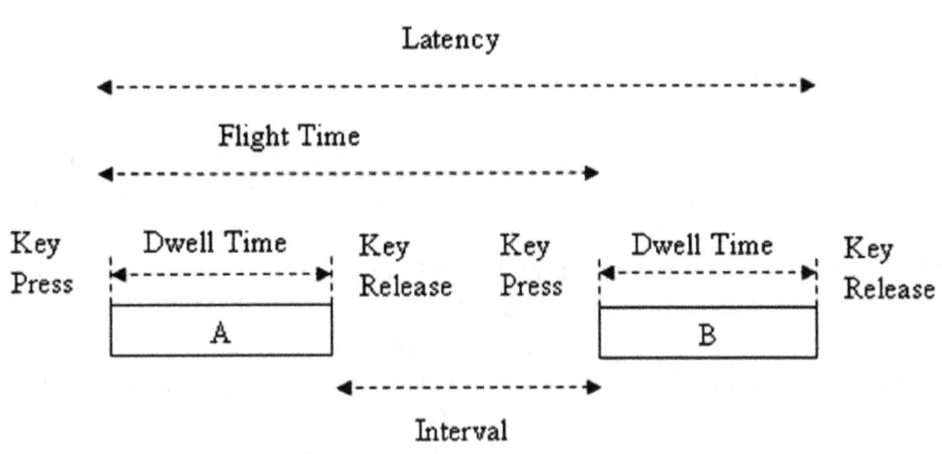

Up-Down Time: (Key Release – Key Press)

It is the key up-key down time i.e the time interval between key release of A and key press of B. This time can be negative too. In figure 2, the up down time is shown as the 'interval'.

Timing data of two back to back keystrokes is known as digraph, three successive keystrokes as trigraph and 'N' continuous keystrokes as N-graph. The timing interval depends on the keyboard layout, environmental factors which affects the users keyboard typing speed etc. Based on the above said features, the word 'timing' would have three digraphs ('ti', 'mi', ng') and two tri-graphs ('tim', 'ing'). Patterns are extracted from these recorded keystroke timing features using statistical approaches like mean, standard deviation, Euclidean distance, Manhattan distance etc. These informations are stored in the database as a template and used for authentication.

Halakou (2013) had used feature selection methods like wrapper methods like Neural Network, K-Nearest Neighbor (KNN) and Ridge Regression (RR) and one filter method. The authors have shown that wrapper method was better than filter method and K-NN was better among the wrapper methods. The features like keydown-keydown time, key holding time, and keyup-keydown time are compared and the best authentication was shown by key hold time.

Shanmugapriya & Padmavathi (2011) had used keystroke features like dwell time, flight time, di-graph, tri- graph and virtual key force of every user. Z-Score method was used in the pre-processing stage. For feature subset selection, Extreme Learning Machine (ELM) algorithm was used together with Particle Swarm Optimization (PSO), Ant Colony Optimization (ACO) and Genetic Algorithm (GA). To find the best method for feature subset selection, experiments were conducted to compare ACO, PSO and GA with ELM and the results had revealed that ACO with ELM was best for feature subset selection.

Karnan et al. (2009) had used Ant Colony Optimization technique (ACO) on keystroke duration, latency and digraph for feature subset selection. Here, ACO is used to reduce the redundancy in the feature values and to reduce the search space. Azevedo et al.(2007) had developed a hybrid system based on Support Vector Machines (SVM) and optimization techniques like Genetic Algorithms (GA) and Particle Swarm Optimization (PSO). These optimization algorithms were analyzed by the authors for feature selection and they have claimed that PSO was better than GA with respect to classification error and processing time, but was inferior regarding the feature reduction rate.

CLASSIFICATION MODELS

Many research works have been done on Keystroke Dynamics using various classification models such as Neural Networks (Schlenker & Sarek, 2013), SVM classifier with a linear kernel and Random Forest Classifier (Bhatia & Hanmandlu, 2017), Artificial Neural Network (ANN) when combining keystroke and mouse biometrics (Mondal & Bours, 2017a), Neural Network to predict the missing digraphs (Ahmed & Traore, 2014), Neural network (Clarke and Furnell 2007, Cho et al. 2000) for mobile and web based applications, Counter-Propagation Artificial Neural Network (CPANN) and the prediction model Support Vector Machine (SVM). Online document verification was done by the authors (Brocardo et al., 2014) using stylometry technique and used SVM for classification. Most widely used classification models like Neural Networks, Support Vector Machines and Decision Trees are discussed below.

Neural Networks (NN)

A Neural Network (NN) is a model motivated by biological model of human brain consisting of millions of neurons. The signals are sent in the form of electrical and chemical signals. NN are connected in such a way that it consists of one input and one output layer, and in between these two layers there are hidden layers consisting of processing units to discover patterns in information received from the previous layer. Artificial Neural Networks(ANN) are computational models used in machine learning. This model is inspired by the working nature of human brain and the model is made to replicate the method of how the individual learns. There are many processing elements that are connected together similar to a human brain and are used to process the received information as human does. Backpropagation is one of the NN techniques that permit networks to fine-tune their hidden layers of neurons when the output does not match with the expected outcome. ANN techniques are applied in many problems for classification, regression, decision making, pattern recognition etc.

Support Vector Machine (SVM)

Classification by using Support Vector Machines (SVM) is performed by separating hyperplanes in a two dimensional space. This hyperplane is a line that divides a plane in two parts where each class label is laid on either side. If the class label cannot be fixed in either side, then the hyperplane is transformed to add one more dimension. The parameters are tuned to adjust the data plots when they overlap. This algorithm can be used for both classification and regression kind of problems like image recognition, classifying the texts in spam detection etc. The performance of this algorithm is better when the samples are limited.

Decision Trees (DT)

A Decision tree is a supervised learning algorithm (class label is fixed) that uses a tree like structure to make decision. Conditional control structures like 'if' statements are used in decision making. The decision rules can be constructed by using association rules. The decision is made by following the edges in the tree representation. Here features are represented by nodes, a rule or a decision is represented by a link, the result is represented by leaf nodes. There are different algorithms to construct a decision tree namely Iterative Dichotomiser 3 (ID3), Classification and Regression Trees (CART), C4.5 (An improved version of ID3) etc. ID3 and C4.5 selects a test using Information Gain whereas CART uses Gini Index as the metric. This algorithm is very easy to understand and can be merged with other techniques to solve classification and regression problems. They can handle both categorical ('Yes' or 'No' type) and continuous (range) types of variables.

Random Forest (RF)

Random forest algorithm is a supervised classification algorithm. This algorithm creates the forest with a number of trees. The number of trees in the forest and the results it can get are directly proportional; i.e, if the number of trees is higher, the result is more accurate. Two stages of Random Forest algorithm are to create random forest and to predict from the random forest classifier. The process of random forest creation is described as 5 steps. i) Randomly select "K" features from total "m" features where k

<< m. ii) Among the "K" features, calculate the node "d" using the best split point. iii) Split the node into daughter nodes using the best split. iv) Repeat the steps i) to iii) until "l" number of nodes has been reached v) Build forest by repeating steps a to d for "n" number times to create "n" number of trees. The process of prediction is as follows. i) Takes the test features and use the rules of each randomly created decision tree to predict the outcome and stores the predicted outcome (target) ii) Calculate the votes for each predicted target iii) Consider the high voted predicted target as the final prediction from the random forest algorithm. This algorithm can be used for both classification and regression.

EVALUATION OF KEYSTROKE DYNAMICS AUTHENTICATION MODEL

The performance of biometrics systems are generally measured by False Acceptance Rate (FAR), False Rejection Rate (FRR), Equal Error Rate (EER).

FAR is the percentage of impostors' attempts classified as valid users

FAR=Total False Acceptance * 100

Total False Attempts

FRR is percentage of valid users' attempts classified as impostors.

FRR=Total False Rejection * 100

Total True Attempts

Equal Error Rate (EER) is the value when the FAR and FRR are equal.

FAR is one, if all impostor patterns are falsely accepted and it is zero, if none of the impostor patterns is accepted. Similarly, if the classification threshold is too high, some of the client patterns are falsely rejected and it is shown as FRR.

FACTORS AFFECTING THE KEYSTROKE PERFORMANCE

The keystroke typing pattern of one person is considerably different from others due to many factors, such as familiarity with the input device, age, primary language, user's typing skills, level of comfort while using a keyboard, type of the keyboard etc. This section describes the various factors that affect the keystroke typing performance of the user.

Data Collection Environment

Data was collected from different users under a controlled environment or in an uncontrolled environment. The difference between these two is, in a controlled environment, data collection was done within a lab, with a predefined text and with a specific configuration of systems (Killourhy & Maxion (2012), Stefan et al. (2012), Goodkind et al. (2017)). During verification, the user may be in a different environment, which may lead to variation in the responses. In uncontrolled environments, keystroke data were collected from the users in different places, with different set of texts or documents, say in test centers, or home or online/kiosk (Ahmed & Traore (2014), Mondal & Bours (2017a)). The data collected using this method may provide real time representation, but the surroundings may be different which can lead to difficulty in analyzing.

Keyboard Type

Keyboards which are used in the data collection process or during the verification process may differ. The keyboards may be a traditional desktop keyboard or laptops or virtual keypads. They differ by their size, brands, space between keys etc. Timing data may get affected due to these factors and this may cause a big difference in typing behaviour of the user. Kang & Cho (2015) had used a soft keyboard typed by means of a stylus pen, conventional PC keyboard, and a touch keyboard typed with one hand and two hands. Experiments were conducted by the authors and they have shown that the highest authentication accuracy was reported on the usage of a PC keyboard, then by a soft keyboard and then by a touch keyboard.

Text Mode

The users may be asked to type a specific fixed text for few times, or they may be asked to type long free text from articles or run various applications like web browsers, word processors etc. These differences in mode of entry may cause difference in the experimental results. Authors like Roth et al. (2013), Alpar (2017), Hu et al. (2008) had used fixed text for authentication. To analyse the keystrokes, the authors (Ahmed & Traore, 2014) had taken free-text and they have combined monograph and digraph analysis for free text. Gunetti & Picardi (2005) performed keystroke analysis on free text by combining A-measure and R-measure. Few other authors like Alshehri et al. (2018), Kim et al.(2018), Goodkind et al. (2017), Bours (2012), Kang & Cho (2015), Matsubara et al. (2015), Tsimperidis et al. (2018) have also worked on long free text.

Text Length

During the enrolment process, few researchers collect text of length 400 to 1500 characters, whereas few others ask their participants to type their username and password multiple times. In some cases, both methods may be employed and the users are monitored at periodic intervals. If the text length is too short, the accuracy of the classifier may be reduced. It could be a irritation, if it is too lengthy (Revett et al., 2007).

User's State of Mind

Every time when the user's typing behaviour is captured, user's physical condition or state of mind plays an important role in the keystroke dynamics authentication process. Different state of mind such as anger, sad or nervous may affect the typing behaviour (Singh et al., 2017).

Monitoring Mode

Along with username and password, the keystroke typing pattern is also captured to authenticate the user. If the username and password matches, the user is permitted to access the documents after which no monitoring is done. This may lead to impersonation after one-time authentication. Few researches were conducted to monitor the users during the entire login process (Flior & Kowalski, 2010).

KEYSTROKE DYNAMICS BASED COMMERCIAL PRODUCTS

Few companies provide software products which capture the users typing behaviour without the knowledge of the user which can be further used for authentication purpose. Two of them are discussed below.

Typing DNA

TypingDNA records Keystroke dynamics behavior of a user and turn them into typing patterns. These are analyzed and verified against previous patterns from the real user. Webcam or Microphone is not required to authenticate the user. This tool can be used to verify the exam taker's identity in e-learning, to prevent the fraudulent attacks during the online financial services, detecting intruders, multi factor authentication etc.

BehavioSec

BehavioSec is a swedish company that provides security to online applications using behavioural biometrics to authenticate the users continuously throughout the session and stop fraudulent behaviours. This can be used in both web and mobile applications.

APPLICATIONS OF KEYSTROKE DYNAMICS

As discussed in the earlier sections, keystroke dynamics can be applied in online learning environment and to authenticate the user during the online exams. Other than online learning, this section describes the applications of keystroke dynamics in different areas.

Additional Layer of Security

The authors (S. D. Abualgasim & Osman, 2011) had proposed that keystroke dynamics acts as an added layer of security together with the traditional username and password type of authentication, PINs used in Point-of-sale, bank ATMs. Experiments were conducted by the authors and they have shown that PINs of length 4-12 digits were encouraging than the PINs with less than 4 digits.

Web Applications

Abikoye et al. (2017) designed a web-based application and also applied keystroke dynamics biometrics on the password to provide the organization (SMEs) a more secure system than using password alone. Cho et al. (2000) had proposed to combine keystroke dynamics and password based system in E-commerce applications by using Neural Network classifier.

Online Banking Transactions

Chourasia (2014) had proposed to use keystroke dynamics for authenticating the users in online banking transactions. This technique has been provided as an additional layer of security to the users. In this paper, based on the variation in the typing pattern of the user, the author is provided with full access or partial access or no access to his account.

Smart Phone Authentication

The viability of using keystroke information was evaluated by using NN classifiers namely Feed forward multi-layered perceptrons (FF-MLPs), Radial basis function (RBF), Generalised regression neural network (GRNN) to classify the users in two handset interactions namely typing telephone numbers and text messages (Clarke & Furnell, 2007). The authors (Tsai et al., 2015) have proposed a novel technique through which the users can modify their PIN codes in their mobile phones at anytime, without any extra retraining of KDA systems. The authors have also claimed that keystroke dynamics can be used to improve the authentication system more effectively.

Emotion Detection

Lee et al. (2015) have proved that the emotions of a user can influence the keystroke typing patterns. Through the experiments, the authors have shown that the keystroke duration and latency are influenced by arousal but not in keyboard typing accuracy rate. Keystroke dynamics has also been applied to detect at least two of the emotions like frustration, anger, stress, relaxation, excitement and tiredness (Epp et al., 2011).

Gender Recognition

Keystroke Dynamics is used in digital forensics to identify the gender of the user during investigation(Tsimperidis et al., 2018). The authors have compared the results by using the classifiers namely SVM, RF, NB, MLP, RBFN. Through the experiments, the authors were able to show that NN, SVM, RBFN model have predicted the gender of an unknown user with more accuracy.

Bot Detection

The robustness of Telling hUman and Bot Apart (TUBA) was evaluated by the authors (Stefan et al., 2012) through experiments which included two series of simulated bots. First-order Markov chains were used for modeling the keystroke sequences forged by the two bots and the authors have used SVM for classification and the authors have shown that Keystroke Dynamics has been robust against the forgery attacks. Hernández et al. (2006) had used mouse and keystroke dynamics to differentiate between human and bot.

Intrusion Detection

Ahmed A. E. A & Issa Traore (2007) have combined keystroke dynamics and mouse dynamics to detect intrusion. Neural network based approach has been used to construct keystroke dynamics and mouse dynamics system.

DATASETS FOR KEYSTROKE DYNAMICS

This section describes the datasets that are available for keystroke dynamics, especially the keystrokes captured during online exams.

The dataset 'Stewart keystroke' was created for students taking online exams using keystroke and stylometry data. Four online short-answer tests were taken by 40 students and each student was given 10 questions, with an interval of 2 weeks. The tests were conducted on classroom Dell desktop computers with associated Dell keyboards. The students who have not completed all 4 online tests, their data were removed from the system. The authors (Stewart et al., 2011) have used 30 students data who have completed all 4 online tests, out of which 17 are male and 13 are female. The authors compared the performances of keystroke, stylometry and by combining keystroke and stylometry. Compared to stylometry, the authors have shown that keystroke was better. (Monaco et al., 2013) have also worked on the same dataset to improve the performance of the stylometry system. The performance of the keystroke system with that of the stylometry system was weaker on test input of 500 and 1000 words, respectively. To investigate the stylometry system further, the authors conducted a separate study on 30 book authors with 5000 and 10000 words. The authors have shown that the performance of the keystroke biometric system is far better than that of the stylometry.

The dataset 'Perez keystroke' was created for the users typing with one hand and two hands. 64 students with atleast 500 keystrokes were recorded on each exam. Exam attempts were conducted in an electronic classroom on desktop computers and laptops. (Perez et al., 2015) has utilized this dataset for the competition with one-handed keystroke biometric evaluation which included typing samples with the labeled and unlabeled portion of the dataset. Labeled portion consists of 500 normally-typed keystroke samples from 64 subjects and the unlabeled portion of the dataset was with 61 subjects which included both hands typing samples of 203 keystrokes, left hand typing samplesof 131 keystrokes, right hand typing samples of 137. Testing data was taken from 1 to 38 samples.

The dataset 'Perez keystroke' have been used by the authors (Mondal & Bours, 2017b) in which Pairwise User Coupling(PUC) method has been used for identifying the persons using keystroke dynamics. Various machine learning algorithms were compared with the PUC and the author had shown that PUC works better than other algorithms. The link to the Stewart keystroke and Perez keystroke are available at http://www.vmonaco.com/datasets.

(Dhakal et al., 2018) have created the dataset which included 136,857,600 keystrokes from 168,960 participants who completed 15 sentences in the questionaire, on an average of about 810 keypress per participant. Experiments were conducted to assess the typing behaviour of letter repetitions, rollover typing of the trained and untrained typists. The link to this dataset is available at http://userinterfaces.aalto.fi/136Mkeystrokes.

CONCLUSION AND FUTURE RESEARCH DIRECTIONS

Online learning continues to gain popularity but serious security threats can compromise the integrity of the Online learning experience. Weak authentication mechanisms and security countermeasures are no longer enough. Hence, a strong authentication method like Keystroke Dynamics is essential which is comparatively inexpensive and the only required hardware is the keyboard. A successful authentication from the beginning till the end of the login session can be done using Keystroke Dynamics and this can overcome the drawbacks of other authentication techniques. This chapter presented the various biometric technologies used for authentication, the methods employed by the students for cheating during the online exams, Keystroke Dynamics as a solution for strong authentication model during the online exams and their applications on different domain.

Commercial products that are available in the market that use keystroke dynamics as the authentication method, datasets that are available to monitor the users during the online exams have also been discussed. Many research works were carried out using Keystroke Dynamics as a major authentication method together with sound, mouse dynamics, stylometry etc., and this has led the researchers a new line of research through Keystroke Dynamics even the position and size of the keys may differ from one type of keyboard to another. In future, the authors have planned to develop a novel authentication mechanism, a multimodal biometric authentication method using username and password together with Keystroke Dynamics in monitoring the users during online exams.

REFERENCES

Abikoye & Sanni. (2017, November). Keystroke Dynamics Authentication for a Web-Based Sales and Stock Solution. *The International Journal of Computing & ICT Research.*

Abualgasim, S. D., & Osman, I. (2011). An Application of the Keystroke Dynamics Biometric for Securing PINs and Passwords. *World of Computer Science and Information Technology Journal, 1*(9), 398–404.

Ahmed, A. A., & Traore, I. (2014). *Biometric Recognition Based on Free-Text Keystroke Dynamics.* Academic Press.

Ahmed Awad, E. Ahmed & Issa Traore. (2007). Anomaly Intrusion Detection Based on Dynamic. *Proceedings of the 2005 IEEE Workshop on Information Assurance United States Military Academy.*

Al-Saleem, S. M., & Ullah, H. (2014). Security Considerations and Recommendations in Computer-Based Testing. *The Scientific World Journal, 2014*, 1–7. doi:10.1155/2014/562787 PMID:25254250

Alpar, O. (2017). Frequency Spectrograms for Biometric Keystroke Authentication Using Neural Network Based Classifier. *Knowledge-Based Systems, 116*, 163–171. doi:10.1016/j.knosys.2016.11.006

Alshehri, A., Coenen, F., & Bollegala, D. (2018). Iterative Keystroke Continuous Authentication: A Time Series Based Approach. *KI - Künstliche Intelligenz,* 1–13. doi:10.100713218-018-0526-z

Asha, S., & Chellappan, C. (2008). Authentication of e-learners using multimodal biometric technology. *Biometrics and Security Technologies, 2008. ISBAST, 2008*, 1–6. doi:10.1109/ISBAST.2008.4547640

Atoum, Y., Chen, L., Liu, A. X., Hsu, S. D. H., & Liu, X. (2017). Automated Online Exam Proctoring. *IEEE Transactions on Multimedia, 19*(7), 1609–1624. doi:10.1109/TMM.2017.2656064

Azevedo, G. L. F. B. G., Cavalcanti, G. D. C., & Carvalho Filho, E. C. B. (2007). Hybrid solution for the feature selection in personal identification problems through keystroke dynamics. *IEEE International Conference on Neural Networks - Conference Proceedings*, 1947–1952. 10.1109/IJCNN.2007.4371256

Bal, A., & Acharya, A. (2011). Biometric authentication and tracking system for online examination system. *2011 International Conference on Recent Trends in Information Systems, ReTIS 2011 - Proceedings*, 209–213. 10.1109/ReTIS.2011.6146869

Bawarith, R., Basuhail, A., Fattouh, A., & Gamalel-Din, S. (2017). E-exam Cheating Detection System. *International Journal of Advanced Computer Science and Applications, 8*(4), 176–181.

Bhagat, V., & Katankar, V. (2014). Novel method for Student Authentication in Online Examination. *International Journal of Research, 1*(7), 974–979.

Bhatia, A., & Hanmandlu, M. (2017). Keystroke Dynamics Based Authentication Using Information Sets. *Journal of Modern Physics, 8*(9), 1557–1583. doi:10.4236/jmp.2017.89094

Bours, P. (2012). Continuous keystroke dynamics: A different perspective towards biometric evaluation. *Information Security Technical Report, 17*(1–2), 36–43. doi:10.1016/j.istr.2012.02.001

Brocardo, M. L., Traore, I., & Woungang, I. (2014). Toward a Framework for Continuous Authentication Using Stylometry. *IEEE Xplore*, 106–115.

Canales, O., Murphy, T., Zych, E., Stewart, J., Tappert, C., Castro, A., … Plains, W. (2011). A Stylometry System for Authenticating Students Taking Online Tests. P. of Student-Faculty Research Day, Ed., CSIS. Pace University.

Cho, S., Han, C., Hee Han, D., & Kim, H.-I. (2000). Web based Keystroke Dynamics Identity Verification using Neural Network. *Journal of Organizational Computing and Electronic Commerce, 10*(4), 295–307. doi:10.1207/S15327744JOCE1004_07

Chourasia, N. (2014). Authentication of the user by keystroke dynamics for banking transaction system. *Proceedings of International Conference on Advances in Engineering & Technology*, 41–45.

Clarke, N. L., & Furnell, S. M. (2007). Authenticating mobile phone users using keystroke analysis. *International Journal of Information Security, 6*(1), 1–14. doi:10.100710207-006-0006-6

Cluskey, G. R., Ehlen, C. R., & Raiborn, M. H. (2011). Thwarting online exam cheating without proctor supervision. *Journal of Academic and Business Ethics, 4*, 1–8.

Dehnavi, M. K., & Fard, N. P. (2011). Presenting a multimodal biometric model for tracking the students in virtual classes. *Procedia: Social and Behavioral Sciences, 15*, 3456–3462. doi:10.1016/j.sbspro.2011.04.318

Dhakal, V., Feit, A. M., Kristensson, P. O., & Oulasvirta, A. (2018). Observations on Typing from 136 Million Keystrokes. *Proceedings of the SIGCHI Conference on Human Factors in Computing Systems - CHI'18*, 1–12. 10.1145/3173574.3174220

Epp, C., Lippold, M., & Mandryk, R. L. (2011). Identifying emotional states using keystroke dynamics. *Proceedings of the 2011 Annual Conference on Human Factors in Computing Systems - CHI '11*, 715. 10.1145/1978942.1979046

Fayyoumi, A., & Zarrad, A. (2014). Novel Solution Based on Face Recognition to Address Identity Theft and Cheating in Online Examination Systems. *Advances in Internet of Things*, *4*, 5–12. doi:10.4236/ait.2014.42002

Flior, E., & Kowalski, K. (2010). Continuous biometric user authentication in online examinations. *ITNG2010 - 7th International Conference on Information Technology: New Generations*, 488–492. 10.1109/ITNG.2010.250

Gao, Q. (2012). Biometric Authentication to Prevent e-Cheating. *International Journal of Instructional Technology and Distance Learning*, *9*(2), 93.

Goodkind, A., Brizan, D. G., & Rosenberg, A. (2017). Utilizing overt and latent linguistic structure to improve keystroke-based authentication. *Image and Vision Computing*, *58*, 230–238. doi:10.1016/j.imavis.2016.06.003

Gunetti, D., & Picardi, C. (2005). Keystroke Analysis of Free Text. *ACM Transactions on Information and System Security*, *8*(3), 312–347. doi:10.1145/1085126.1085129

Halakou, F. (2013). Feature Selection in Keystroke Dynamics. *Norwegian Information Security conference(NISK)*.

Hernández, J. A., Ochoa, A., Muñoz, J., & Burlaka, G. (2006). Detecting cheats in online student assessments using Data Mining. *Conference on Data Mining*, 6, 205.

Hu, J., Gingrich, D., & Sentosa, A. (2008). Authentication through Biometric Keystroke Dynamics. *IEEE International Conference on Communications*, 1556–1560. 10.1109/ICC.2008.301

Jung, I. Y., & Yeom, H. Y. (2009). Enhanced security for online exams using group cryptography. *IEEE Transactions on Education*, *52*(3), 340–349. doi:10.1109/TE.2008.928909

Kang, P., & Cho, S. (2015). Keystroke Dynamics-based User Authentication Using Long and Free Text Strings From Various Input Devices Keystroke Dynamics-based User Authentication Using Long and Free Text Strings From Various Input Devices. *Journal of Information Science*, *308*, 72–93. doi:10.1016/j.ins.2014.08.070

Karim, N. A., & Shukur, Z. (2015). Review of user authentication methods in online examination. *Asian Journal of Information Technology*. doi:10.3923/ajit.2015.166-175

Karnan, M., Akila, M., & Kalamani, A. (2009). Feature subset selection in keystroke dynamics using ant colony optimization. *Journal of Engineering and Technology Research*, *1*(5), 72–80.

Khlifi, Y., & El-Sabagh, H. A. (2017). A novel authentication scheme for E-assessments based on student behavior over e-learning platform. *International Journal of Emerging Technologies in Learning*, *12*(4), 62–89. doi:10.3991/ijet.v12i04.6478

Killourhy, K. S., & Maxion, R. A. (2012). Free vs. transcribed text for keystroke-dynamics evaluations. *Proceedings of the 2012 Workshop on Learning from Authoritative Security Experiment Results - LASER '12*, 1–8. 10.1145/2379616.2379617

Kim, J., Kim, H., & Kang, P. (2018). Keystroke dynamics-based user authentication using freely typed text based on user-adaptive feature extraction and novelty detection. *Applied Soft Computing*, *62*, 1077–1087. doi:10.1016/j.asoc.2017.09.045

Krause, M. (2014). A behavioral biometrics based authentication method for MOOC's that is robust against imitation attempts. *Proceedings of the First ACM Conference on Learning @ Scale Conference - L@S '14*, 201–202. 10.1145/2556325.2567881

Kumlander, D. (2008). Soft biometrical students identification method for e-learning. *Advances in Computer and Information Sciences and Engineering*, *114–118*, 114–118. doi:10.1007/978-1-4020-8741-7_21

Lee, P. M., Tsui, W. H., & Hsiao, T. C. (2015). The influence of emotion on keyboard typing: An experimental study using auditory stimuli. *PLoS One*, *10*(6), 1–16. doi:10.1371/journal.pone.0129056 PMID:26065902

Lee-Post, A., & Hapke, H. (2017). Online Learning Integrity Approaches: Current Practices and Future Solutions. *Online Learning*, *21*(1), 135–145. doi:10.24059/olj.v21i1.843

Matsubara, Y., Samura, T., & Nishimura, H. (2015). *Keyboard Dependency of Personal Identification Performance by Keystroke Dynamics in Free Text Typing*. Academic Press.

Monaco, J. V., Stewart, J. C., Cha, S. H., & Tappert, C. C. (2013). Behavioral biometric verification of student identity in online course assessment and authentication of authors in literary works. *IEEE 6th International Conference on Biometrics: Theory, Applications and Systems, BTAS 2013*. 10.1109/BTAS.2013.6712743

Mondal, S., & Bours, P. (2017a). A study on continuous authentication using a combination of keystroke and mouse biometrics. *Neurocomputing, 230*, 1–22. doi:10.1016/j.neucom.2016.11.031

Mondal, S., & Bours, P. (2017b). Person Identification by Keystroke Dynamics Using Pairwise User Coupling. *IEEE Transactions on Information Forensics and Security*, *12*(6), 1319–1329. doi:10.1109/TIFS.2017.2658539

Moten, J., Fitterer, A., Brazier, E., Leonard, J., & Brown, A. (2013). Examining online college cyber cheating methods and prevention measures. *Electronic Journal of E-Learning*, *11*(2), 139–146.

Perez, G., Tappert, C. C., Bours, P., Mondal, S., Morales, A., Fierrez, J., & Ortega-garcia, J. (2015). *One-handed Keystroke Biometric Identification Competition*. Academic Press.

Ramu, T., & Arivoli, T. (2013). a Framework of Secure Biometric Based Online Exam Authentication: An Alternative To Traditional Exam. *International Journal of Scientific and Engineering Research*, *4*(11), 52–60.

Revett, K., Gorunescu, F., Gorunescu, M., Ene, M., Tenreiro De Magalhaes, S., Santos, D., ... Santos, H. (2007). A machine learning approach to keystroke dynamics based user authentication. *International Journal of Electronic Security and Digital Forensics*, *1*(1), 55. doi:10.1504/IJESDF.2007.013592

Rodchua, S. (2017). Effective Tools and Strategies to Promote Academic Integrity in e-Learning. *International Journal of E-Education, E-Business, E-Management and E-Learning*, *7*(3), 168–179. doi:10.17706/ijeeee.2017.7.3.168-179

Rodchua, S., Yiadom-Boakye, G., & Woolsey, R. (2011). Student Verification System for Online Assessments: Bolstering Quality and Integrity of Distance Learning. *Journal of Information Technology*, *27*(3), 1–8.

Roth, J., Liu, X., Ross, A., & Metaxas, D. (2013). Biometric authentication via keystroke sound. *Proceedings - 2013 International Conference on Biometrics, ICB 2013*. 10.1109/ICB.2013.6613015

Rudrapal, D., Das, S., Debbarma, S., Kar, N., & Debbarma, N. (2012). Voice Recognition and Authentication as a Proficient Biometric Tool and its Application in Online Exam for P. H People. *International Journal of Computers and Applications*, *39*(12), 975–8887. doi:10.5120/4870-7297

Schlenker, A., & Sarek, M. (2013). Neural Networks in Keystroke Dynamics for Multi-Factor Authentication in. *Measurement*, 109–112.

Shanmugapriya, D., & Padmavathi, G. (2011). An Efficient Feature Selection Technique for User Authentication using Keystroke Dynamics. *IJCSNS*, *11*(10), 191.

Singh, D., Jaggi, B., Nayyar, H., & Kumar, A. (2017). *Presskey- A Keystrokes Dynamics Based Authentication System*. Academic Press.

Stefan, D., Shu, X., & Yao, D. (2012). Robustness of keystroke-dynamics based biometrics against synthetic forgeries. *Computers & Security*, *31*(1), 109–121. doi:10.1016/j.cose.2011.10.001

Stewart, J. C., Monaco, J. V., Cha, S. H., & Tappert, C. C. (2011). An investigation of keystroke and stylometry traits for authenticating online test takers. *2011 International Joint Conference on Biometrics, IJCB 2011*. 10.1109/IJCB.2011.6117480

Syed Idrus, S. Z., Cherrier, E., Rosenberger, C., & Bours, P. (2014). Soft biometrics for keystroke dynamics: Profiling individuals while typing passwords. *Computers & Security*, *45*, 147–155. doi:10.1016/j.cose.2014.05.008

Traoré, I., Awad, A., & Woungang, I. (2017). *Ensuring Online Exam Integrity Through Continuous Biometric Authentication. In Information Security Practices: Emerging Threats and Perspectives* (pp. 73–81). Cham: Springer; doi:10.1007/978-3-319-48947-6

Tsai, C. J., Peng, C. C., Chiang, M. L., Chang, T. Y., Tsai, W. J., & Wu, H. S. (2015). Work in progress: A new approach of changeable password for keystroke dynamics authentication system on smart phones. *Proceedings of the 2014 9th International Conference on Communications and Networking in China, CHINACOM 2014*, 353–356. 10.1109/CHINACOM.2014.7054316

Tsimperidis, I., Arampatzis, A., & Karakos, A. (2018). Keystroke dynamics features for gender recognition. *Digital Investigation*, *24*, 4–10. doi:10.1016/j.diin.2018.01.018

Chapter 9

A Framework for Improving the Accuracy of Keystroke Dynamics-Based Biometric Authentication Using Soft Computing Techniques

Shanmugapriya D.
Avinashilingam Institute for Home Science and Higher Education for Women, India

Padmavathi Ganapathi
Avinashilingam Institute for Home Science and Higher Education for Women, India

ABSTRACT

The global access of information and resources from anywhere has increased the chance of intrusion and hacking of confidential data. Username with password is the commonly used authentication mechanism which is used for almost all online applications from net banking to online examinations. However, advanced safeguard mechanisms are sought against unauthorized access as the passwords can be hacked easily. To strengthen the password, it can be combined with biometric technology. Keystroke biometrics, a strong behavioral biometric, can be considered as a secure method compared to other methods even if the imposter knows the username and password as it is based on user habitual typing rhythm patterns. For any online application the accuracy level plays a vital role. But the accuracy of keystroke authentication when compared with other biometric authentication mechanisms is low. To improve the accuracy and minimize the training and testing time, this chapter proposes a wrapper-based classification using PSO-ELM-ANP algorithm which gives 99.92% accuracy.

DOI: 10.4018/978-1-5225-7724-9.ch009

INTRODUCTION

Unauthorized access of vital information threatens the computer security. User authentication is the only way to prevent the unauthorized access. User Authentication is classified into three categories namely: Knowledge based authentication, Object based authentication and Biometric based authentication. The Figure 1 shows the classification of User Authentication mechanisms.

Knowledge Based Authentication

Knowledge based authentication is based on "something the user knows". Passwords or PINs come under this category. Passwords are widely used because they are simple, inexpensive, and portable.

Even though the passwords are strong or weak or medium, they are one of the very poor forms of authentication because they can be easily cracked or hacked by Guessing, Brute force attacks, Dictionary attacks, Shoulder surfing, Key logger attack. Apart from the above mentioned, user may have multiple passwords or a password that needs to be continuously changed, which are very hard to remember by the user. Passwords are classified as Strong passwords, Weak Passwords and Medium Passwords according to password generators. Strong passwords are hard to remember which may lead to dictionary attacks. Medium and Weak passwords can easily be hacked or cracked.

Knowledge based Authentication can be categorized as Static and Dynamic Knowledge based authentication methods. Static method is commonly used by banks, financial services companies and e-mail providers where the user has to authenticate with the password before accessing the information. The user has to provide the answer for a shared secret question during the registration phase. The user is allowed to reset the password using shared secret question by providing answer for the question. If the shared secret question is hacked or the imposter is able to guess the answer for the question, he can

Figure 1. User authentication methods

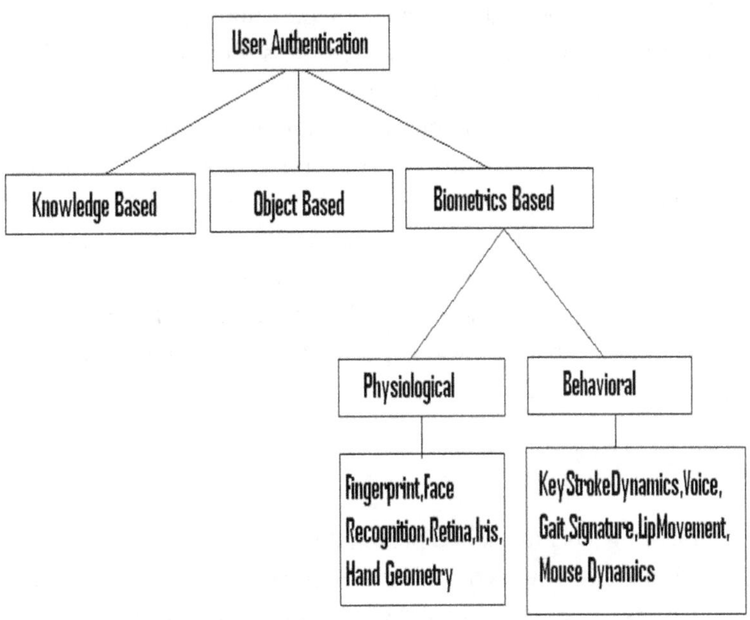

easily authenticate and access the information. The Dynamic knowledge based authentication does not require the user to provide the secret questions and answers during registration phase. The questions will be generated by the system based on the records of the individuals. A time limit will be given for the user to respond to the Questions and numbers of attempts are also limited to answer the question. This method provides high level of authentication than the static method.

Object Based Authentication

Something the user has or possesses is called Object based authentication. Tokens like ATM cards, credit/ debit cards, ID cards etc., comes under this category of authentication through “something the user has”. This authentication method is popular because tokens are portable and does not require any special software or equipment. Tokens prevent the stolen password from accessing the website by the unauthorized user as he/she must steal the physical token as well as the password.

However, they are expensive when compared with Knowledge based authentication as it requires replacement of tokens every few years. Object based authentications are not convenient for the user since he/she must manually enter the token value as well as the password. Tokens can also be stolen by strange party which leads to the stealing of information. In this type of authentication, user can either forget the password or lose the card.

Biometric Based Authentication

“Something whom you are” is biometrics. Biometrics based authentication authenticates a person based on physical or behavioral characteristics of the claimed person. Fingerprints, Iris, face recognition, typing rhymes are popular biometric authentication methods. The advantages are: it is difficult to duplicate biometric and cannot be stolen, shared, guessed, lost or forgotten. Biometric methods authenticate a person without the necessity to carry or to memorize anything as the person’s body itself is the password. It too has some limitations like: special equipment and software are required for recognition, expensive, leaks the personal information of the body when spoofed, cannot change or replace the biometric data that has been stolen and finally some biometric applications are not portable. The biometric methods that are not portable cannot be used for remote authentication.

The following Table 1shows the comparison of user authentication methods.

The traditional method of providing authentication for remote access application such as online examination, Internet banking, payment gateways etc., depends on knowledge or object-based authentication mechanisms. Almost all the online applications use password or PIN for authenticating the user. However, these methods are not secured and vulnerable to various attacks.

One of the ways to strengthen the password or PIN is to combine Biometric technology. Biometric authentications are secured and consider various parameters for authentication which are not possible by knowledge or object-based methods.

Types of Biometrics

Biometrics involves something a person is or a person does. Biometrics systems are classified as Physiological biometrics and Behavioral biometrics. Physiological biometrics refers to what the person is or they measure physical parameters of certain parts of the body. Fingerprints, iris, retina, hand geometry,

Table 1. Comparison of user authentication methods

Authentication Methods	Examples	Advantages	Limitations
Knowledge based Authentication Based on "something the user knows"	Username and Passwords	• Does not require any special software for incorporating. • Portable as the users memorize them. • Authenticate the user directly because the user only knows the password	• Vulnerable to hackers. • Recovery is either expensive or insecure.
Object based Authentication Based on "something the user has"	PIN, Tokens	• Prevent accounts from being shared • Portable. • Does not require any special software on the user's computer.	• Expensive and must be replaced every few years. • Easy to misplace or damage. • Not convenient for the user since he/she must manually enter the token value as well as the password.
Biometric based Authentication Based on "something the user is"	Keystroke, Gait, Fingerprint, Voice, IRIS, Signature, Hand Geometry etc.,	• Cannot be stolen, shared, guessed, lost or forgotten unlike passwords and tokens. • Very strong because they are difficult to steal. • Authenticate a person without the necessity to carry or to memorize anything.	• Appropriate biometric sensor and software must be installed on the users' computers which are tending to be expensive. • It leaks the personal information of the body.

face are some of the examples of Physiological biometrics. Behavioral biometrics shows how the person is using the body for authentication. For Example keystroke dynamics, mouse dynamics, signature, gait, voice are certain behavioral biometrics. The table 2 analyses the different types of biometrics.

Behavioral based biometrics must frame over its physiological counterpart on the capacity to work in secrecy mode verification. The way people do things such as speaking like voice, typing rhythm like keystroke dynamics, writing like signature, and walking style as gait recognition are known to be behavioral biometrics. Behavioral based biometrics such as the way one signs his/her name or type his/her password are unique as well and have much lower impact on privacy and costs. The approach and the method in which one types on the computer keyboard varies from individual to individual and is considered to be a unique as far as behavioral biometric is concerned.

Table 2. Analysis of different types of biometrics

Physiological Biometrics	Behavioral Biometrics
What the person is or measure physical parameters of a certain part of the body	What a person does or how the person uses the body
Examples: Fingerprint, Iris, Face Recognition, etc.,	Examples: Voice, Handwriting, walking gait, Typing rhythm or Keystroke Dynamics
Limitations: • Can be spoofed • Difficulties in implementing • Expensive • Requires special Hardware • More Intrusive	Limitations: • Voice Recognition, the most promising technology but it is difficult to distinguish a live user from a recorded one • Handwriting, gait requires a special device, which can be cost prohibitive and signatures can be forged

Behavioral biometrics finds its applications in many areas like financial institutions, businesses, government facilities, retail Point of Sale (PoS) and in other environments for secure authentication. Unlike many types of physical biometrics, behavioral biometrics can be embedded with existing hardware and requires only software for analysis. This capacity makes behavioral biometrics simpler and less costly to implement compared to other biometric and user authentication mechanisms. The Figure 2 shows some of the examples of Physiological and Behavioral biometrics.

Keystroke Dynamics

Keystroke dynamics is one of the behavioral biometric approaches which is cheaper and does not require any sophisticated hardware other than the keyboard. It is easier to implement and has many advantages over other biometric approaches. Keystroke dynamics is a strong behavioral biometric which does not consider what password the user is typing but it analyzes how the user is typing (typing pattern of user). Keystroke biometrics can be considered as a secure method compared to other methods even if the imposter knows the username and password as it is based on user habitual typing rhythm patterns. The following table 3 shows the advantages of keystroke dynamics over other biometric authentication systems.

Keystroke Approaches

There are two keystroke authentication modes namely: Static and Dynamic modes (Anil Jain, Lin Hong and Sharath Pankanti, 2000). Static mode (Fixed text mode) is based on password typing rhythm and the authentication is done only at login time. Dynamic mode (Free text mode) is based on pattern regardless of the typed text. The authentication is done continuously even after the login and in this mode it is not required to memorize a predetermined text (username and password). The research work proposed in this chapter uses static mode verification of keystroke dynamics.

Figure 2. Different types of biometrics

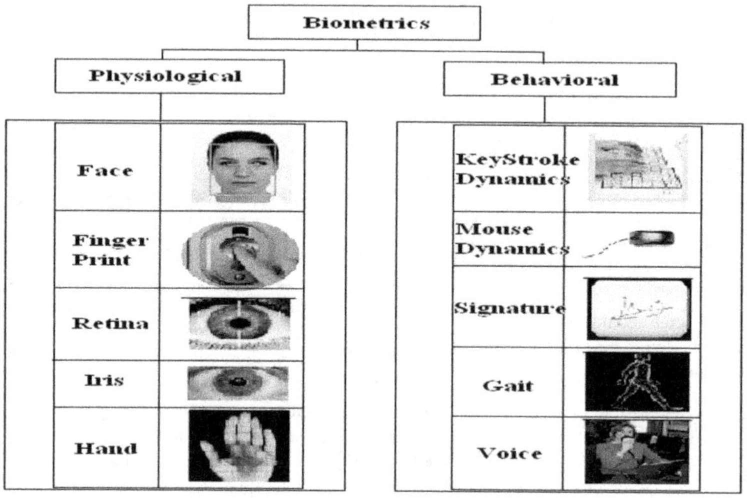

Table 3. Advantages of keystroke dynamics over other biometric authentication systems

S.No	Keystroke Authentication Method	Other Biometric Authentication Methods
1.	Cheap	Expensive
2.	No special Hardware Required other than keyboard/keypad	Requires Special Hardware
3.	Provides one-time and Continuous authentication	Provides only one-time authentication
4.	Remote authentication is possible	Remote authentication is not possible
5.	Collection of sample is possible from anywhere using Internet.	Collection of samples not possible from remote place
6.	No personal information leaked	Authentication systems like fingerprint, Iris, palm leaks the personal information of a person
7.	If stolen, possible to create new sample.	If stolen, not possible to create new sample. User loses his/her unique identity.

For any online application the accuracy level plays a vital role. But the accuracy of keystroke authentication when compared with other biometric authentication mechanisms is low. So to put this strong biometric authentication, which can very well be adapted for conducting online examination, into practice, a higher level of accuracy is desired. Hence, it is highly required to identify the parameters that optimize the accuracy. To improve the accuracy and minimize the training and testing time, this chapter proposes a wrapper-based classification using PSO-ELM-ANP algorithm.

The rest of the chapter is organized as follows: the next section discusses the background works done in research area which is followed by methodology, discussion of the results and the comparing the obtained results with the previous works. Final section concludes the work with future directions.

BACKGROUND

Collecting keystroke data, feature extraction, normalization, feature subset selection and classification are the major steps involved in keystroke authentication. This section discusses the related works done in the data collection or various benchmark datasets available, features extracted, feature subset selection and neural network-based classification areas.

Literature on Data Collection

Data collection is the process of collecting the user data and storing in the template. The sensor unit senses the key press timing of the user. When the user is interacting with the system for the first time, all the required credentials are stored as a user profile in the database. User information includes the username and password. When user is typing the information to be authenticated, his/ her typing behavior is captured, processed and stored as a reference keystroke template for future comparisons during identification process. Authentic Datasets are very important for reliable experimentation and derivations.

There are only few public benchmark datasets available for testing Keystroke dynamics authentication. The datasets available are very less in number when compare to other biometrics systems. This section elaborates the publically available benchmark datasets for Keystroke dynamics.

The dataset GREYC was proposed by (Giot et al., 2009) which contains 133 users, and 100 of them provided samples of, at least, 5 distinct sessions. The password "greyc laboratory" was typed by each user 12 times during each session. Raw data and hold time and latencies are the features extracted.

Giot et al., (2012) proposed another dataset named WEBGREYC{A,B}.The dataset contains two kinds of biometric samples: (i) imposed login and password -WEBGREYCA; (ii) login and password chosen by the user - WEBGREYCB.

Killourhy and Maxion (2009) proposed DSN 2009 dataset. In this dataset, 51 users provides 400 samples captured in 8 sessions. 50 inputs per session are captured. Each user is asked to type the password ".tieRoanl". The dataset contains hold time, interval between two pressures, interval between the release of a key, and the pressure of the next one.

The dataset PRESSURE{0,1,2}created by (Allen, 2010) contains key code, time when pressed, time when released, pressure force. Three different passwords have been typed: "pr7q1z" (PRESSURE2), "jeffreyallen" (PRESSURE1) and "drizzle" (PRESSURE3). 104 users participated in the sample collection.

The dataset BIOCHAVES{A,B,C} is used in many papers (Filho and Freire, 2006; Montalvao Filho and Freire, 2006; Montalvao et al., 2006). The number of users in the dataset is 15 and each user provided 10 samples. The dataset contains pressed key and the elapsed time since the last key down event.

Loy et al. (2007) created Keystroke100 dataset which contains keystroke pressure information. 100 users provided 10 samples each.

Syed Idrus et al. (2013) have released a dataset GREYC-NISLAB {A, B, C, D, E}. 100 users provided 10 inputs of 5 different passwords.

The Table 4 shows the literature on benchmark datasets used for study in keystroke dynamics.

The majority of the authors, in their studies, have used their own datasets. Only few benchmark dataset are available publically. Different datasets are used by different authors. This chapter uses 2 benchmark datasets and one real time dataset for testing the efficiency of the proposed algorithms. The next step in the keystroke Text Phase is feature extraction.

Literature on Feature Extraction

When a user is typing via the keyboard, key press time and release time are recorded as the raw data. After obtaining the raw data, the other features such as Duration and Flight time or Latency can be extracted. This section discusses about the various features and feature extraction techniques available so far in the literature.

Akila and Suresh Kumar (2011) measured Mean, Median and Standard deviation of feature values such as Latency, Duration and Digraph and compared the performance. Digraph with median as the feature gives good result when compared to other features.

Alaa Darabseh and Akbar Siami Namin (2015) analyzed the features such as key duration, flight time latency, digraph time latency, and word total time duration. The authors captured the logged experimental data for 28 users and found that key duration time offers the best performance result among all four keystroke features, followed by word total time.

Table 4. Literature on benchmark datasets used in keystroke dynamics research

S.No	Authors(s)	Dataset	No of Users	Samples	Password	File Format
1.	Killourhy and Maxion (2009)	DSN2009	51	400	.tie5Roanl	text, csv and Excel files.
2.	Giot et al., (2009)	GREYC	133	Each user 12 times	greyc laboratory	sqlite
3.	Giot et al (2012)	WEBGREYC {A,B}	118	-	WEBGREYCA - imposed login and password WEBGREYC B - login and password chosen by the user	text
4.	Allen (2010)	PRESSURE {0,1,2}	104	97 users provided complete samples. Each user provided between 3 and 15 samples .	3 passwords used 1."pr7q1z" 2."jeffrey allen" and 3."drizzle"	csv or sql files
5.	Freire (2006a, b,c)	BIOCHAVES {A,B,C}	15	10 samples each user.	-	text
6.	Loy et al., (2007)	Keystroke100	100	10 samples for each user(100*10)=1000 samples	-	text
7.	Syed Idrus et al.,(2013).	GREYC-NISLAB{A,B,C,D,E}	110	10 samples each user.	5 different passwords	excel

Balagani et al. (2011) classified the keystroke feature vectors as homogeneous if it has either key hold latencies only or key interval latencies only as its features, a heterogeneous vector if it has both key hold and key interval latencies as features, an aggregate vector if its features are derived by aggregating key hold and key interval latencies. They measured all the above-mentioned vectors and concluded that a heterogeneous vector has higher discriminability than an aggregate vector, especially when the reference text is short. However, as the length of the text increases, the difference in the discriminability between heterogeneous and aggregate vectors tends to decrease.

Eesa Al Solami et al. (2011) suggested four statistical-based feature extraction techniques. The techniques, most frequently typed n-graph selection, quickly-typed n-graph selection, Time-stability typed n-graph selection, and Time-variant typed n-graph selection are tested with k-means clustering algorithm and found that the most-frequent 2-graphs technique can represent user's typing patterns effectively because of their high statistical significance.

Giot., et al., (2015) measured Mean and Standard deviation of feature values such as hold time and latencies interval between two pressures, interval between the release of a key, and the pressure.

Giovanni Saggio et al. (2011) considered cumulative and ratio timings (sum and ratio of dwell and flight times), which do not change even if the user's typing style evolves with practice.

Gissel Zamonsky Pedernera et al. (2010) measured R-Distance, A- Distance and weighted Distance (Combination of R and A Distance) and evaluated the results using Adapted k-Means (AKM), Adapted Subtractive Clustering (ASC) algorithms. The Best classification results are given by the Weighted A-distance using the median.

Juan Liu et al. (2014) had investigated feature extraction based on FFT, DCT and Gabor wavelet.

A force-sensing keypad is constructed by Liang Tang et al. (2010) by adding the Force Sensitive Resistor (FSR) sensors and extracted applied force and key-press timings. Compared to the traditional dynamics identification methods, the Force time domain information provided more appropriate user input information. Preliminary experimental results show that the force is more difficult to imitate than rhythm.

Margit Antal et al. (2015) examined the effect of touch screen features though their dataset of 42 users. Both time and touch screen features were studied.

Monaco, J.V., et al., (2013) measured Mean and Standard deviation of feature values such as Key press duration (dwell) and key transition (latency) and accuracy obtained is between 88.2 and 91.5.

Morales. et al., (2015) measured the difference between timestamps of press and release events of the same key difference between time, Time stamps of press and press events of consecutive keys.

Rohit A.Patil and AmarL.Renke (2016) measured extracted Digraph, trigraph, dwell time and Flight time and compared performance of Static and continuous approaches.

Sonali B. Wankhede and ShilpaVerma (2014) measured how long a key is pressed, how long it takes to move from one key to another and feature value such as Dwell time, Flight time and key code and obtained FAR:4.8%, FRR: 3.1% and ERR: 5%.

Wong et al. (2011) extracted the keystroke latencies using a box plot algorithm to create typing patterns for individuals.

Y. Deng and Y. Zhong (2013) measured Mean and Standard deviation of feature values such as Latency Timings and Dwell time and compared the performance. Here EER is reduced to 0.035

Zhang et al. (2011) analyzed the existing Press Release – Release Press (PR-RP) Model in order to provide a new improved PR-RP model for identity authentication of keystroke rhythm. Press - Press (P-P) and Release – Release (R-R) relay datasets are rebuilt, with the improved PR-RP model. The authors tested the result with four critical datasets and obtained improved recognition rate.

Zhong, Y. Deng and Anil K.Jain (2012) measured Mean and Standard deviation of feature values such as Latency Timings and Dwell time and compared the performance. Here Equal Error Rate (EER) is reduced from 8.7% to 8.4%.

From the literature available for feature extraction, it is observed that almost all the authors extracted Dwell time and Flight time. No new features were extracted. In this chapter during feature extraction phase commonly extracted features like Duration, Latency, Digraph and trigraph are extracted. A new feature – Virtual Key Force is proposed which plays vital role in improving the accuracy of the proposed system.

Literature on Normalization Methods

Keystroke data may be preprocessed before being used. Such pre-processing allows for better performance using a normalization function. Very few normalization techniques are available in the literature for keystroke and some of the significant works are tabulated in this section. The following table 5 gives the Normalization techniques available in the literature.

From the above table 5, it is observed that only few authors applied normalization techniques and not much importance is given for normalization. The chapter applies Z-Score Normalization to preprocess the extracted features.

Table 5. Literature on normalization techniques

Author(s)	Normalization Techniques Followed
Anil Jain et al., (2005)	Different normalization techniques for different biometric applications.
Azevedo et al., (2007)	Sample standard deviation method.
Campisi et al., (2009)	User dependent and independent normalization.
Filho and Freire (2006)	Log- Normal.
Francesco et al., (2002)	Divided degree of disorder by the value of the maximum disorder.
Hai-Rong Lv at al., (2006)	Normalized the mean value of pressure to zero and the standard deviation of the pressure to value one.
Hu et al., (2008)	Divided the distance calculated from trigraph by $N^2 / 2$.
Mariusz Rybnik et al., (2008)	Counting average and standard deviation for dwell and flight values.
Wong et al., (2001)	Standard normal bell curve algorithm.

Literature on Feature Subset Selection Methods

Feature subset selection is applied to select the most optimal features (Singhi and Liu, 2006). It reduces the number of features and results in speeding up the searching process thus reducing the computational time. There are two feature subset selection approaches (Sung and Cho, 2006): Filter and Wrapper approach. Filter approach filters out the discriminating features and passes only the relevant features to the classifiers. On the other hand, the wrapper approach tries different sets of features using a learning algorithm, and then chooses the best one. This method has the computational overhead of evaluating the feature subset using the learning algorithm. However, if the learning algorithm used to train the classifier is computationally efficient, it yields better results than the filter approach. This section discusses the major feature subset selection methods applied in keystroke dynamics.

Akila M., Suresh Kumar V., Anusheela N. and Sugumar K. (2012) measured mean, median and standard deviation of features such as latency, duration and digraph during feature extraction. The authors used artificial bee colony algorithm for feature selection and neural network for classification.

Bermejo et. al (2014) used the Naïve Bayes classifier as induction algorithm for wrapper-based method and observed that the wrapper feature subset selection with Naïve Bayes classifier is better than filter based FSS.

Giot.,et al (2014) used methods like PSO, SVM and Genetic Algorithm and it is observed that PSO gives better results than GA.

Optimization techniques such as Genetic Algorithm (GA) and Particle Swarm Optimization (PSO) are used for feature subset selection and their performance is compared by Akila and Suresh Kumar (2011) and Karnan and balagan (2009a). Particle Swarm Optimization (PSO) gives better performance than Genetic Algorithm with regard to feature reduction rate.

Particle Swarm Optimization (PSO), Genetic Algorithm (GA) and Ant Colony Optimization (ACO) are used by Karnan and Akila (2010) for feature subset selection. Backpropagation Neural Network (BPN) has been used for classification. ACO gives better performance than PSO and GA with regard to feature reduction rate and classification accuracy.

Muthuramalingam A., Gnanamanickam J., and Muhammad R (2018) proposed Firefly algorithm (FA) for feature selection. The authors computed and compared the performance efficiency of FA with existing techniques and found that the convergence rate and iteration generations to reach the optimum solution is 41% and 18% less respectively.

Rohit A.Patil and AmarL.Renke(2016) used the three methods Filter, Wrapper and hybrid model and compared.

Shivam Sharma (2015) used methods like Multi –SVM, KNN and Ant Colony Optimization and it is observed that accuracy of feature selection technique using Ant Colony Optimization on Keystroke Dynamics on Benchmark dataset gives desirable results.

V. Chandrasekar, S. Suresh Kumar and T. Maheswari (2016) measured the Hausdorff timing values, mean, standard deviation, and median of keystroke features, such as latency, duration, digraph, and their combinations, and compared their performance. The authors used stochastic diffusion search for feature subset selection.

V. Chandrasekar and S. Suresh Kumar (2016) evaluated the Keystroke features, dwell time and flight time of every user by preprocessing techniques such as Hausdroff timing, mean, median, and standard deviation. The authors applied artificial immune system is used for feature selection.

Many researches are being done in keystroke feature subset selection. This chapter proposes a novel wrapper based feature subset selection using PSO-ELM-ANP method which greatly reduces the number of features.

Literature on Neural Network Based Classification Approaches

The next step after feature subset selection is classifying the selected features. Classification aims to find the best class that is closest to the classified pattern (Karnan and Akila, 2010). A number of studies have been performed in the area of classification since its inception. Five different classification techniques namely statistical classification, neural network-based classification, classification using pattern recognition, hybrid classification techniques and other techniques are being followed in the area of keystroke dynamics.

The proposed research focuses on Neural Network based classification techniques. This section discusses the various Neural Network based classifications followed in the area of keystroke dynamics so far.

A Multilayer Perceptron (MLP) neural network with a Back Propagation (BP) learning algorithm is used by Harun et al. (2010) to train and validates the features. The simulation results revealed that MLP with BP network is suitable to discriminate and classify a nonlinear keystroke database as high as 98% in correct acceptance rate.

Ahmed (2014) used ANN/MLP method and tested for 17 users.

Alaa Darabseh and Akbar Siami Namin (2015) employed four machine learning techniques for classification such as two-class support vector machine (TC) SVM, one-class support vector machine (OC) SVM, k-nearest neighbor classifier (K-NN), and Naive Bayes classifier (NB). The authors found that TC-SVM and KNN perform the best among the four classifiers.

Idrus et.al (2013) used SVM and used the datasets that are composed of several soft biometrics information of users for classification. It consists of information on the user's way of typing by defining the number of hands used to type (one or two), gender, age and handedness.

J. Yadav, K. Pandey, S. Gupta and R. Sharma (2017) proposed keystroke dynamics based authentication system using fuzzy logic and the proposed fuzzy logic model gives 98% authentication accuracy and 2% false rejection rate.

Kaganovet et.al (2015) used fuzzy sets kernel and it is observed that FRR is equal to FAR.

Karnan and Akila (2009b and 2010) obtained 92.8% classification accuracy using Back Propagation Neural Network (BPN). The authors applied feature subset selection methods before classification.

Margit Antal et al. (2015) used machine learning classification algorithms like Random Forest, Bayseian Nets and SVM. The author concluded with an increase of 10% increase in the accuracy using touch screen features.

MdLiakat Ali et al, (2016) used methods like SVM, Decision tree and ANN and it is observed that overall accuracy is lower than other biometrics.

Orcan Alpar (2014) introduced authentication and discrimination system using artificial neural networks with RGB histograms. Keycodes and interkeytimes for passwords were used as feature. The system has granted 90% of real user attempts and rejected 90% of frauds.

R.Abinaya and An. Sigappi (2018) used random forest classifier and compared their methods with other classifiers such as Support Vector Machine and Random tree classifier and obtained 90% accuracy for random forest classifier.

Rohit A.Patil and AmarL.Renke(2016) used the classification methods of ANN, Pattern recognition and applied for Human behavior-based password protection for web based applications.

Syed Idrus et al., (2014) studied static passwords and free text. The authors combined machine learning and fusion process to authenticate users using keystroke dynamics.

Yohan Mulionoa, Hanry Hamb and Dion Darmawanb (2018) used deep learning using optimizer for classification and obtained 92.60% accuracy.

Yunbin Deng and Yu Zhong (2015) adopted Gaussian Mixture Model with Universal Background Model (GMM-UBM), identity vector (i-vector) approach to user modelling, deep machine learning approach to perform keystroke dynamics based user authentication.

Keystroke for Online Examination Authentication

Very few works were done which uses Keystroke authentication for securing Online Examination. This section discusses the various works on securing Online Examination through keystroke dynamics.

P. K. Mungai and R. Huang (2017) proposed 3-leveled architecture which comprised of a statistical level, machine learning level, and a logical comparison level in hierarchical order to continuously authenticate the user while taking an online examination.

Ramu . T and Arivoli. T (2013) proposed a multimodal authentication framework using Keystroke Dynamics to secure online with two layers of authentication.

Traore et al.(2017) used multimodal continuous biometric framework which involves mouse dynamics, keystroke dynamics, and face biometrics to ensure the authenticity of remote test takers during online exams.

This chapter proposes a novel wrapper based classification approach using PSO-ELM-ANP method and tested on all the datasets.

From the background study of all the phases, it is obvious that there is no single comprehensive mechanism that is available to give increased performance for better keystroke recognition. Moreover, many evolving soft computing algorithms are not applied to improve the accuracy rate. Not much importance is given for wrapper based approaches which reduces the computation time. The accuracy obtained using neural network classification can further be improved. Hence an attempt has been made in this chapter to develop a comprehensive methodology using the wrapper based approaches that improves the accuracy and reduces the computation time. The next section discusses the methodology.

METHODOLOGY

The entire methodology is divided into four phases namely, feature extraction phase, normalization phase, feature subset selection phase and classification phase after obtaining the raw keystroke data (press time and release time). This section explains each step in detail.

Feature Extraction

The features that can be extracted from the raw typing data forms the template for further processing. While typing, the computer can record the key pressed and the time at which it is pressed. In addition to this, it can also record the time at which the same key is released and for how long it is pressed.

There are many features that can be extracted from the recorded keystroke timestamps. The figure 3 shows the various features that can be measured from keystrokes.

The following table 6 shows the definition of features extracted from the raw keystroke data.

Figure 3. Timing features from the keystrokes

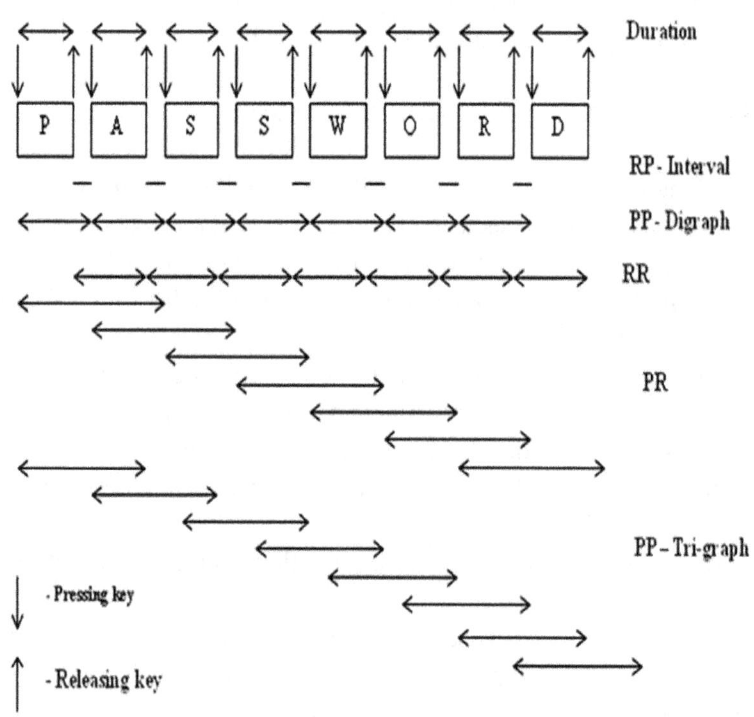

Table 6. Definition of timing features of keystrokes

S.No	Features	Definition	Formula
1.	Dwell Time or Duration (DT)	How long a key is pressed until it is released	DT=R1-P1
2.	Flight Time (FT) or latencies	Interval time between key press and release of different keys	
3.	a. Release-Press (RP)	Interval between a key release and next key press time	RP=P2-R1
4.	b. Press-Press (PP) (Digraph)	Interval between a key Press and next key Press	PP=P2-P1
5.	c. Release-Release (RR)	Interval between two successive key Releases	RR=R2-R1
6.	d. Press - Release (PR)	Elapsed time between Pressing of first key and releasing of next key	PR=R2-P1
7.	Trigraph(TRI)	Elapsed time between the first key press and the third key press	TRI=P3-P1

Force of key typed is one of the features which can be obtained using a special force sensitive keyboard which is expensive and also may not be available commonly (Liang Tang et al, 2010). In addition to the entire timing features such as duration, latency, digraph and trigraph, a new proposed Feature called Virtual Key Force (VKF) has been introduced. The Virtual Key Force is measured without using any special key board which also improves the accuracy when the feature is used for classification. The Virtual Key Force is calculated based on the typing speed and the behavior of the user at the keyboard. It measures the time taken by the user between releasing one key and pressing another key. It is based on the fact that each user has different typing speed and each user takes his/her own time to release and press another key. Virtual Key Force can be determined from the Complexity Label (CL), Average time taken to press the keys and Key Complexity (KC).

The Complexity Label (CL) calculation is fully user dependent and also depends on the hand used by the user. After calculating the complexity label, the average time is calculated. It is the total time taken divided by time taken to press a key. If the actual time is less than the average time, then the key complexity is assumed as 0 (i.e Shorter Interval between Keys) otherwise key complexity is assumed as 1 (i.e longer Interval between keys). Based on the key complexity and the complexity label, the following algorithm which is shown in Table 7 has been formulated:

The three levels of force namely High, Medium and Low are identified. If the key distance is nearer or longer and time interval taken to hit the key is below average then the key is assumed to be hit with low force. If the key distance is nearer and time interval taken to hit the key is above average then the key is assumed to be hit with medium force. If the key distance is longer and time interval taken to hit the key is above average then the key is assumed to be hit with high force. The Virtual Key Force for each character is stored as feature for the passwords.

Table 7. Virtual key force algorithm

```
if (key distance is nearer || longer && time interval is below average)
VKF=3 (low)
else if (key distance is nearer && time interval is above the average)
VKF=2(medium)
else if (keys are longer and the average time interval is above the average)
VKF=1 (high)
End
```

Normalization

After feature extraction, the next step is pre-processing. This section focuses on the pre-processing using normalization techniques. The typing pattern of the user varies from time to time even for the same user. There will also be a significant difference in the keystroke patterns exhibited from person to person, even when asked to type the same words (Chen and Chang, 2004). Therefore, data normalization is done between 0 and 1 to pre-process the data. Min-Max, Z-Score, Zero Mean and Standard Deviation, Median - MAD, and Tanh normalizations are done, and Z-score normalization is found to be effective compared to other normalization methods.

Feature Subset Selection

Feature subset selection is a Subset selection method that forms a group of features based on certain evaluation. When the dimensionality of a dataset grows significantly there is an increasing difficulty in processing the data to prove that the results are statistically significant. Large datasets lead to the problem of over fitting. An over fitted model can mistake small fluctuations for important variance in the data which can lead to classification errors. As the dimensionality increases the computational cost also increases, usually exponentially. To overcome this problem, it is necessary to find a way to reduce the number of features in consideration. The technique used is Feature Subset selection.

Feature Subset Selection Methods can classified as follows.

- Wrapper Method.
- Filter Method and
- Embedded Method or Hybrid Method

Feature selection techniques are divided into two types namely, feature ranking and feature subset selection. Feature ranking ranks the features by a metric and eliminates all features that do not achieve an adequate score. Subset selection searches for the set of possible features to construct the optimal subset. To carry out this search, a starting point, a strategy to traverse the space of subsets, an evaluation function and a stopping criterion are to be specified. Although there are varieties of subset selection techniques developed, usually two methods namely filter, and wrapper approaches are considered. In Filter approach, the feature subset selection is done independently of the learning algorithm. The subset selection procedure in this case be a preprocessing step.

In the wrapper approach, the feature subset selection is done by induction algorithm. The feature subset selection algorithm conducts a search for a good subset using the induction algorithm as a part of the evaluation function. The purpose of the induction algorithm is to induce from training data a classifier that will be useful in classifying future cases. The accuracy of induced classifiers is estimated by accuracy estimation technique. The classification algorithm itself is used to determine the attribute subset. Since the wrapper approach optimizes the evaluation measure of the classification algorithm while removing features, it mostly leads to greater accuracy than the filter approaches such as X^2Statistic, Information gain, Symmetrical uncertainty, ReliefF and Correlation based feature selection. Wrapper methods generally result in better performance than filter methods because the feature selection process is optimized by the classification algorithm to be used. Hence a wrapper based feature subset selection approach is proposed.

The Following Table 8 shows Advantages and Limitations of filter, wrapper and embedded method.

After pre-processing with Z-Score normalization, feature subset selection is done for reducing the features for further processing. Particle Swam Optimization (PSO) is wrapped with Extreme Learning Machine (ELM) – Analytic Network Process (ANP) which routinely selects an appropriate subset of features and the rest will not be considered, thus concluding in a more comprehensive model.

Particle Swam Optimization (PSO)

The authors of Sidhartha Panda and Padhy (2007), and Kennedy and Eberhart (1995) defines that Particle Swarm Optimization (PSO) is a stochastic search technique that aims to optimize an objective function, motivated by social activities of birds gathering or fish schooling. PSO is a population dependent search algorithm in which each individual is indicated as a particle and represents a candidate solution. All the particles in PSO moves through the search space with an adjustable velocity that is dynamically transformed based on its individual moving experience and also the moving experience of the other particles. In PSO each particles attempt to enhance themselves by imitating traits from their successful peers. The position in proportion to the best fitness is known as pbest (Particle best) and the overall best out of all the particles in the population is called gbest (Global best) (Sidhartha Panda and Padhy, 2007). The velocity $v_i(t)$ a nd the positions $x_i(t)$ of the particles are updated with the following equations.

$$v_i\left(t+1\right) = w * v_i\left(t\right) + c_1 r_1\left(y_i\left(t\right) - x_i\left(t\right)\right) + c_2 r_2\left(\hat{y}\left(t\right) - x_i\left(t\right)\right) \tag{1}$$

Table 8. Advantages and limitations of filter, wrapper and embedded methods

S.No	Model	Advantages	Limitations	Methods
1.	Filter	• Fast • Scalable • Independent of classifiers • Better computational complexity	Ignores relations with classifier	• Chi-square, • Euclidean distance, • Information gain, • Correlation based feature selection, • Markov blanket filter, • Fast correlation based feature selection
2.	Wrapper	• Simple • Interacts with classifier • Models feature dependencies, • Good classification accuracy, • Minimizes computational cost	Computationally exhaustive	• Sequential forward selection, • Sequential backward selection, • Randomized hill climbing, • Genetic Algorithms
3.	Embedded/ Hybrid	• Similar to wrapper model • computationally expensive • Less prone to over fitting	Computationally exhaustive	• L1 (LASSO) regularization • decision tree

$$x_i(t+1) = x_i(t) + v_i(t+1) \quad (2)$$

where i=1, 2, …, n

$v_i(t)$: Velocity of agent i at iteration t and must lie in the range

$$V_{min} \leq V_i(t) \leq V_{max} \quad (3)$$

w: inertia weight factor
c_1, c_2: Cognitive and social acceleration factors respectively,
$r_{1,} r_2$: Uniformly distributed random number between 0 and 1,
$x_i(t)$: Current position of agent i at iteration t,
$y_i(t)$: Personal best (pbest) and is updated using the following equation:

$$y_i(t+1) = \begin{cases} y_i(t), & f(x_i(t+1)) \geq f(y_i(t)) \\ x_i(t+1), & f(x_i(t+1)) < f(y_i(t)) \end{cases} \quad (4)$$

$\hat{y}(t)$: Global best (gbest) and is updated using the following equation:

$$\hat{y}_i(t+1) = \arg\min\left\{f(y_i(t+1))\right\}, i \leq i \leq s \quad (5)$$

where s = number of particles in the swarm.

Extreme Machine Learning (ELM)-Analytic Network Process(ANP)

Computational intelligence techniques are widely being used in variety of applications. Neural Networks like Back Propagation Neural Network (BPN) and Support Vector Machines (SVMs) are the most popular computational intelligence techniques that have been playing the leading roles for classification. However, both neural networks and SVMs requires trivial human intervene which slows the learning speed of the algorithm. The learning speed of Extreme Learning Machine (ELM) is faster and with least human intervene it gives better performance when compared with the above mentioned computational intelligent techniques.

The following Table 9 summarizes the ELM Algorithm (Huang et al., 2004 and 2006).

The input weights and the hidden bias are randomly chosen in Extreme Learning machine. The output weights are analytically determined by Moore Penrose inverse. ELM overcomes the problems such as slow learning speed, number of epochs etc., which are being faced by the computational intelligence techniques such as BPN and SVMs and has faster learning rate. However, ELM requires more number

Table 9. Extreme learning machine algorithm

Given a training set $N = (x_i, t_i), x_i \in R^n, t_i \in R^m$, an activation function $f : R \mapsto R$ and the number of hidden nodes H.
• Randomly assign input weights wi and bias $b_i \in [1, H]$ • Calculate the hidden layer output matrix H • Calculate the weight matrix $\beta = H^{\dagger}Y$

of hidden neurons than that of human intervene based algorithms that are mentioned above. The input weights are generated using Analytic Network Process (ANP).

Wrapper Based Feature Subset Selection Using PSO-ELM-ANP

In the proposed PSO-ELM-ANP Wrapper approach, Duration, Latency, Trigraph, Digraph and the proposed Virtual Key Force are given as input features. Feature subset selection is done and the selected features are evaluated byPSO- ELM-ANP in order to find the fitness value. The process is repeated until the best solution is obtained. The fitness function of the PSO is evaluated using ELM-ANP.

The algorithm of the proposed PSO-ELM Wrapper approach is given below:

```
Input: Duration, Latency, Trigraph, Digraph, Virtual Key Force
Output: Subset feature values.
Step 1: Initialize the number of Iterations, Number of particles, Weight,c1,
c2, r1,r2, vi(t)
 Step 2: Compute the feature values of xi(t) (Duration, Latency, Digraph, Tri-
graph, and Virtual Key Force).
Step 3: Evaluate fitness for each feature value using ELM-ANP.
Step 4: The following is repeated for number of iterations:
1.        Check if p>=pbest then, pbest=p else pbest= pbest.
2.        If pbest >=gbest then, gbest= pbest else gbest=gbest.
3.        Update velocity by (1) and position is updated by (2), pbest and
gbest position are updated using (4) and (5).
Step 4 is repeated until gbest is optimum value.
```

CLASSIFICATION

The selected features by PSO are given as input to ELM-ANP, ELM and Back Propagation Neural Network (BPN). The wrapper based classification approach using PSO-ELM-ANP yields the average accuracy of 99.92% and takes less training and testing time when compared to wrapper based PSO-BPN and PSO-ELM. The Wrapper based classification approach is explained below.

The training algorithm for ELM is Levenberg-Marquardt (LM) and the learning parameter μ is a constant number in standard Levenberg-Marquardt algorithm. During the classification phase the learning parameter μ is modified as follows.

$$\mu = 0.01e^{T} \tag{6}$$

Analytic Network Process (ANP)

The input weights and the hidden bias are randomly chosen in Extreme Learning machine. The output weights are analytically determined by Moore Penrose inverse. ELM overcomes the problems such as slow learning speed, number of epochs etc., which are being faced by the computational intelligence techniques such as BPN and SVMs and has faster learning rate. However, ELM requires more number of hidden neurons than that of human intervene based algorithms that are mentioned above. The input weights are generated using Analytic Network Process (ANP).

Analytic Network Process (ANP) proposed by (Saaty. T. L, 1996) facilitates for more complex interrelationships among the decision levels and attributes and does not require any strict hierarchical structure. The ANP approach handles interdependent relationships among the elements by obtaining the composite weights developed by the supermatrix.

Wrapper Based Classification Using PSO-ELM-ANP

In the proposed wrapper based classification approach, the features are selected by PSO feature subset selection method and are given as input vectors to the input layers. The input weights and hidden bias are generated by Analytic Network Process. Extreme Learning Machine analytically determines the output weights and the weights and bias are updated using Modified Levenberg–Marquardt (LM) algorithm which is used to train the network. The Extreme Learning Machine with Analytic Network Process is used as the induction algorithm and wrapped with PSO for feature selection and classification.

The process for the Extreme Learning Machine with ANP is described below:

Given a training set $\aleph = \left(x_i, t_i\right)$, $x_i \in R^n$, $t_i \in R^m$, an activation function f1(x) and f2(x), $\breve{N}$ & K, the number of nodes hidden in first and second layer.

Step 1: The input weight vectors w_1 are randomly initialized with the help of ANP technique and bias vectors without using the ANP technique.

Step 2: Compute the hidden first layer output matrix a_1. Using ELM algorithm, determine the output weight.

$$w_2 = a_1^{-1}.t \tag{7}$$

Step 3: Determine the hidden second layer output matrix a_2 errors

$$e_1 = t - a_2 \tag{8}$$

and determine the sum of squared errors of overall input.

Step 3: Compute the Jacobian matrix J as follows:

Computer the sensitivities with the recurrence relations S_q^m, after initializing S_q^m

$$S_q^m = f^m\left(\eta_q^m\right)\left(w^{m+1}\right)^T .S_q^{m+1} \tag{9}$$

$$S_q^M = -f^m\left(\eta_q^m\right) \tag{10}$$

Augment the individual matrices into the Marquardt sensitivities using the following equation.

$$S^m = \left[S_1^m, S_2^m, \ldots, S_Q^m\right] \tag{11}$$

Calculate the elements of the Jacobian matrix with the equations

$$\left[J\right]_{h,l} = S_{i,h}^m \times S_{j,k}^{m-1} \tag{12}$$

and

$$\left[J\right]_{h,l} = S_{i,h}^m \tag{13}$$

Step 4: Determine increments of weights, Δw_k by the following equation

$$\Delta w_k = \left[J^T\left(w_k\right).J\left(w_k\right) + \mu.I\right]^{-1} .J^T\left(w_k\right).e\left(w_k\right) \tag{14}$$

where $\mu = 0.01\ e^T e$.

Step 5: The sum of the squared errors is recomputed using $w + \Delta w$. If this new sum of squared errors is lesser than the sum of squared errors calculated in step 2, then

$$\mu = \mu.\beta\left(\beta = .1\right) \tag{15}$$

$$w_{k+1} = w_k + \Delta w_k \tag{16}$$

and continue from step 4.

Otherwise compute μ with the following equation and continue step 5.

$$\mu = \frac{\mu}{\beta} \tag{17}$$

Step 6: Update weight vectors and bias.

Step 7: Recompute the sum of the squared errors using $w + \Delta w$. If this new sum of squared is lesser than the sum of squared errors calculated in step 3, then calculate μ with the equation (15) and w_{k+1} with the equation (16) and continue from step 5. Otherwise compute μ with the equation (17) and continue the step 7.

Thus, by using the ANP technique, the Extreme Learning Machine is modified as ELM-ANP which has the advantage of quick learning.

Feature subset selection is done for the original extracted features using wrapper based PSO-ELM-ANP approach in which the performance is evaluated using the classifier. The selected feature set acts as training set and testing set (70% as training set and 30% as test set). Classification is done using wrapper based PSO-ELM-ANP algorithms in which the learning parameter is modified and the final accuracy of the system is estimated.

EXPERIMENTAL RESULTS

The methodology is tested with data from real time users and bench mark datasets. The datasets used are described as follows.

Dataset Description

Three different types of dataset are used to test the methodology. KSP Dataset -Pressure sensitive keystroke dynamics dataset (2010), which represents the typing of 103 individuals on three different words ('drizzle', 'jeffrey allen', 'pr7q1z'). From the dataset, the following passwords are considered for the study. drizzle – Weak password. jeffrey allen– Strong password. DSL Dataset - Keystroke Dynamics - Benchmark Data Set (2009) consists of keystroke-timing information from 51 subjects each typing a password (.tie5Roanl – Medium password) 400 times. Real time dataset with 100 users in the age group of 18-60 each typing the following 3 different passwords for 10 times. 'pass_tie.R' – Strong password, '.tie5Roanl' – Medium password, 'nopassword' – Weak password is also considered for the study. Three different strength passwords (weak, medium and strong) have been chosen for study. The reason for choosing different strength passwords is that the weak and medium passwords can be easily guessed and vulnerable to hacker attacks. The strong passwords are complex to remember and the user

Table 10. Percentage of accuracy obtained without VKF, preprocessing and feature subset selection

Classifier	Accuracy					
	Dataset					
	KSP Weak Password	KSP Strong Password	DSL Medium Password	Real time weak password	Real time Medium password	Real time Strong password
BPN	50	51.45	54.80	51.48	53.69	52.56
ELM	54.31	55.50	58.00	54.28	57.07	57.94
ELM-ANP	58.12	56.53	59.63	59.74	59.94	59.69

may store or write it which may lead to Dictionary attacks. For all such types of passwords, security can be more tightened using Keystroke dynamics.

All the above passwords are checked for strength using publicly available strength checker (Microsoft, 2008) and the checker determines the strength of a password depending on the different types of characters that is used, the overall length of the password (it should be 8 or more characters long), and whether the password can be found in a dictionary. The strength of the password also varies from software to software used for checking

RESULTS AND DISCUSSION

The Particle Swarm Optimization Algorithm for wrapper based feature subset selection is implemented with 30 particles, 100 iterations, coefficients of positive acceleration (c_1,c_2) as 1 and random Variables (r_1, r_2) as 0.9. The Extreme Learning Machine is implemented with 5 and 20 input and hidden neurons, Sigmoid as activation function and Levenberg-Marquardt as training algorithm. The Extreme Learning Machine- Analytic Network Process is implemented with 5 and 15 input and hidden neurons, Sigmoid as activation function and Modified Levenberg-Marquardt as training algorithm.

The Table 10 shows the accuracy obtained without the new feature VKF, without Preprocessing and without Feature Subset Selection for all the datasets and using BPN, ELM and ELM-ANP classifiers.

From the above table 10, it is observed that the classification accuracy obtained using ELM-ANP is slightly higher than other methods.

New feature Virtual Key Force is added to the existing features and preprocessing is done by applying Z-Score Normalization technique to all the databases. The results are measured with and without the Virtual Key Force.

From the above table 11, it is observed that after preprocessing using Z-score method and even with the newly added feature Virtual Key Force(VKF), the classification accuracy has improved for ELM-ANP method.

After preprocessing using Z-Score method, a PSO wrapper-based feature selection is applied to all the databases to select the optimal features.

From the above table 12, it is observed that PSO-ELM-ANP has reduced the existing features. The proposed PSO-ELM-ANP wrapper-based approach has selected less number of features for further classification which is shown in the following table 13.

Table 11. Percentage of accuracy obtained with normalization, with and without VKF and without feature subset selection methods

Classifier	Dataset											
	KSP Weak Password		KSP Strong Password		DSL Medium Password		Real time weak password		Real time Medium password		Real time Strong password	
	W.O. VKF	W. VKF	W.O. VKF	W. VKF	W.O. VKF	W. VKF	W.O. VKF	W. VKF	W.O. VKF	W. VKF	W.O. VKF	W. VKF
BPN	51.46	54.34	53.3	59.23	60.45	62.00	52.5	55.34	54.5	56.20	55.33	59.45
ELM	58.54	59.27	59.54	62.99	61.44	63.90	55.34	57.42	60.76	61.44	62.41	66.93
ELM-ANP	62.55	64.82	61.45	65.67	64.93	67.34	62.5	65.83	62.12	65.67	64.96	67.34

W.O.VKF – Without VKF, W. VKF – With VKF

Table 12. Features selected with virtual key force

Classifier	Dataset											
	KSP Weak Password		KSP Strong Password		DSL Medium Password		Real time weak password		Real time Medium password		Real time Strong password	
	F	FS	F	FS	F	FS	F	FS	F	FS	F	FS
PSO-BPN	43	31	85	67	71	60	64	56	71	55	78	64
PSO-ELM	43	30	85	59	71	57	64	50	71	53	78	58
PSO-ELM-ANP	43	28	85	58	71	55	64	46	71	51	78	50

F- Number of Features, FS – Features selected.

Classification is done after Feature selection. Table 14 shows the average accuracy obtained using the wrapper-based classification by wrapping PSO with BPN, ELM and ELM-ANP classifiers.

It is observed that PSO-ELM-ANP wrapper-based classification approach gives an average accuracy of 99.92% for real time medium password. The proposed wrapper-based classification gives more accuracy for all types of passwords and for all datasets compared to that of BPN and ELM wrapper-based approaches.

Table 13. Feature selection percentage with virtual key force

Classifier	Dataset											
	KSP Weak Password		KSP Strong Password		DSL Medium Password		Real time weak password		Real time Medium password		Real time Strong password	
	FS	S%	FS	S%	FS	S%	FS	S%	FS	S%	FS	S%
PSO-BPN	31	72.09	67	78.82	60	84.51	56	87.5	55	77.5	64	82.1
PSO-ELM	30	69.77	59	69.41	57	80.28	50	78.1	53	74.7	58	74.4
PSO-ELM-ANP	28	65.12	58	68.24	55	77.46	46	71.9	51	71.8	50	64.1

F- Number of Features, S%- Selection percentage

Table 14. Percentage of Accuracy (%) obtained with wrapper-based classification

Classifier	Dataset					
	KSP Weak Password	**KSP Strong Password**	**DSL Medium Password**	**Real time weak password**	**Real time Medium password**	**Real time Strong password**
PSO-BPN	90.42	90.04	90.66	92.96	94.31	94.77
PSO-ELM	96.90	96.46	95.21	96.64	97.92	96.3
PSO-ELM-ANP	98.56	98.01	98.95	98.55	99.92	98.3

The training time and testing time (in sec) taken by the wrapper based classification approaches are given in the Table 15 and Table 16. It is observed that the PSO – ELM-ANP wrapper approach takes less time for training and testing when compared with the other wrapper-based methods.

COMPARISON WITH PREVIOUS WORKS

From the above result tables, it is clearly seen that the proposed PSO-ELP-ANP method obtains the accuracy of more than 98% which outperforms the other existing methods. The maximum accuracy of 99.92% for real time medium password is obtained. Whereas the works which were previously done in keystroke authentication had resulted in 92% (Karnan, M. and Akila, M, 2010) which used soft computing methods and 98% (J. Yadav, K. Pandey, S. Gupta and R. Sharma, 2017) which used fuzzy logic. Our proposed work also outperforms the previous works done in keystroke authentication.

Table 15. Training time (in sec)

Classifier	Dataset					
	KSP Weak Password	**KSP Strong Password**	**DSL Medium Password**	**Real time weak password**	**Real time Medium password**	**Real time Strong password**
PSO-BPN	8.61082	16.76341	17.66666	18.09596	11.67424	19.66666
PSO-ELM	0.11009	0.09615	0.03421	0.20628	0.12500	0.03421
PSO-ELM-ANP	**0.01081**	0.08001	0.00002	0.10121	0.01210	0.03002

Table 16. Testing time (in sec)

Classifier	Dataset					
	KSP Weak Password	**KSP Strong Password**	**DSL Medium Password**	**Real time weak password**	**Real time Medium password**	**Real time Strong password**
PSO-BPN	0.02389	0.01892	0.03381	0.03427	0.03381	0.03381
PSO-ELM	0.02186	0.01241	0.01560	0.02502	0.01561	0.01561
PSO-ELM-ANP	0.00230	0.00152	0.00560	0.00302	0.00188	0.00564

CONCLUSION AND FUTURE DIRECTIONS

Feature extraction, normalization, feature subset selection and classification are done to identify the users using keystroke dynamics. The extraction of features can be done from the press and release time measured from the users. The features such as duration, latency, digraph and trigraph can be extracted from the users' press and release time. This research work proposes a new feature called Virtual Key Force. The obtained experimental results show that even though a new additional feature is added, when the proposed feature is used with wrapper-based approaches, the accuracy is improved. Feature subset selection is done using Particle Swarm Optimization wrapped with Extreme Learning Machine and Anlytic Network Process. A wrapper-based classification technique is proposed using PSO-ELM-ANP method and compared with Back Propagation Neural Network and ELM. An average accuracy of 99.92% is obtained and proposed method takes less training and testing times. This chapter proposes a wrapper-based classification technique using PSO-ELM-ANP method for keystrokes obtained from laptop. The same proposed wrapper-based classification technique using PSO-ELM-ANP method can be tested for touch screen gadgets and mobile environments.

REFERENCES

Abinaya, R., & Sigappi, A. (2018). Biometric Identification of a Genuine User/Imposter from Keystroke Dynamics Dataset. *International Journal of Chemtech Research*, *11*(08), 147–160. doi:10.20902/IJCTR.2018.110816

Akila, M., & Suresh Kumar, S. (2011). Improving Feature Extraction in Keystroke Dynamics using Optimization Techniques and Neural Network. *Proceedings of the International Conference on Sustainable Energy and Intelligent Systems (SEISCON 2011)*, 891–898. 10.1049/cp.2011.0493

Akila, M., Suresh Kumar, V., Anusheela, N., & Sugumar, K. (2012). A Novel Feature Subset Selection Algorithm Using Artificial Bee Colony in Keystroke Dynamics. In *Proceedings of the International Conference on Soft Computing for Problem Solving (SocProS 2011)*. Springer. 10.1007/978-81-322-0491-6_74

Al Solami, E., Boyd, C., Clark, A., & Ahmed, I. (2011). User-Representative Feature Selection for Keystroke Dynamics. *Proceedings on 5th International Conference on Network and System Security (NSS)*, 229-233.

Allen, J. D. (2010). *An analysis of pressure-based keystroke dynamics algorithms* (Master's thesis). Dallas, TX: Southern Methodist University.

Alpar, O. (2014). Keystroke recognition in user authentication using ANN based RGB histogram technique. *Engineering Applications of Artificial Intelligence*, *32*, 213–217. doi:10.1016/j.engappai.2013.11.009

Antal. (2015). Keystroke Dynamics on Android Platform. Procedia Technology, 820-826.

Awad, Ahmed, & Traore. (2005). Anomaly Intrusion Detection based on Biometrics. *Proceedings of 6th IEEE Information Assurance Workshop*, 452- 453.

Chandrasekar, V., & Suresh Kumar, S. (2016). A dexterous feature selection artificial immune system algorithm for keystroke dynamics. *Stochastic Analysis and Applications*, *34*(1), 147–154. doi:10.1080/07362994.2015.1110707

Chandrasekar, V., Suresh Kumar, S., & Maheswari, T. (2016). Authentication based on keystroke dynamics using stochastic diffusion algorithm. *Stochastic Analysis and Applications*, *34*(1), 155–164. doi:10.1080/07362994.2015.1112291

Chen, W., & Chang, W. (2004). Applying Hidden Markov Models to Keystroke Pattern Analysis for Password Verification. *Proceedings of IEEE International Conference on Information Reuse and Integration*, 467–474. 10.1109/IRI.2004.1431505

Darabseh, A., & Namin, S. (2015), On Accuracy of Keystroke Authentications Based on Commonly Used English Words. *2015 International Conference of the Biometrics Special Interest Group (BIOSIG)*, 1-8. 10.1109/BIOSIG.2015.7314612

Deng, Y., & Zhong, Y. (2013). Keystroke dynamics user authentication based on Gaussian mixture model and deep belief nets. *ISRN Signal Processing*, *2013*, 1–7. doi:10.1155/2013/565183

Deng, Y., & Zhong, Y. (2015). Keystroke Dynamics User Authentication Using Advanced Machine Learning Methods. In Y. Zhong & Y. Deng (Eds.), *Recent Advances in User Authentication Using Keystroke Dynamics Biometrics* (Vol. 2, pp. 23–40). GCSR.

Filho, J. M. R., & Freire, E. O. (2006). On the equalization of keystroke timing histograms. *Pattern Recognition Letters*, *27*(13), 1440–1446. doi:10.1016/j.patrec.2006.01.010

Giot, R., & Rosenberger, C. (2012). A new soft biometric approach for keystroke dynamics based on gender recognition. *International Journal of Information Technology and Management*, *11*(1/2), 35–49. doi:10.1504/IJITM.2012.044062

Huang, G. B., Zhu, Q. Y., & Siew, C. K. (2004). Extreme Learning Machine: A New Learning Scheme of Feedforward Neural Networks. *Proceedings of 2004 International Joint Conference on Neural Networks*. 10.1109/IJCNN.2004.1380068

Huang, G. B., Zhu, Q. Y., & Siew, C. K. (2006). Extreme Learning Machine: Theory and Applications. *Neurocomputing. Elsevier*, *70*(1-3), 489–501. doi:10.1016/j.neucom.2005.12.126

Idrus, S., Zulkarnain, S., Cherrier, E., Rosenberger, C., & Bours, P. (2013). Soft Biometrics For Keystroke Dynamics. *International Conference on Image Analysis and Recognition (ICIAR)*, 11-18. 10.1007/978-3-642-39094-4_2

Idrus, S., Zulkarnain, S., Cherrier, E., Rosenberger, C., & Bours, P. (2014). Soft biometrics for keystroke dynamics: Profiling individuals while typing passwords. *Computers & Security*, *45*, 147–155. doi:10.1016/j.cose.2014.05.008

Idrus, S. Z. S., Cherrier, E., Rosenberger, C., & Bours, P. (2013).Soft biometrics database: a benchmark for keystroke dynamics Biometric methods. *2013 International conference of the biometrics special interest group (BIOSIG)*, 1-8.

Jain, Hong, & Pankanti. (2000). Biometric Identification. *Communications of the ACM, 43*(2), 91-98.

Jain, L., Monaco, J. V., Coakley, M. J., & Tappert, C. C. (2014). Passcode keystroke biometric performance on touch screen is superior to that on hardware keyboards. *International Journal of Research in Computer Applications and Information Technology*, *2*(4), 29–33.

Karnan, M., & Akila, M. (2009). Personal Authentication based on Keystroke Dynamics using Ant Colony Optimization and Back Propagation Neural Network. *International Journal of Computer and Network Security*, *1*(2), 8–15.

Karnan, M., & Akila, M. (2009). Identity Authentication based on Keystroke Dynamics using Genetic Algorithm and Particle Swarm Optimization. *Proceedings of 2nd IEEE International Conference on Computer Science and Information Technology*, 203–207. 10.1109/ICCSIT.2009.5234420

Karnan, M., & Akila, M. (2010). Personal Authentication based on Keystroke Dynamics using Soft Computing Techniques. *Second International Conference on Communication Software and Networks*, 334–338. 10.1109/ICCSN.2010.50

Kennedy, J., & Eberhart, R. (1995). Particle Swarm Optimization. *Proceedings of International Conference on Neural Networks*, 1942-1948. 10.1109/ICNN.1995.488968

Killourhy, K., & Maxion, R. (2009) Comparing anomaly-detection algorithms for keystroke dynamics. *IEEE/IFIP International Conference on Dependable Systems& Networks.*

Liu, J. (2014). The BeiHang Keystroke Dynamics and baselines. *Neurocomputing*, *144*, 271–281. doi:10.1016/j.neucom.2014.04.044

Loy, L., W. K., & Lim, C. P. (2007). Keystroke patterns classification using the ARTMAP-FD neural network. *International Conference on Intelligent Information Hiding and Multimedia Signal Processing*, 61–4. 10.1109/IIH-MSP.2007.218

Montalvao, Almeida, & Freire. (2006). Equalization of keystroke timing histograms for improved identification performance. *Telecommunications Symposium, 2006 International*, 560–565.

Montalvao Filho, J., & Freire, E. (2000). Multimodal biometric fusion–joint typist (keystroke) and speaker verification. *Telecommunications Symposium, 2006 International*, 609–614.

Mulionoa, Y., Hamb, H., & Darmawanb, D. (2018). Keystroke Dynamic Classification using Machine Learning for Password Authorization. *Procedia Computer Science*, *135*, 564–569. doi:10.1016/j.procs.2018.08.209

Mungai, P. K., & Huang, R. (2017). Using keystroke dynamics in a multi-level architecture to protect online examinations from impersonation. *2017 IEEE 2nd International Conference on Big Data Analysis (ICBDA)*, 622-627.

Muthuramalingam, A., Gnanamanickam, J., & Muhammad, R. (2018). Optimum Feature Selection Using Firefly Algorithm for Keystroke Dynamics. In A. Abraham, P. Muhuri, A. Muda, & N. Gandhi (Eds.), *Intelligent Systems Design and Applications. ISDA 2017. Advances in Intelligent Systems and Computing* (Vol. 736). Cham: Springer. doi:10.1007/978-3-319-76348-4_39

Panda & Padhy. (2007). Comparison of Particle Swarm Optimization and Genetic Algorithm for TCSC-based Controller Design. *World Academy of Science, Engineering and Technology*, *27*, 497–505.

Patil, R. A., & Renke, A. L. (2016). Keystroke Dynamics for User Authentication Identification by using Typing Rhythm. *International Journal of Computers and Applications*.

Ramu, T., & Arivoli, T. (2013). A Framework of Secure Biometric Based Online Exam Authentication: An Alternative to Traditional Exam. *International Journal of Scientific & Engineering Research*, *4*(11), 52–60.

Romain, Dorizzi, & Rosenberger. (2015). A review on the public benchmark databases for static keystroke dynamics. *Computers & Security*, 46–61.

Saaty, T. L. (1996). *Decision Making with Dependence and Feedback- The Analytic Network Process*. Pittsburgh, PA: RWS Publications.

Saggio, G., Costantini, G., & Todisco, M. (2011). Cumulative and Ratio Time Evaluations in Keystroke Dynamics to Improve the Password Security Mechanism. *Journal of Computer and Information Technology*, *1*(2), 2–11.

Sharma. (2015). Feature Selection Technique Using Ant Colony Optimization on Keystroke Dynamics. *International Journal of Scientific & Engineering Research*, 668-678.

Tang, L., Yang, C., & Wu, Z. (2010). User Authentication based on Force-Sensing Keypad by Fuzzy C-Means Clustering Method. *Journal of Computer Information Systems*, *6*(11), 3659–3667.

Wankhede & Verma. (2014). Keystroke Dynamics Authentication System Using Neural Network. *International Journal of Innovative Research & Development*.

Wong, F., Wong, M.H., Supian, A.S.M., Ismail, A.F., Kin, L.W., & Soon, O.C. (2011). Enhanced User Authentication through Typing Biometrics with Artificial Neural Networks and k-Nearest Neighbor Algorithm. *Conference Record of the Thirty-Fifth Asilomar Conference on Signals, Systems and Computers, 2,* 911–915.

Yadav, J., Pandey, K., Gupta, S., & Sharma, R. (2017). Keystroke dynamics based authentication using fuzzy logic. *2017 Tenth International Conference on Contemporary Computing (IC3)*, 1-6. 10.1109/IC3.2017.8284338

Zamonsky, G., & Pedernera et al. (2010). Revisiting Clustering Methods to their Application on Keystroke Dynamics for Intruder Classification. *Proceedings of 2010 IEEE Workshop on Biometric Measurements and Systems for Security and Medical Applications (BIOMS)*, 36-40. 10.1109/BIOMS.2010.5610443

Zhang, Z.-Y., Chuang-bai, X., Zhao, D.-Q., Hong, S., Kai, X., & Tian, G.-Z. (2011). Identity Authentication System based on Improved PR-RP Model. *Proceedings of 2nd International Conference on Advanced Computer Control (ICACC)*, 65-69.

Zhong, Y., Deng, Y., & Jain, A. K. (2012). Keystroke dynamics for user authentication. *Proc. IEEE Comput. Soc. Conf. Comput. Vis.PatternRecognit. Workshops (CVPRW)*, 117–123.

Chapter 10
Biometric Authentication Techniques and E-Learning

Ramgopal Kashyap
https://orcid.org/0000-0002-5352-1286
Amity University Chhattisgarh, India

ABSTRACT

The primary goal of this chapter is to answer online exam frameworks by utilizing face acknowledgment to verify students for going to an online exam. A strategy in light of the utilization of neural systems to validate individuals' computerized unique mark framework for e-learning is present. This chapter centers around breaking down and contrasting the distinctive facial verification frameworks to confirm the understudies when they utilize e-learning stages, itemizing the expenses and the highlights of each structure recorded. Biometrics is a sensible verification used as a type of distinguishing proof and access control. It is additionally used to distinguish people in bunches that are under observation. Biometric identifiers are then particular quantifiable qualities used to mark and portray people. Biometric authenticators are as often as possible named as conduct and additionally physiological attributes. Physiological qualities are identified with the state of the body. In this chapter, the essential focus is on the distinctive biometrics and their applications.

INTRODUCTION

Information these days, e-learning frameworks have turned out to be imperative parts in the instruction and preparing areas a few nations are endeavoring to conquer the knowledge division through instruction and preparing, nations are capable of building up the aptitudes of their natives. Thusly crossing over the knowledge divide inside the nation and with more created ones. Accomplishment in the knowledge economy depends intensely on a qualified and talented populace; consequently, viable instruction and preparing frameworks are required. At the same time, Information and Communication Technologies (ICTs) keep on growing at a fast pace and have changed the way individuals live, work and learn. The coordination of ICT apparatuses in instruction and preparing has made better approaches for conveying, getting to, and handling helpful information, and has offered help to information sharing between vari-

DOI: 10.4018/978-1-5225-7724-9.ch010

ous performing artists and too long-lasting learning (Daniel, 2017). Likewise, mechanical advancement, what's more, the development of the Internet has brought about the rise of e-learning as an essential learning approach. E-learning gives creative strategies to teaching individuals. In addition, the e-learning market is extending as a result of its numerous preferences over conventional instruction. E-learning is additionally profoundly adaptable, versatile, utilizes a fast learning strategy, more affordable, and turned out to be viable contrasted and customary training (Khairnar, 2015). Specifically, the accompanying is the three fundamental drivers for the expanding worldwide significance of e-learning:

- The movement toward a learning-based economy;
- Paradigm moves in instruction conveyance;
- Technological advancements and Internet development.

The advancement of e-learning and online evaluation frameworks is expanding quickly, both all around and locally, with numerous colleges and companies putting critical capital in e-learning projects and activities. This development is additionally reflected in the report by Ambient Insight, which was distributed in 2010, showing that the e-learning market has achieved US$ 27.1 billion of every 2009 and will outperform $49.6 billion by 2014. The development of the e-learning industry requires new administrations to guarantee unwavering quality and viability of its frameworks, particularly amid the examinations procedure, by tending to the issue of duping in online examinations and data fraud.

E-learning is flourishing on worldwide and neighborhood levels. In Saudi Arabia, the legislature is centering to the instruction area by and large and to e-learning specifically in reacting to the expanding number of male and female understudies selected in instructive foundations. Numerous colleges in Saudi Arabia have effectively actualized e-learning frameworks and are putting forth separate learning courses and degrees. In this manner, guaranteeing the unwavering quality of e-learning frameworks, particularly amid examinations, is exceptionally basic. Online examination deceiving and personality robbery ought to be considered, while the security of students' information and all the more critically, their pictures is ensured. Common distinguishing proof strategies in view of what we have an attractive card, personality card; chip (Cross & Palese, 2015). The ongoing fear monger acts and dangers too numerous nations, combined with the developing advancement of cybercrime in reality as we know it where all errands have a tendency to turn into programmed distinguishing proof/validation of immaculate characteristic people have turned into an issue pivotal for security reasons. Every one of these issues has energized the facilitate improvement of biometrics as a distinguishing proof and validation device. The utilization of fingerprints is one of the most seasoned and best strategies for recognizable proof/validation biometrics. Concerning a unique mark demonstrate the presence of the papillary illustrations from birth to death; this specific game plan of the papillary lines shapes focuses called particulars, which are the reason for uniqueness and permanence of the unique finger impression.

The unique computerized mark shapes a particular class of example with a specific peculiarity and measurements attributes known. At that point, unique mark acknowledgment is by all accounts more compelling than the other issue of example acknowledgment like manually written character acknowledgment where the neural system has been all around connected. The primary inquiry we along these lines consider here is: the manner by which to naturally verify faultless individuals who wish to get to a PC framework with the most minimal conceivable blunder rate? The approach comprises in the utilization of neural systems to grow such a framework. The manufactured model will at that point be connected to E-learning with a generally stable database ("Handwritten English Character Recognition Using Logistic Regression and Neural Network", 2016). Truth be told, the neural system has demonstrated their capacity to manage numerous spaces. Their eccentricity is the capacity to adjust to the information to be handled that they offer, and the capacity to perform estimations in parallel, enabling them to work in different fields of use. In this chapter, we exhibit a model of verification by fingerprints, in view of the utilization of a neural system having a multilayer perceptron structure and extraction calculation and arrangement particulars at two levels.

The main level comprises of the unique finger impression characterization at the info layer thinking about its general geology as indicated by the characterization and is partitioned into two phases: the principal arrange grants the order of the info unique finger impression as per the models of the top of the line of built information by straightforward figuring of probabilities the second stage fills in as a control of the main grouping also, decreases the danger of inaccurate answers. This second control can diminish the rate of false dismissal, notwithstanding when the quantity of identified particulars is low. The second level which is given to correspondence or validation is to look at in the system preparing stage, the component vector of the information unique mark which is invariant to interpretation and turn to all those having a place with the class victory after the two past groupings, by a straightforward closeness measure the Euclidean standard for instance (Zhong & Peng, 2016). Whatever remains of the archive is composed into three areas. The primary segment presents calculations of the neural system for unique mark acknowledgment. The second segment depicts the proposed technique in light of multilayer perceptron with two concealed layers. The third area displays the utilization of contemplated strategies and got comes about; we wrap up by a conclusion.

BIOMETRIC SYSTEMS

Biometrics is a continually advancing field that is ending up more far-reaching in the business the term biometrics is depicted as a programmed individual acknowledgment framework in view of physiological or social attributes. Biometrics utilizes organic properties of a human, for example, fingerprints, iris, and voice acknowledgment confront acknowledgment, and hand geometry to distinguish people (Gottschlich & Huckemann, 2014). Figure 1 shows cases of the natural properties utilized for biometrics.

You can see the three different methods as shown in figure 2.

Which the biometric innovation works or the genuine biometric coordinating capacities is more perplexing. The accompanying is the fundamental procedure stream of biometric check and recognizable proof (see Figure 3).

A client at first selects in a biometric framework by giving biometric information, which is changed over into a format (≈ digitized data). Templates are put away in the biometric framework with the end goal of consequent examinations. In a request to be checked or distinguished after enlistment, the client gives biometric information, which is then changed over into a format. The check layout is contrasted and at least one enlistment formats. The consequence of a correlation among biometric formats is rendered as a score or certainty level, which is contrasted with a limit utilized for a particular technology, framework, client, or exchange. Table 1 is showing feature of the distinctive biometric frameworks if the score surpasses the edge, the correlation is a match, and that outcome is transmitted. If the score does not meet the edge, the examination is a non-coordinate, and that outcome isn't transmitted.

Figure 1. Biological properties for biometric

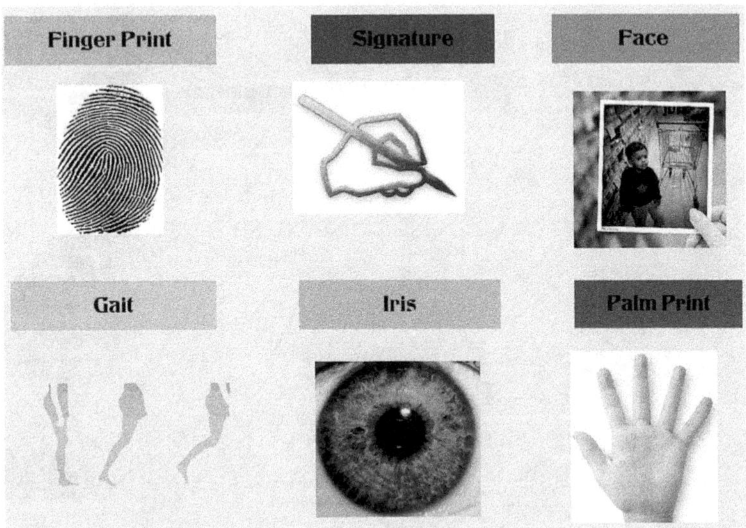

Figure 2. Increases in security and comfort

E-LEARNING SYSTEMS

The Web is anything but difficult to utilize, simple to refresh, and is accessible around the world. The Web is the driver of the learning economy and is subsequently a characteristic vehicle for learning. Corridor and Snider characterized e-learning as synonymous to all PC related applications, devices, and procedures of learning and educating. E-learning offers adaptable learning whenever and anyplace. The expanding pace of the Internet, the development of the World Wide Web, and the development of rapid PCs add to the accessibility of e-learning all day, every day and around the world. Additionally, e-students can get to materials whenever and put advantageous for them (Berglund, 2013). E-learning has different points of interest for students, instructors, and instructional engineers. Students, for instance,

Figure 3. Biometric enrollments and storage process flow

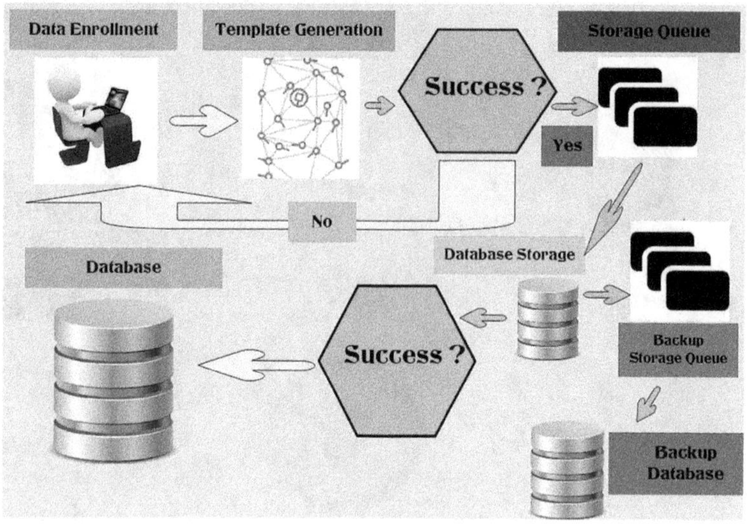

Table 1. A feature of the distinctive biometric frameworks

S.N.	Biometric System	Description	Price
1	Secure exam Remote Proctor	Utilization of fingerprints for the client's recognizable proof and requires a video reconnaissance framework	150 dollars per year
2	ProctorU	Requires the username, secret word and a photo to recognize the understudy and the ID of the client progressively done by a man of the organization.	It is around 20-30 dollars for a two hours' exam.
3	Webassessor	A facial acknowledgment programming and keystrokes beat designs are required and video reconnaissance framework utilizing a webcam.	50-80 dollars for the webcam in addition to the expenses for utilizing the application
4	Smowl	Webcam of the understudy's terminal to catch pictures and sounds and App that can be downloaded from Smowltech server and can be executed in the client's program with no sort of establishment process.	4,99€ every hour.

will advantage from their connection with different understudies and with their teachers, and they can learn at their favored pace. Students can likewise choose the material they need or be coordinated to the substance that meets their level of learning, premium, and requirements. Moreover, students wind up in charge of their learning. In the meantime, instructors can create materials utilizing on the web assets, and afterward distribute them in a wide range of routes, for example, content, pictures, video, sound, recreations, and amusements. Instructors will along these lines pick up fulfillment through quality understudy interest. At last, designers can create point by point and institutionalized courses. E-learning enables engineers to plan the course and utilize it for various circumstances by utilizing learning objects. E-learning gives much improvement to the numerous models, markets, premium gatherings, and diverse degrees of fulfillment in the instructive procedure. E-learning innovations offer potential for astounding developmental appraisal. Online evaluation gives dynamic visuals, sound, client intelligence, adjustment to singular students, and constant score detailing, in this way extending examination alternatives past the restrictions of customary tests. Student appraisals are without a doubt basic in the instructive procedure. Examination scores advise the educator on whether an understudy's advance is acceptable or not. Online evaluation frameworks give teachers with numerous points of interest, for example, the production of online examinations inside a brief timeframe, organization of the examination through the PC, simple checking of answers amid the examination, quick access to examination comes about without investing energy in assessment and rectification, and simple computation of the patterns of the examination comes about (Chen, Bastedo & Howard, 2018). Online examinations advantage students by enabling them to not be physically present at an offered area to take an examination and the outcomes can be influenced accessible to them to instantly. In online examinations, students should enlist their names and passwords the examination things will at that point be produced by the test bank as indicated by the parameters set by the teacher. The things in online appraisals are generally valid/false and different decision questions, and in addition include reordering/revamp coordinating, arrangement, positioning, and others, fulfillment, idea maps, and expositions. Examination inquiries will show up on the screen, and after that, every student will begin noting them through his/her PC. Toward the finish of the time restraint, the examination will stop and the score will show up. The student will then acquire the test outcomes instantly.

E-Learning Activities

Build up an individual website, profile, or blog then take an interest in content talk sessions official with the teacher. Take part in content talk sessions informal with different understudies and partake in live voice visit sessions official with an educator. Take an interest in live voice-visit sessions informal with different understudies. Post another dialog gathering message authority to the teacher and post another exchange gathering message informal to different understudies. The answer to discourse gathering messages authority to the teacher then answer to exchange gathering messages informal to different understudies. Send messages to the teacher and different understudies. Therefore offer assignments with different understudies by means of dialog discussion and also offer assignments with alternate understudies through email. Submit assignments online and take exams online after this submit tests online and ungraded practice tests online (Menekse, Stump, Krause & Chi, 2013).

Online Assessment Reliability

Security issues of e-learning frameworks have been examined by numerous specialists online examination is a challenge for e-learning security right now. Online evaluations are generally led at particular examination focuses and require supervision chiefly in light of the fact that, if regulated in unsupervised areas, students may get help from others to enhance their examination comes about or have another individual take the examination for them. In such cases, educators will wind up dubious on who addressed the examination questions, which clashes the adaptability favorable position of the online instruction.

In this manner, notwithstanding the extension of the e-learning market locally and universally, an issue remains, particularly with off-site examinations. Inability to confirm the students going to an examination is a noteworthy test in internet learning conditions (Kim, 2013). Next to no consideration has been given in taking care of the issue of students' deceptive lead. In addition, numerous scientists censured current e-learning frameworks for not concentrating on the verification of students, especially the individuals who take part in online tests and examinations. The biometrics has supplanted the ordinary secret word frameworks another investigation featured the significance of recognition instruments past the beginning access to the e-learning framework. In this manner, a framework must be created to guarantee that the individual taking an examination is the understudy selected in the course featured numerous advances related to security that have not been utilized as a part of e-learning. These advancements incorporate biometrics advances that are progressively getting to be fundamental in various applications. Biometric verification is the programmed acknowledgment and recognizable proof of students by utilizing their physiological qualities, for example, voice, hand geometry, fingerprints, and facial pictures. For the most part, biometric verification requires correlation of the put-away information against the caught information (Thiemann, 2016). No flawless biometric framework that fits all needs has so far been made. Every known framework have their preferences and impediments a couple of studies had concentrated on enhancing e-learning security utilizing biometric frameworks, be that as it may, a set number of them tended to the issue of constant client verification. In an ongoing report, biometric client verification in online examinations by means of keystroke elements, keystroke biometrics has its inconveniences, for example, the significant contrasts that can happen over the long haul because of changes in composing design, tiredness of the hands after a time of writing, furthermore, change of abilities.

In accordance with the beforehand said idea, scientists are searching for the best biometric confirmation strategy that will help approve the recognizable proof of the student going to the examination and that will guarantee that he/she is an indistinguishable individual from the one enlisted in the course without bargaining his/her protection (George & B., 2017). Face acknowledgment frameworks are human-accommodating on the grounds that they require no contact and no extra equipment (given that most PCs and PCs accompany a camera). All the more vitally, confront acknowledgment frameworks can be utilized for the consistent verification of the student amid the whole examination time frame.

Development Model for Authentication

Face acknowledgment advances work by checking the individual's face and coordinating it with the put-away picture. The confront acknowledgment biometric framework is a framework that records recognized facial highlights and stores the format in a server. In examining the face, the camera recognizes facial highlights and transmits the flag to the server where the examined highlights are prepared for coordinating (See Figure 4). Facial acknowledgment recognizes key highlights from the facial picture. The framework distinguishes the face and catches images of facial highlights that don't change over time, while maintaining a strategic distance from those that change, for example, outward appearances or hair (Liu, Li, Ma & Song, 2017). The initial phase in confronts acknowledgment is the location of the face in the image.

Yang and his associates assessed the fundamental techniques utilized for confronting identification to be specific, information-based strategy, appearance-based approach, include invariant technique, and format coordinating technique. The second step is the direction of a few ways to deal with demonstrating and perceive the facial picture, such as immediate connection, versatile chart coordinating, neural sys-

Figure 4. Standard authentication process

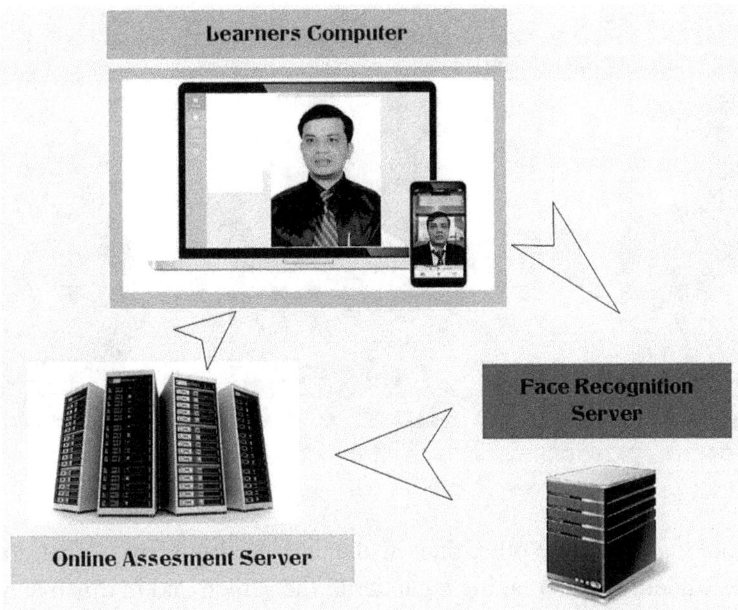

tems, main segment investigation and multiresolution examination (Höynälänmaa, 2015). Examination framework incorporates the advancement of test put money on a particular field. The test bank contains an assortment of question things e.g., coordinating, positioning, article, classification, numerous decision things, and genuine/false inquiries. These inquiries are arranged into three trouble levels: simple, direct, and troublesome. In outlining the test, the educator can determine the number of test questions, kinds of inquiries, and the trouble related to each inquiry. The framework will then naturally create an irregular arrangement of inquiries in view of the criteria indicated by the teacher. In this manner, every student will get an alternate arrangement of inquiries yet with the same trouble level. The educator is additionally permitted to outline another inquiry and indicate its trouble recently detailed inquiries will be put away in the database and after that added to the test bank. Toward the finish of the examination time or in the wake of finishing the test, the score and time spent by every student amid the examination will show up (Menke, 2014). The database will likewise spare other information identified with the enrolled clients or students, including profiles what's more, pictures. The face acknowledgment framework is incorporated in the online evaluation instrument to recognize and confirm the students permitted to get to the exam and to constantly approve the student's personality until the finish of the examination. In particular, amid enlistment in the course when utilized as a major aspect of an e-learning framework or for an examination, pictures of students, notwithstanding other required information, are caught and put away in the database. Caught pictures are encoded to secure students' protection. Amid examination time, the student's character is checked for participation in the examination and is observed by contrasting the caught pictures and the one put away in the database.

Likewise, to address the issue of conning in online examination frameworks, ceaseless checking is executed, as appeared in Figure 5. In the said figure, the student is caught taking a gander at the screen (left), perusing (center), and taking a gander at an adjoining student's PC (right). In the two-second video taken amid the examination period, the pictures in the video are contrasted with each other with the check

Figure 5. Images of learners in different positions

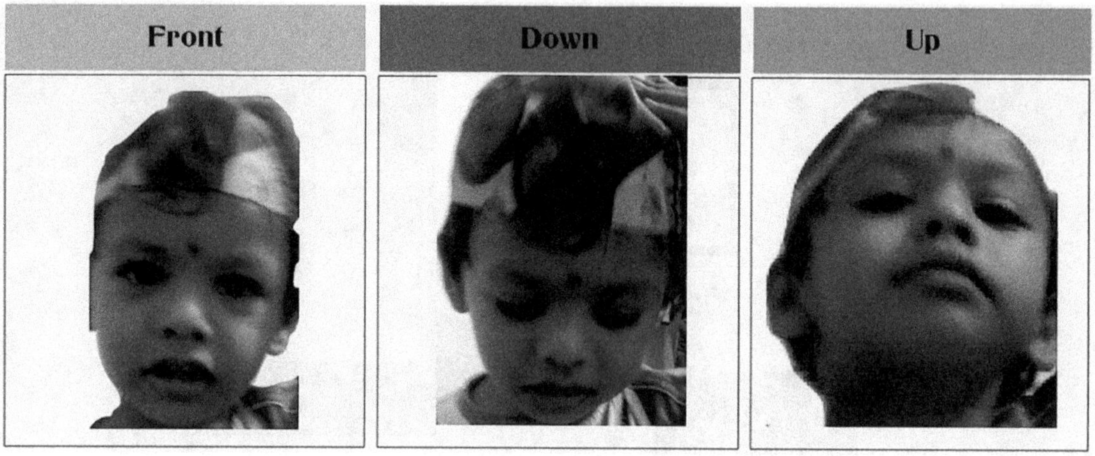

if the student was looking elsewhere other than his/her screen. In the event that, for all pictures inside the two seconds, the student was not taking a gander at the screen and in this manner was not engaged in understanding the examination questions, he/she will be cautioned by an adjustment in foundation shading, as appeared in Figure 5. Fizzled validation will likewise be made unmistakable to the students by the change in foundation shading (Omar Zughoul, 2013).

In the event that the confirmation disappointment proceeds for in excess of a couple of moments, the framework will stop and perform the collective check. In this stage, the framework will request that the client put his/her face in a fitting position to catch another picture. On the off chance that the mistake is more than once experienced, the examination won't be controlled for suspected bamboozling. Students not to be irritated and to limit coordinating disappointments a two-second video of the student is recorded and the best picture as far as outward appearances, lighting, and determination will be chosen by an application already introduced on the customer's side of the framework. This picture is sent to the server and will be contrasted and the put-away picture. The face is utilized as a contribution for key age sees Figure 6 the entire plan incorporates four principal steps. 1. Picture pre-preparing finding and situating of the face in the picture, Hambridge Framing and diminishment of information size and sensibility against varieties, 2. Biometric Eigen analyzes 3. Making of discrete biometric information with a Token biometric data is changed to somewhat set {0,1} and a bit string is created, 4. The intersection of the Key with a Token.

Client Verification

Desktop Vs Mobile condition from the underlying time of cell phones, the nature of client confirmation has remained generally unaltered. A Personal Identification Number (PIN) is utilized limitlessly as a point-of-section security in cell phones. Comparative circumstance is likewise found in work areas where confirmation approaches, for the most part, depend on mastery learning. Be that as it may, clients of cell phones may utilize numerous instruments to bolt different parts of usefulness. For instance, Windows Mobile handsets bolster two unmistakable confirmation instruments, for example, one to secure cell phone and other to ensure client's SIM (Subscriber Identification Module). The forefront insurance

Figure 6. A solution for key generation using face imaging

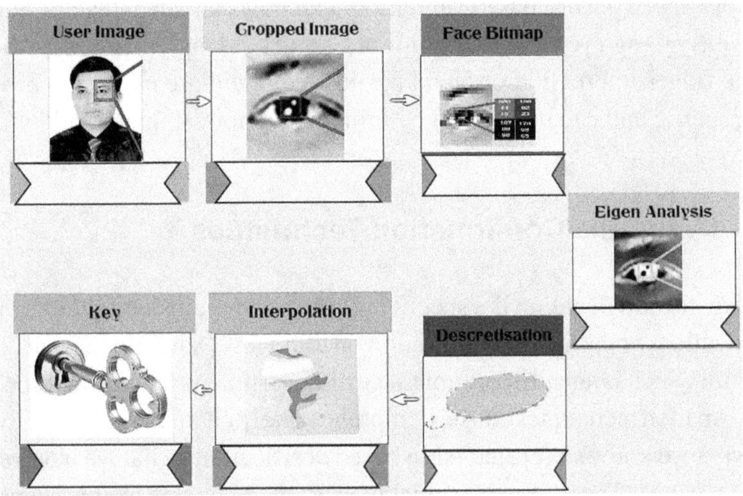

of the handset is guaranteed by gadget level validation, especially when the gadget is exchanged on. Therefore gadget level validation monitors access to applications and client's put away information in the telephone. Though SIM-level confirmation gives the security of the substance of SIM and cell arrange an account. Something else, SIM card can be expelled from the approved gadget and can be utilized in any unapproved gadget. In this way, SIM level confirmation successfully oversees the utilization of cell information and confines from making voice calls. One of the one of a kind distinction among cell phones and workstations is that cell phones are exclusively close to the home gadget, regularly utilized by one client. The little-estimated cell phones are defenseless against unapproved use by misfortune and burglary. Therefore, cell phone confirmation and ID is not the same as work area verification techniques (Botha et al. 2009). Potential assailants get a controlled and operational gadget if there should be an occurrence of misfortune or robbery as cell phones are commonly dependably on gadgets. The client should be confirmed to the telephone before utilization to counteract unapproved use and to guarantee the assurance of put away information. The telephone ought to be bolted after the client finishes his/her assignments. Consequently, cell phone working frameworks, for the most part, give short yet visit sessions. Subsequently, a critical effect on the convenience of cell phones depends on the picked validation technique by the client. The speed and solace of utilization are the key two elements for the vast majority of the clients while choosing about the confirmation strategies (Luca and München 2015).

As cell phones are extremely adaptable for developments, these are frequently utilized in suspicious spots with an immense number of potential onlookers. An easy to use and simple methodology of verification needs the autonomy of the condition the telephone is utilized in, other than guaranteeing security. For the validation on cell phones a couple of prerequisites are recorded underneath from (Luca and München 2015): • A client needs to verify before every single session • A passing eyewitness ought not master anything about the mystery utilized for confirmation • Authentication ought to be secure and quick with least client exertion Authentication techniques accessible by and by these days are not ready to satisfy every one of these necessities. Stick/Password validation is generally basic technique which requires tedious client cooperation's (composing over and over for each time while opening the telephone). Hence, a large portion of the clients like to utilize simple to type, short Password/PIN and in the long run, less anchored particularly when utilized in swarmed open spots. A graphical example is one of the choices of PIN/Password based validation which has an advantage of quicker info yet nearly simpler to learn by easygoing eyewitnesses and it is a security danger. Biometric-based confirmations, for example, unique finger impression or facial acknowledgment are as yet not broadly introduced because of the extra expense and can be avoided by utilizing a phony unique mark or a basic photo. (Luca and München 2015)

Classification of Versatile Confirmation Techniques

"Validation is the way toward deciding if a specific individual or gadget ought to be permitted to get to a framework, an application, or particular information on a gadget" (Sametinger et al. 2012). Confirmation strategy is a procedure which is an indispensable security instrument. There are three essential classifications in which confirmation techniques can be comprehensively characterized. For example, information based validation (what we know), proprietorship based verification (what we have) and inherence based confirmation (what we are). These sorts are quickly examined beneath (Sametinger et al. 2012)

Knowledge-Based Authentication

PIN/Password based verification is a case of learning based confirmation or "what we know". Graphical examples/passwords other than inquiries and answers are another case of this sort of verification. The gadget is being verified by the utilization of learning. Here, there is a plausibility that mystery learning goes to unapproved hands. Any individual who has the learning of mystery for the gadget verification is able to utilize it for validation. Test reaction verification relies on learning as well. Secret word validation is the most straightforward case of test reaction verification where requesting the secret key is the test and the substantial main reaction is simply the secret phrase (Ahmed, 2017). Graphical passwords are an approach to keep away from numerical/alphanumerical passwords which are difficult to recall as human cerebrum is fit for recollecting noticeable pictures as opposed to complex series of characters. There are review based and acknowledgment based graphical passwords. The client needs to recollect pictures in review based frameworks. Acknowledgment based frameworks expect clients to recognize pictures whether they have seen a picture previously. In this framework pictures that are seen as of now must be perceived as opposed to producing from memory (Sametinger et al. 2012).

Ownership Based Validation

Ownership based confirmation ('what we have') incorporates the case of smart cards and electronic tokens. A key which opens an entryway is a case of genuine possession based confirmation. Anybody can enter the entryway that has the proprietorship of the key. It is likewise conceivable to have duplicates of the way to enable numerous individuals to enter the entryway. Radio Frequency-Identification (RFID) labels or attractive strip cards are computerized cases of this sort of verification. For short separation correspondence, up to 10 cm Near Field Communication (NFC) is a standard for radio correspondence. Proprietorship confirmation based NFC labels should behold intently to the NFC label perusing gadget. The distinguishing proof number of the NFC tag can be utilized for examination. Age of secure passwords is leverage of proprietorship based confirmation; however, to stay in the responsibility for a thing a sheltered consideration must be taken. On the off chance that the thing is lost or stolen then someone else can effectively validate himself/herself. In the event that duplicates of the thing made and left unnoticed then it turns into an extraordinary risk.

Security Factors for Portable Validation

Security is one of the greatest issues which it should have been considered while confirming on a cell phone. The key center territory of verification strategies is on latent and persistent validation of clients as clients have the power over their gadgets while in customary techniques talked about above, confirmation is, for the most part, performed toward the start of framework utilization. For instance, a number of various qualities of clients, for example, a client's unmistakable composing design, use of a cursor, intellectual preparing time can be utilized to screen clients ceaselessly and to validate them, which can be alluded as persistent verification. It is necessitated that clients build up a profile first by associating with the framework they need to use in nonstop confirmation frameworks and afterward exercises amid the use of telephone are contrasted and the client's known profile. Barely any models of persistent validation techniques are quickly talked about underneath which are really not to be utilized like or supplant a conventional confirmation conspire, rather help other verification instruments: Keystroke dynamics is conceivable to distinguish a client for verification by utilizing his/her time interims and weight of console presses. It tends to be utilized in cell phones however commonly it is connected to customary consoles (Choong et al. 2016). On-Body Detection if accelerometer of a cell phone is dynamic then this component keeps the gadget opened (when the gadget is connected with a moving client) and when the accelerometer is latent the gadget is bolted (development isn't distinguishing). (Choong et al. 2016) Location-Based Awareness: A client's area can be recognized through gadget's Global Positioning System (GPS) area, IP deliver or vicinity to a particular remote system and this area can be utilized for the help of confirmation of a client (Choong et al. 2016). Some of the normal security factors for portable validation are depicted beneath:

Social Designing

It is an exceptionally outstanding security factor which takes in the control of individuals to reveal private proof like a PIN or secret key. Social designing may likewise result in the disclosure of inconceivable image employments of Face Unlock. NFC labels are likewise available if the assailant can come neighboring adequately to peruse the tag. Motion Puzzle is thought to be difficult to social designing as various passwords are utilized and a grouping of pictures in addition to the coordinating the motion would need to be uncovered for every secret word (Sametinger et al. 2012).

Shoulder Surfing

It is less demanding to distinguish a PIN or an open example for somebody who is viewing a client. Along the secret phrase is genuinely harder to perceive in light of its length while open examples are helpless to bear surfing as they are drawn on the screen and can be perceived even from a separation. The various systems don't post danger while being viewed by others (Sametinger et al. 2012).

Malware

Malware is the most established and most regular security danger. It can show up in different structures. For occasion, spyware or phony applications can keep running out of sight, log client input and send it to a server controlled by the intruder. The assailant may endeavor to get physical access to the gadget on the off chance that the validation information goes to the server and when a client is entering a PIN or a secret phrase it is anything but difficult to log a client's information. If there should be an occurrence of NFC, the main data required is a picture or the recognizable proof number of the NFC. It is additionally simple to make a UI like the screen discourse to take the PIN or secret word from the client by a malware. Signal Puzzle guarantees some assurance on the grounds that the info relies upon pictures appeared to clients (Sametinger et al. 2012).

Speculating

It is conceivable to create arbitrary PINs or passwords utilizing an application by an aggressor, which can be attempted after certain interim. In the event that the interim is short, the gadget may deactivate itself because of the expanded number of unsuccessful attempts. As fingers leave an oily leftover portion on the touch screen, it is conceivable to follow the example utilized to Unlock design. Face Unlock, NFC labels and Secure Lock don't allow any type of speculating (Sametinger et al. 2012).

Copies

It is conceivable to copy the Face Unlock pictures and NFC labels. It is likewise conceivable to sidestep face unlock utilizing a photograph of the real client. If there should be an occurrence of burglary, it is conceivable to get to the gadget proprietors photograph which can be later used to validate on the gadget (Sametinger et al. 2012).

Dumpster Jumping

Dumpster jumping is an issue if a client records his/her confirmation certifications on paper what's more, arrange them later. An assailant can get hold of the paper and thus, the data and can go into the gadget. This can be conceivable if there should arise an occurrence of a PIN or secret phrase while NFC labels are saved from such assaults as it is impossible that clients will discard their labels (Sametinger et al. 2012).

Ignorance

Ignorance of a client is a security danger by and large; numerous clients don't think it important to monitor their gadgets with a bolted screen. They think their gadget is secure as they continuously convey it with them. Ignorance is likewise an issue if there should be an occurrence of picking appropriate PIN or the secret key for confirmation. They may utilize a feeble PIN or secret word which is anything but difficult to figure or to bear surf. Open examples and Gesture Puzzle may likewise endure designs that are not precisely picked and simple to make sense of. Face Unlock, NFC labels, and Secure lock have fewer issues with unconscious clients too (Sametinger et al. 2012).

BIOMETRIC PERFORMANCE MEASUREMENTS

The execution of biometric frameworks is tried for the most part as far as false rejection rate (FRR), false acceptance rate (FAR), failure to enroll rate (FER), enlistment time, and confirmation time. The false acceptance rate is most imperative when security is a need though low false dismissal rates are favored when comfort is the need. The biometric system used in the flight deck must have a low false acknowledgment rate since security is the need. In case the false affirmation rate is as low as possible then we have a prevalent shot of not allowing unapproved subjects into the system. The time when the FAR and FRR meet or cross breed is known as the identical error rate. This rate gives a more sensible measure of the execution of the biometric system rather than using either the FAR or FRR freely. The inability to select rate is the rate which a subject can't familiarize his or her biometric with the system which is palatable to the structure (Kukula, Sutton and Elliott, 2010). For example, if there is a one of a kind check sifting device which is greatly fragile to the photos acquainted with it and a subject can't give an undeniable picture then he or she won't have the ability to choose into the system. Regularly, there are systems that will empower the subject a couple of undertakings to choose biometric information into the structure. Both the selection and live presentation times are essential factors in choosing or testing structure execution. The enrollment time is that course of occasions amidst and including the getting of the biometric test and making the set away design of that case. The affirmation time is an estimation of the system of live presentation. This method consolidates the catch of the unrefined data, a live design taking care of, examination of the set away organization to the live arrangement and the time it takes for the structure to give a decision i.e. facilitate or non-arrange. To give the persevering approval segment needed for the flight deck, the check time must be close constant for a productive flight deck biometric confirmation system.

MULTI-BIOMETRICS

Biometric systems need to fight with noisy data, bound degrees of chance, and powerlessness to choose issues, spoof ambushes, and unsatisfactory bungle rates. In a couple of conditions, it may be achievable to pass on a biometric structure that endeavors more than one technique for conspicuous verification or affirmation to beat these issues. A biometric contraption can either be joined with non-biometric sorts of approval or with various kinds of biometric affirmation devices. Exactly when a biometric contraption is joined with various sorts of biometric confirmation devices, it can be portrayed as a multi-biometric system. Multi-biometric structures may be stronger and give higher affirmation rates in light of the closeness of different, free bits of verification. Multi-biometric systems address the issue of non-completeness, since various qualities ensure sufficient people extension, and give against ridiculing measures by making it troublesome for an intruder to take various biometric traits of a true blue customer (Neves, Moreno and Proença, 2017). If there is a deficiency in one procedure for biometrics, by then going along with it with a biometric technique that is more grounded in regards to that weakness will help that issue. For instance, it may be conceivable to pass on a biometric system in the flight deck that contains both extraordinary finger impressions separating and voice affirmation devices. Additionally, a multi-biometric structure may diminish the false reject rate and the failure to enroll issue.

Figure 7. Portrays a multi-biometric framework utilizing and arrangement

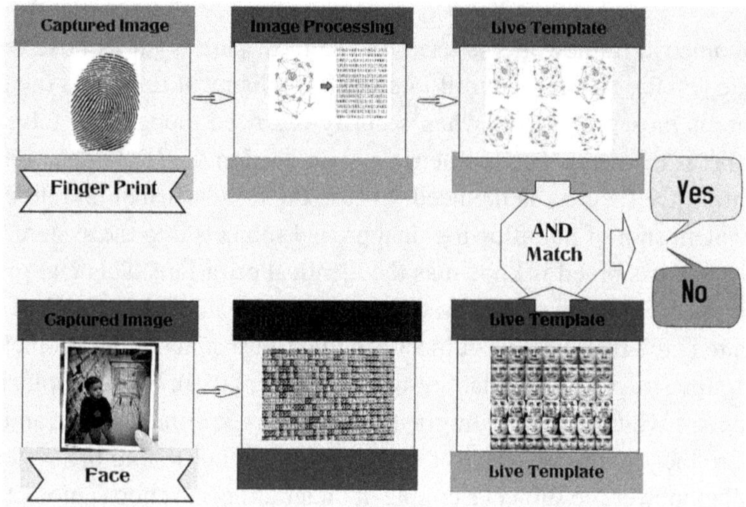

In this plan as appeared in figure 7, it is principal that both of the biometric methodologies achieve an organizing score proportional to the affirmation score set for the structure which is set up at first. This system would give high assurance that the person who is familiarizing their biometric information with the structure is who he says he is. Ridiculing is more troublesome in light of the way that two biometric characteristics are used. It is possible to set individual biometric limits for each system used or to weight one biometric method more than the other all through the structure by and large. In this couple of formulas are displayed for the false recognize and false reject rates similar to probabilities while using the AND outline. The condition, as given and portrayed is by all accounts:

$$PA(FR) = 1-[1-P1(FR)][1-P2\ (FR)] = P1(FR) + P2(FR) - P1(FR)P2(FR)$$

This condition shows that the probability of conveying a false reject would lessen in case one used a singular biometric system alone, instead of joining diverse biometric techniques, especially in case one is essentially more grounded than the other. Conditions for the OR setup are practically identical besides that a false reject can simply happen if both biometric systems make a false reject. "OR" Configuration a multi-biometric structure using the OR game plan, in this outline as sown in figure 8, general affirmation by the structure can be defined either by both biometric procedures having a planning score proportional to the affirmation score set for the system at first or by either biometric technique having an organizing score identical to the affirmation score set for the system at first. This plan does not give the conviction that the individual is who they say they are and what's more the AND course of action does.

This outline may reduce the false rejection rate general in light of the fact that the customer will be recognized into the structure by for example, either their remarkable check design organizing the as of now set away one of a kind finger impression picture or by their facial format planning the in advance set away facial picture or both (Gawande, Zaveri and Kapur, 2011). Since using this outline may lessen the false rejection rate, the false affirmation rate will extend, which is unquestionably not a keen idea for significantly tied down zones.

Figure 8. Portrays a multi-biometric framework utilizing or arrangement

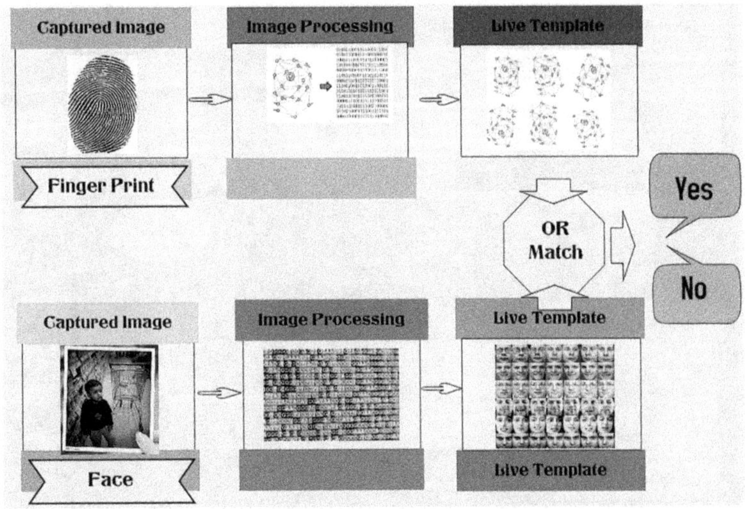

BIOMETRIC SYSTEM PROCESS

All biometric structures in a general sense take after a comparative game plan of systems for biometric feature organizing addressed in Figure 9. Biometric get occurs at the biometric contraption i.e. exceptional stamp scanner. The photo of the biometric is taken care of using specific figurings redid for that biometric system to convey a live organization. The live organization of the biometric is a numerical depiction of the starting at now acquired biometric. From the limit device, the design of the biometric which was secured as a noteworthy part of customer enrollment is recuperated and should facilitate the motivator from the live configuration (Kashyap, Anderson, 2018). Exactly when this happens a biometric facilitate is picked up. Biometric get occurs at the biometric contraption (i.e. extraordinary check scanner). The photo of the biometric appears in figure 9 that is taken care of using specific counts uniquely designed for that biometric procedure to convey a live arrangement. The live design of the biometric is a numerical depiction at introducing acquired biometric.

From the limit device, the design of the biometric which was secured as a noteworthy part of customer enrollment is recuperated and should arrange the motivation from the live configuration. Exactly when this happens a biometric facilitate is acquired. Figure 9 addresses the strategy that occurs between biometric data selection and DVD amassing of the configuration. This methodology begins with the biometric data enrollment. Once the contraption has gotten the biometric data, a configuration is delivered. This arrangement will be used later for connection at the period of affirmation/organizing. In case the biometric data isn't successfully changed over into its proportionate format then the biometric data must be re-chosen into the structure. After the biometric data has been changed over into its corresponding design, it will be secured into the database, close by the biometric configurations of other endorsed customers (Biggio, fumera, Russu, Didaci and Roli, 2015). In case there is an issue with the database accumulating, the path toward getting biometric from the customer and limit is attempted again until the point that accomplishment has been come to. Right when the database amassing of the design is compelling, the biometric organization will be secured into the fortification database at the enrollment office and thereafter set away into a trusted in DVD.

Figure 9. Standards biometric system process

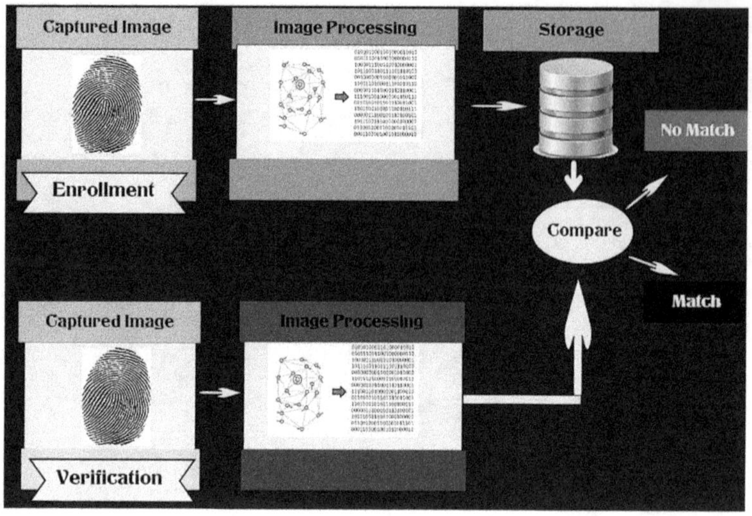

OPERATIONAL MODE

The approval technique incorporates two phases to be particular, enrolment arranges and conspicuous confirmation organizes. In the midst of enrolment arrange, biometric tests are gotten from individuals and set away as biometric designs in databases. In recognizable proof stage, the biometric tests are expelled from the general population and considered against the one set away in the database. If there is a match, the individual is successfully confirmed and the client is allowed to utilize the advantage or organization s and client is honest to goodness person. It there is no match, the individual isn't successfully affirmed and the client is kept from utilizing the advantage or organization s and a client is an illegal person. Figure 4 shows the two particular times of individual affirmation (Kashyap, Tiwari, 2018). In the midst of enrolment arrange the biometric tests removed from the customer is secured in the database as biometric formats. Designs reveal singular information about a man and are slanted to ambushes. Subsequently, it is pressing for any biometric system to ensure the security of the biometric positions. Biometric systems substantiate the identity of a man using his natural or lead traits, for instance, face one of a kind finger impression, iris, palm print, hand geometry, voice, stamp, and walk et cetera. Which are generally called biometric modalities or qualities (Kashyap and Gautam, 2017)? Presumably, the most by and large used biometric traits by the identity organization systems are showed up in Figure 10.

The prime favored viewpoint of biometric structures stood out from common conspicuous evidence procedures is supplanting what the individual passes on and what the individual reviews perfect models with which the individual is, along these lines balancing identity distortion by using biometrics plans that are attested to be uncommon, enduring and hard to produce. Since, the individual is required to be accessible at the period of the check by the biometric structure, along these lines also turning away false disavowal claims (Pointner, 2017). In order to fight to extend security perils in information time, governments, organizations, and foundations have, all things considered, got and enabled explore on biometric check structures.

Figure 10. Biometric template matching process

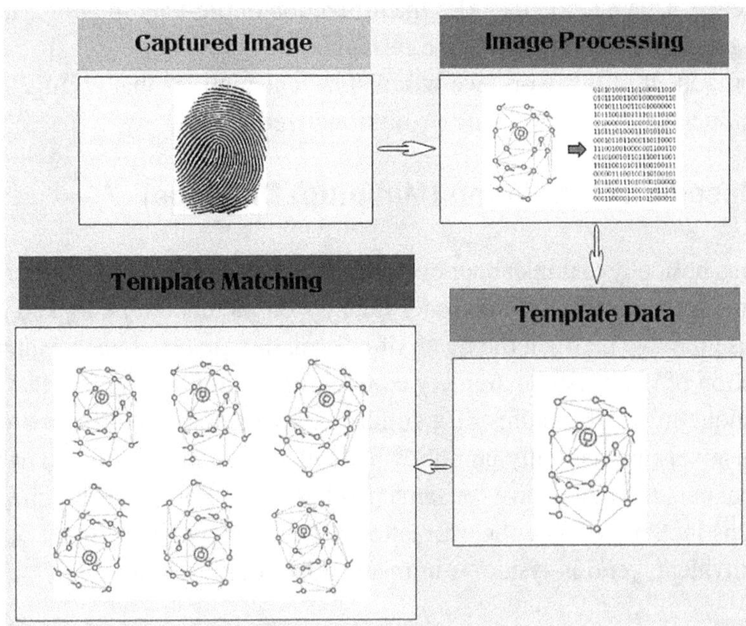

STUDY RESULTS

Two studies were led the main overview focused on online educators, and the second focused on potential clients/understudies. The principal examination was led to distinguish the rate of picture catch and the circumstances the suspicious conduct is acknowledged without influencing students' focus amid the examination. Additionally, e-learning teachers and specialists (n = 8) were given the portrayal of the proposed arrangement and were requested to rate the accompanying articulations utilizing a five-point Likert scale (5—Strongly Agree to 1—Strongly Disagree): the framework will give solid outcomes that mirror students' accomplishments, the framework will adequately approve the students' personality the likelihood of swindling amid examinations will be limited, I will prescribe this framework to different teachers, and the understudies will consider harder. What's more, the teachers what's more; specialists were requested to indicate the accompanying data: the catch rate of pictures for nearness checks, what's more, the adequate number of times of suspicious conduct admonitions amid the examination.

Unique Mark Recognition and Neural Network

From a specialized perspective, there are two unique issues in the field of unique mark investigation, to be specific: characterization and correspondence. The order comprises in assembling a few fingerprints that have formats that are comparable in predefined subclasses. Today, there are five classes of fingerprints characterized by Henry. Correspondence suggests verification and recognizable proof, depends on other more unpretentious components details, streaks' introductions and requires a substantially more nitty-gritty investigation of the unique mark picture (Yang, Li & Liao, 2014). The entire picture is never again thought about here, however, an arrangement of components which all in all constitutes the engraving format or on the other hand the element vector: This is the details and their rules when making the correspondence. This method includes the utilization of pre-handling binarization, diminishing, separating, and so forth that will empower the refinement of the picture to extricate the most helpful highlights. As a component of this work, we will display calculations that arrangement solely with the presence of correspondence, the order being institutionalized.

Examination Based on Correlation (Minimum Distance)

This technique points basically to think about pixels cluster of two unique mark pictures and to figure the relationship of that pixels. For two unique finger impression pictures to be looked at, this strategy just thinks about the base remove between the mean of information pixels of each example. The method in light of the connection of pixels isn't extremely compelling, in light of the fact that we can get various pictures for one unique mark. If pictures of a similar unique finger impression are extraordinary, the estimations of their pixels are naturally unique; what's more two pictures taken at various resolutions from two unique scanners might not have the same pixel esteems. For this technique, we can give a limit for instance if the distinction between the estimations of the two pixels is under 3, it is viewed as that these pixels are equivalent, yet this system is unmistakably not thorough.

Correlation Based on Minutiae (Image Mapping)

Correlation in light of particulars Image mapping it's the most broadly utilized strategy since it depends on singular highlights in every individual, the particulars. Details are separated from two prints and are spoken to as an arrangement of focuses in two-dimensional plans relying upon the model directions kind of details, organize, and point. The examination is to locate a legitimate arrangement of the two unique mark details (F1 and F2) which creates a greatest of sets of comparative particulars (Gu & Liu, 2011). Calculation number increment with vector number officially characterized; Classification result relies upon a basic level (two fingerprints are indistinguishable when a number t of indistinguishable particulars is achieved, this number can shift starting with one framework then onto the next relying upon the level of security picked by the architect).

Probabilistic Neural Network

The calculation depicted only manages the part of correspondence comprises of two fundamental advances, to be specific: a preprocessing step and a choice advance. The preprocessing stage essentially adjusts the two pictures and concentrates, from every single one of them, a focal locale. The two focal districts are sustained to the choice stage, which is the best possible neural system part of the calculation and subject to preparing from illustrations. Though the preprocessing stage is genuinely standard, the choice stage is imaginative and in light of a neural organize that executes a probabilistic Bayesian way to deal with the gauge of the likelihood p of a match (Savchenko, 2013). It is shaped by the plunge of the slope utilizing a preparation set of n sets of pictures from a few extraordinary fingers. Adding some extra fingers to make utilizing stage reinforces the vigor of the neural system made and tried in advance.

Points of Interest

- Use of a focal area of the picture with arrangement and information pressure;
- Calculation of probabilities to make the correspondence between two examples gave at the system section after a few progressive sifting;
- The calculation is adjusted for validation of a man who wishes to get to a framework in which its unique finger impression is beforehand known;
- The calculation is appropriate for frameworks utilizing a little database.

Constraints

- Use complex recipes for the count of probabilities;
- Usable for a set number of unique mark's photos including the learning stage.

LOCALLY CONNECTED NEURAL NETWORK

For this calculation, design acknowledgment frameworks normally comprise of two phases: development of highlight vectors, what's more, characterization of example. In the initial step, a unique finger impression is passed in the framework section and is diminished to a component vector which is invariant to interpretation and turns; this tosses tasks of change computing an initiation for each neuron of the system and a worldwide vitality. In the wake of having extricated highlight vectors, model vectors are shaped by a subclass of information. In the second step, a grouping or direct examination which prompted the acknowledgment or not of the example is performed between the info include vector and the models of the database (Azhar & Anderson, 2012). This technique in view of the locally associated neural system is more fitting for check applications, security, and recognizable proof.

Points of Interest

- Use of models (include vectors of unique finger impression by a subclass of information) invariants to interpretation/pivot to make correspondence;
- This calculation is appropriate for recognizable proof of individual in an extraordinary arrangement of information;
- Use of closest neighbors to play out the order.

Constraints

- The learning time is high;
- The relatively high rate of false dismissals;
- Using limits and actuation computations to play out the correspondence.

A TWO HIDDEN LAYERS PERCEPTRON FOR FINGERPRINT RECOGNITION

By utilizing the advantages of the already specified strategies, we have built up another system in light of the utilization of a multilayer perceptron. The structure of the neural system will be that of a privately associated neural system, which utilizes the computation of probabilities to arrange the info picture sub-blocks thinking about the henry's grouping. We additionally utilize a limit (to decide if there is coordinating or not between two examples) and the retro-proliferation calculation to prepare the system. The neural system utilized here is a multilayer perceptron with an info layer, a concealed layer which itself contains three layers (two characterization layers and a correspondence layer) lastly a yield layer. The rule is similar like a standard acknowledgment technique in light of the utilization of the neural system and depends on two primary advances: preparing and acknowledgment arrange (Abdullah, Rahman, Abas & Saad, 2016). Before applying the preparation arrange, we order the whole test of information into four primary classes curve, whorl, left and right circle as indicated by Henry's grouping also, we decide two models for each class of information. After this strict order of the database and the assurance of K1 models for the first and second characterization layer, the preparation stage can be continued as take after: the three layers of our system are emphatically associated among its. Information layer is associated with the first arrangement layer, the first characterization layer is associated with the second grouping layer, the second order layer and correspondence layer are associated with the correspondence layer is associated with the yielding layer by association weights. The N input unique mark pictures to be perceived (after pretreatment tasks), is connected at the info layer of our neural arrange.

Association weights of the first and second grouping layer are separately initialized by K1 here K1 = 4 on account of four principle classes models of each datum class definitely known by the unique mark class made. At that point, association weights of the correspondence layer are initialized by all the K2 expected unique mark picture of the database for the preparation organizes for each class of information decided previously. So K2 is the quantity of example of an information class, for instance, is the victor neuron after the two grouping has a place with the class of whorl at that point K2 will be equivalent to the quantity of example of the whorl class.

The info vector is contrasted with all model of each datum class of the first and second order layers, and it's appointed to the class of victor neuron the second order fill in as control of the first and help to lessen mistake rate. After this, the information vector is contrasted with all information of the class of the victor neuron at the correspondence layer, and it's relegated the victor neuron as well. On the off chance that the champ neuron isn't the model of the right class, conferred mistakes are back proliferated. The yield layer gives us the initiated victor neuron. Scientifically, the proposed calculation can be connected this way:

1. Choose the underlying models two by two for every datum class, as indicated by the characterization of Henry.
2. Initialize neuron of the main characterization layer by the principal model of every datum class.
3. Initialize neuron of the second order layer constantly model of every datum class.
4. Initialize each correspondence layer neuron weight with the database design class by class.
5. Apply an info vector.
6. Apply the closeness measure on the principal grouping layer by likelihood computation.

SECURITY FOR ONLINE EXAM

Jung and Yeom proposed an upgraded secure online exam administration condition intervened by amass cryptography utilizing remote observing and control of ports and information. Since the security of online examinations remains an issue, the individual written work the exam on an organized PC is observed by a delegate at some foreordained area (Jung & Yeom, 2009). Be that as it may, the necessity for an exam area conflicts with the openness, the significant fascination of e-learning or separation learning. The prerequisite may likewise refute the cost reserve funds produced by learning or stance deterrents for remote understudies. Rearrangements and robotization of instructive procedures are different advantages of online training, and online exams acquire these points of interest. To expel the prerequisite for human intercession in secure online exam administration in order to profit by the benefits of online procedures, answer for the issue of security and deceiving for online exams. This arrangement utilizes an upgraded Security Control framework in the Online Exam which depends on gather cryptography with an e-observing plan. The cryptography underpins upgraded security control for the online exam process, and verification and trustworthiness. The e-observing gives a delegate capacity to remote examinees to avert bamboozling and hence evacuates the necessity of going to a settled area. The proposed framework (Jung and Yeom 2009), directs an online exam at a settled time with similar inquiries for all examinees, much the same as a disconnected exam, yet without limiting the physical area of the examinees. As the SeCOnE framework empowers numerous sorts of tests to be given on the web, it can give instructors better assessment measures for understudies and may add to enhancing the nature of training. The framework depended on a protected exam convention with an all-powerful focal administrator who controlled all the data for understudies, instructors, issue sheets, answer sheets and grades. The shortcoming of this framework was that the director was thought cutting straight to the chase. Additionally, a limited room was required for the exam, to forestall conning. The security issues identified with online exams incorporates unapproved access to the issue sheets before the exams, yet additionally adjustment of the inquiries, the appropriate responses, and the evaluations. Moreover, extraordinary tricking designs exist the Blackboard Northern Illinois Univ. Web-based), including replicating the appropriate responses of others, trading answers, hunting the Internet down answers, utilizing the information and programming saved money on the understudy's nearby PC, and examining the exam by email, telephone, or texting. A few strategies for fighting this incorporate giving an alternate issue set to every understudy, confining the exam room or restricting the number of answer entries to one (Okada, Scott & Mendonça, 2015). Research has concentrated on techniques to check understudy personalities and to impart safely amongst educators and understudies instead of on countermeasures against undermining on the web exams. Undermining disconnected exams is likewise a major issue. Agreeing the investigations as the level of correspondence amongst educators and understudies diminishes, the inclination to swindle increments. This impact directly affects online exams, when understudies may have little contact with their instructors. The most current online training utilizes Web-based business course administration programming. Past Web-based ways to deal with online exams have featured simple availability and streamlined exam administration. Notwithstanding, validation through just a client name and watchword can be the powerless point in the security of online exams. The simple condition in which understudies can utilize a Web program and the Internet empowers them to look through the Internet and to speak with others for help amid the exam. In any case, a few soundless photos of an understudy don't demonstrate what that understudy is doing or why he or she is doing it, or regardless of whether bamboozling is occurring through Web seeking, the utilization of spared information, or visiting. The significant dialog had occurred on aggregate conventions and gathering interceded correspondences to guarantee anchor interchanges among bunch individuals. In this way have incorporated the thought of secure gathering arrangement, secure intergroup correspondence utilizing an open key and secure group correspondence utilizing the symmetric key through the Diffie-Hellman key trade adopts two gatherings for secure correspondence between disseminated substances in the online exam framework. The intergroup correspondence is ensured through an open key framework, while intragroup correspondence utilizes a few symmetric Diffie Hellman keys. All elements in the SeCOnE framework play out their parts as individuals from either gathering or get the issues and the correct answers from, and after that disperses the issues and gathers the appropriate response sheets from. A delegate screens the examinees through utilizing the screen information in. Through, an examinee having a place with and overseen by can take the online exam. The gathering specialists and make an arrangement of open and private keys for each gathering. They disseminate this arrangement of keys to their gathering individuals at every exam, and trade people in general keys with each other. General society key of each gathering is utilized for secure intergroup correspondences. For secure interchanges among assemble individuals, they utilize the symmetric keys made by the Diffie-Hellman key exchange. The examinees' PCs ought to be outfitted with Webcams and amplifiers. Top notch Webcams are promptly accessible now and are continually making strides. The SeCOnE framework programming is separated into two sections relying upon the part, that is, regardless of whether it is on the customer side, or server side. Bamboozling is any conduct that places the decency of the exam in question. The kinds of bamboozling methods secured in spite of the fact that not every single conning strategy might be secured; the plan proposed is an approach to maintain a strategic distance from essentially depending on the examinee's feeling of respect not to swindle. The SeCOnE framework receives five strategies to avoid and recognize tricking. To start with, the characters of substances in the framework are confirmed by a Webcam, and the reference photographs taken amid the confirmation procedure are put something aside for verification amid the exam. Second, the screen information for the examinees is recorded and spared amid exam. With a persistent account of video and sound amid the exam instead of secluded pictures, a delegate can better comprehend the examinee's circumstance and lessen the possibility of false positives or negatives in the assurance of conning, even after the exam. Third, through the screenshots spared in parallel with recordings of an examiner, a delegate can better figure out what the examiner is really doing with his or her PC. Fourth, all interchanges by the examinees, aside from those required for the online exam, are incapacitated through port control. All ports aside from those required for the online exam are debilitated and the ports utilized can be picked haphazardly for every examinee; the ports to be utilized have just to be sent to the exam regulatory gathering with the IP of the exam customer. Subsequently, tricking through a settled port can be uncommon (King, Guyette & Piotrowski, 2009). Fifth, every other program with the exception of the online exam customer is deactivated by controlling the contributions of the examinees. By cutting off electronic interchanges and impairing other PC projects or contributions on the examinees' PCs, the examinees can be restricted from conning utilizing their nearby PC or the Internet. The primary downsides in the framework are, age or in excess of one inquiry set, as there is no limitation on the season of taking the examination.

Fundamental Stream

This stream begins when a performer wishes to enlist a record: a. the framework requires that the client gives an email address, a secret key, and a telephone number, b. the client enters the data required, and c. the framework approves the entered email address, secret word, and telephone number and registers the client into the database.

Elective Stream

Invalid mail address/secret word/telephone number In the event that in the Basic Flow the client gives an invalid email address, secret key, or telephone number, a blunder message is shown and the client can choose whether to attempt again or go back to the underlying page. In the event that the data gave by the client is right, at that point, the client is enrolled and another page is shown and this page contains a connection to the login page.

Login of the Client

After characterizing the enrollment procedure for a client, we composed the login framework for our execution. Figure 5-1 demonstrates a utilization case chart that portrays how a client signs into the framework. At login, the client needs to give their email address and the secret key gave amid enlistment. The principal login subsequent to making a record is a SFA login where a right mix of username and secret key are adequate to access the record. We made the Two-factor authentication (2FA) alternative client empowered in light of the fact that we trust that DigiExam's obligation is to furnish the client with a 2FA alternative, however, it is up to the client to pick whether to utilize this choice or not. Thusly, clients can empower or impair 2FA for their record. At the point when 2FA is incapacitated, at that point, our login execution begins similarly as DigiExam's present usage, i.e. the entered username and secret word are checked in the database and if there is a match, at that point the client is allowed get to (Chen & Lee, 2014). After login, an appreciated page is shown where data around 2FA is shown and also the QR-code for OTP login. By means of this page, the client can proceed with SFA (the default choice) or empower 2FA login. For effortlessness, the client can't do whatever else on the appreciated page other than log out. Our framework handles fizzled login endeavors in the same was as DigiExam does, i.e. the client can neglect to login ten times amid a 3 minutes significant lot before the record is hindered for 3 minutes. On the off chance that 2FA is empowered, 'at that point the client needs to give the right blend of username and secret key and is then diverted to a page through which the code got by SMS or the code created on their cell phone can be entered and if approved access is allowed.

DISCUSSION

Clients' primary need is the convenience users don't care to type, in any event amid verification; users don't want to retain something hard despite the fact that it is more anchored. Pattern-based validation is more favored than PIN/secret word based confirmation because of its usability, notwithstanding any sorts of verification technique, the greater part of the appropriate responses gathered from clients was comfort concerned. Despite the fact that secret key got a most extreme rating for security, 45% of secret phrase clients believe that it is anything but difficult to recall and 27% don't think that it's difficult to type. 82% of secret phrase clients recognized the primary issue of utilizing secret word is that composing both letter set and numbers are hard amid validation. Along these lines, it very well may be said that the individuals who are utilizing a secret key for portable verification they don't think the secret key is badly arranged to utilize and they are exceedingly worried about security. For every single other strategy we dissected, it is observed that clients' primary reason for inclination is the comfort of utilization. Their inclination was predominantly for a quick confirmation component with least composing troubles and effortlessly of retaining mystery or no remembering by any means.

Clients' inclination is for the most part for a quick validation instrument with least or no composing challenges and effortlessly of remembering mystery or no retaining by any stretch of the imagination. Security comes after accommodation as a need to a large portion of the clients. Some clients are greater security concerned and they trade off the troubles they look amid validation to guarantee better security. Normally, cell phones are not utilized for a long constant period like work areas or PCs. Clients need access to their telephone for periodical occasions, generally ordinarily in multi-day. Each time clients utilize their gadget they have to confirm them to the gadget, notwithstanding when they keep it appended to them (e.g., in pocket). For this, the validation process ought to be quick and should offer the greatest conceivable ease of use for clients other than guaranteeing the wellbeing of their information and gadget situated highlights. Cell phone confirmation techniques are contemplated and talked about altogether to distinguish the majority of its classes, confirmation techniques that are being used for all intents and purposes, confirmation techniques that are not achievable for cell phones and the strategies that can be the potential for future advancement of the validation process. In the ongoing pattern of cell phone verification, unique finger impression, a biometric confirmation strategy has picked up clients' inclination for the most part and chiefly because of accommodation. In spite of the fact that unique mark does not guarantee the most grounded security, despite the fact that clients lean toward it in the wake of making the exchange off among security and convenience as indicated by their comprehension (broad piece of respondents favored, in light of the fact that it is a quick procedure). Once more, Fingerprint isn't accessible broadly in all cell phones because of its extra equipment cost.

CONCLUSION

Results demonstrate that all e-learning educators concur with the given proclamations. In any case, a few teachers communicated questions with respect to the compelling validation of students and the diminished likelihood of tricking amid examinations. Then, the understudies' study demonstrated a high level of unbiased cases, predominantly with respect to the advantages that the framework will give to understudies. Two inquiries were asked to both e-learning educators and understudies. These inquiries were whether the likelihood of deceiving will diminish and whether the framework will urge understudies to examine harder. The two gatherings trusted that swindling will be decreased with the execution of the facial acknowledgment framework and will motivate understudies to ponder harder. The facial acknowledgment framework ought to consider the picture catch rate of 30 seconds for nearness checks. This adjusted esteem is proper to reduce the heap on the framework and to ensure the level of unwavering quality. Also, the student with suspicious conduct ought to be cautioned twice. Different utilization of the facial acknowledgment framework incorporates observing the participation in virtual classrooms. In estimating the participation rate, if the consequence of the understudy's check is sure, the framework will enlist a participation stamp for him/her. Else, he/she will be checked truant. The proposed strategy, in light of the utilization of a perceptron with three concealed layers, offers the best outcomes as far as acknowledgment rate and false dismissal rate as appeared by Table 3 and Table 4. This can be clarified by the way that the proposed organize utilizes three concealed layers to play out the order, utilizing estimation of the probabilities on highlights vectors (which are invariant to interpretation and pivot). Truth be told, the utilization of the second arrangement. The point of this chapter was to give a unique finger impression acknowledgment calculation in view of fake neural systems for verification. The outcomes acquired with various strategies considered demonstrate that utilizing squares of handled picture empowers an acknowledgment with low false acknowledgment rate in the scope of ±0.5% by and large. For against, when the acknowledgment framework utilizes the mean classes of whole pictures, these false acknowledgment rates are high and reflect the subjectivity of this strategy.

Numerous fingerprints' acknowledgment frameworks for verification have just been set up in the writing; however the strategy we propose is very creative and considers both the estimation of probabilities and the correlation of layouts invariants to interpretation/revolution to perform characterization, an unexplored path till now as far as anyone is concerned. An uncommon exertion has been utilized to consider different singularities of unique finger impression lines for the instance of pictures from a similar person. Tests directed with this strategy have driven us to a few empowering comes about, albeit upgradeable. In fact, we constrained ourselves to perform tests with the greatest set at squared blunder between the normal yield and that got; on the grounds that past this esteem which is as of now high, the outcomes are to some degree one-sided. One little-specialized issue that we propose is that it doesn't deal with extremely well the instance of inert pictures because of the target that we set toward the begin which was to limit blunder rate specifically false dismissal rates. The way things are, the execution of various methodologies is viewed as acceptable for pictures of good characteristics. Anyway, further investigation on the decision and number of times to be connected pretreatment tasks on the unique mark pictures would have all the more intense calculations more qualified to the instance of dormant pictures or unmistakable half. Future work will center on the foundation of a confirmation framework that utilizations more fingerprints, other biometrics, for example, confront shape, the elements of the plot of the mark or potentially flow keystrokes. The blend of these methods will feature a principal component in the acknowledgment previously and amid utilization of the web-based preparing framework, in particular, the marking of the client. In reality, the utilization of marking as opposed to the format is one of a kind that it enables recording notwithstanding the unique mark includes, an arrangement of different components particular to the singular conduct of every client and it underpins verification previously and amid framework.

REFERENCES

Abdullah, S., Rahman, A., Abas, Z., & Saad, W. (2016). Multilayer Perceptron Neural Network in Classifying Gender using Fingerprint Global Level Features. *Indian Journal of Science and Technology*, *9*(9). doi:10.17485/ijst/2016/v9i9/84889

Ahmed, I. (2017). *Smartphone Authentication: User experience, expectation and satisfaction*. Lappeenranta University of Technology.

Azhar, F., & Anderson, W. (2012). Predicting Single-Neuron Activity in Locally Connected Networks. *Neural Computation*, *24*(10), 2655–2677. doi:10.1162/NECO_a_00343

Berglund, A. (2013). A framework to study learning in a complex learning environment. *Research in Learning Technology*, *12*(1). doi:10.3402/rlt.v12i1.21457

Biggio, B., fumera, Russu, P., Didaci, L., & Roli, F. (2015). Adversarial Biometric Recognition: A review on biometric system security from the adversarial machine-learning perspective. *IEEE Signal Processing Magazine*, *32*(5), 31–41. doi:10.1109/MSP.2015.2426728

Botha, R. A., Furnell, S. M., & Clarke, N. L. (2009). From desktop to mobile: Examining the security experience. *Computers & Security*, *28*(3–4), 130–137. doi:10.1016/j.cose.2008.11.001

Chen, B., Bastedo, K., & Howard, W. (2018). Exploring Best Practices for Online STEM Courses: Active Learning, Interaction & Assessment Design. *Online Learning*, *22*(2). doi:10.24059/olj.v22i2.1369

Chen, C., & Lee, C. (2014). A two-factor authentication scheme with anonymity for multi-server environments. *Security and Communication Networks*, *8*(8), 1608–1625. doi:10.1002ec.1109

Choong, Y.-Y., Franklin, J.M. & Greene, K.K., 2016. Usability and Security Considerations for Public Safety Mobile Authentication. *National Institute of Standards and Technology Interagency Report 8080*. Available at: . doi:10.6028/NIST.IR.8080

Cross, T., & Palese, K. (2015). Increasing Learning: Classroom Assessment Techniques in the Online Classroom. *American Journal of Distance Education*, *29*(2), 98–108. doi:10.1080/08923647.2015.1023594

Daniel, B. (2017). Enterprise Lecture Capture Technologies and Value to Student Learning. *International Journal of Information and Communication Technology Education*, *6*(2). doi:10.1515/ijicte-2017-0009

Gawande, U., Zaveri, M., & Kapur, A. (2011). Multi-algorithmic approaches to iris recognition. *Biometric Technology Today*, *2011*(4), 8-10. doi:10.10160969-4765(11)70075-3

Gottschlich, C., & Huckemann, S. (2014). Separating the real from the synthetic: Minutiae histograms as fingerprints of fingerprints. *IET Biometrics*, *3*(4), 291–301. doi:10.1049/iet-bmt.2013.0065

Gu, G., & Liu, F. (2011). Contourlet domain image encryption based on chaos mapping. *Jisuanji Yingyong*, *31*(3), 771–773. doi:10.3724/SP.J.1087.2011.00771

Handwritten English Character Recognition Using Logistic Regression and Neural Network. (2016). *International Journal Of Science And Research*, *5*(6), 750–754. doi:10.21275/v5i6.nov164228

Höynälänmaa, T. (2015). Multiresolution analysis for compactly supported interpolating tensor product wavelets. *International Journal of Wavelets, Multresolution, and Information Processing*, *13*(02), 1550010. doi:10.1142/S0219691315500101

Internet of Things Service. (n.d.). *ISPRS International Journal of Geo-Information*, *5*(7), 116. doi:10.3390/ijgi5070116

Jung, I., & Yeom, H. (2009). Enhanced Security for Online Exams Using Group Cryptography. *IEEE Transactions on Education*, *52*(3), 340–349. doi:10.1109/TE.2008.928909

Kashyap, R., & Gautam, P. (2017). Fast Medical Image Segmentation Using Energy-Based Method. *Biometrics, Concepts, Methodologies, Tools, and Applications*, *3*(1), 1017–1042. doi:10.4018/978-1-5225-0983-7.ch040

Kashyap, R., & Piersson, A. (2018). *Impact of Big Data on Security. In Handbook of Research on Network Forensics and Analysis Techniques* (pp. 283–299). IGI Global. doi:10.4018/978-1-5225-4100-4.ch015

Kashyap, R., & Tiwari, V. (2017). Energy-based active contour method for image segmentation. *International Journal of Electronic Healthcare*, *9*(2–3), 210–225. doi:10.1504/IJEH.2017.083165

Khairnar, C. (2015). Advance Pedagogy: Innovative Methods of Teaching and Learning. *International Journal of Information and Education Technology (IJIET)*, *5*(11), 869–872. doi:10.7763/IJIET.2015.V5.629

Kim, H. (2013). E-learning Ongoing Sesion Authentication Security. *Journal Of Security Engineering*, *10*(5), 601–608. doi:10.14257/jse.2013.10.5.08

King, C., Guyette, R., & Piotrowski, C. (2009). Online Exams and Cheating: An Empirical Analysis of Business Students' Views. *The Journal Of Educators Online*, *6*(1). doi:10.9743/JEO.2009.1.5

Kukula, E., Sutton, M., & Elliott, S. (2010). The Human–Biometric-Sensor Interaction Evaluation Method: Biometric Performance and Usability Measurements. *IEEE Transactions on Instrumentation and Measurement*, *59*(4), 784–791. doi:10.1109/TIM.2009.2037878

Liu, Y., Li, Y., Ma, X., & Song, R. (2017). Facial Expression Recognition with Fusion Features Extracted from Salient Facial Areas. *Sensors (Basel)*, *17*(4), 712. doi:10.339017040712

Menekse, M., Stump, G., Krause, S., & Chi, M. (2013). Differentiated Overt Learning Activities for Effective Instruction in Engineering Classrooms. *Journal Of Engineering Education*, *102*(3), 346–374. doi:10.1002/jee.20021

Menke, J. (2014). Implementation of Online Poster Sessions in Online and Face-to-Face Classrooms as a Unique Assessment Tool. *Journal of Chemical Education*, *91*(3), 414–416. doi:10.1021/ed400665n

Neves, J., Moreno, J., & Proença, H. (2017). *QUIS-CAMPI: an annotated multi-biometrics data feed from surveillance scenarios*. IET Biometrics. doi:10.1049/iet-bmt.2016.0178

Okada, A., Scott, P., & Mendonça, M. (2015). Effective web videoconferencing for proctoring online oral exams: A case study at scale in Brazil. *Open Praxis*, *7*(3). doi:10.5944/openpraxis.7.3.215

Omar Zughoul, O. (2013). Privacy and Security in Online Examination Systems. *IOSR Journal Of Computer Engineering*, *10*(4), 63–70. doi:10.9790/0661-01046370

Pointner, P. (2017). Hybrid tech: the future of biometric ID verification? *Biometric Technology Today*, *2017*(8), 8-10. doi:10.10160969-4765(17)30183-2

Sametinger, J., Schlöglhofer, R. & Sametinger, J. (2012). *Secure and usable authentication on mobile devices Secure and Usable Authentication on Mobile Devices*. Academic Press.

Savchenko, A. (2013). Probabilistic neural network with homogeneity testing in recognition of discrete patterns set. *Neural Networks*, *46*, 227–241. doi:10.1016/j.neunet.2013.06.003

Thiemann, T. (2016). Picking the right path to mobile biometric authentication. *Biometric Technology Today*, *2016*(2), 5-8. doi:10.10160969-4765(16)30034-0

Yang, W., Li, Y., & Liao, Q. (2014). Fast and Robust Personal Identification by Fusion of Finger Vein and Finger-Knuckle-Print Images. *Applied Mechanics and Materials*, *556-562*, 5085-5088. doi: 10.4028/www.scientific.net/amm.556-562.5085

Zhong, Y., & Peng, X. (2016). Multi-level fingerprint continuous classification for large-scale fingerprint database using fractal analysis. *International Journal of Biometrics*, *8*(2), 115. doi:10.1504/IJBM.2016.077829

KEY TERMS AND DEFINITIONS

Artificial Intelligence: Computerized thinking is understanding appeared by machines, as opposed to the trademark learning appeared by individuals and changed animals. In programming designing AI asks about is described as the examination of "sharp masters": any device that sees its condition and goes for broke exercises that open up its danger of viably achieving its objectives. Casually, the articulation "synthetic mental ability" is associated when a machine imitates "subjective" limits that individuals interface with other human identities, for instance, "learning" and "basic reasoning."

Artificial Neural Network: An artificial neural network (ANN) is information dealing with the perspective that is awakened by the way natural tactile frameworks, for instance, the cerebrum, process information. The key part of this perspective is the novel structure of the information taking care of the system. It is made out of a broad number of extraordinarily interconnected planning segments (neurons) filling in as one to deal with specific issues. ANNs, like people, learn by case. An ANN is intended for a specific application, for instance, outline affirmation or data gathering, through a learning technique. Learning in characteristic systems incorporates changes as per the synaptic affiliations that exist between the neurons.

EHR: An electronic health record (EHR) is a propelled adjustment of a patient's paper graph. EHRs are consistent, tolerant centered records that make information open instantly and securely to affirmed customers. While an EHR contains the remedial and treatment accounts of patients, an EHR structure is attempted to go past standard clinical data accumulated in a provider's office and can be thorough of a broader point of view of a patient's care. EHRs can: Contain a patient's helpful history, dissect, remedies, treatment outlines, inoculation dates, hypersensitivities, radiology pictures, and research focus and test results. Empower access to affirm based gadgets that providers can use to settle on decisions about a patient's care k process. One of the key features of an EHR is that prosperity information can be made and regulated by affirmed providers in an automated mastermind fit for being granted to various providers across finished more than one therapeutic administration's affiliation. EHRs are attempted to grant information to other human administrations providers and relationship, for instance, inquire about offices, experts, and helpful imaging workplaces, medicate stores, emergency workplaces, and school and workspace focus so they contain information from all clinicians drew in with a patient's care.

ICT: Information and communications technology (ICT) suggests all the advancement used to manage to communicate correspondences, convey media, savvy building organization structures, changing media taking care of and transmission systems, and framework-based control and watching limits. Notwithstanding the way that ICT is routinely seen as an extended identical word for information development (IT), its degree is more extensive. ICT has all the more starting late been used to depict the association of a couple of advances and the usage of fundamental transmission lines passing on to a great degree contrasting data and correspondence composes and outlines. Joining developments that embody ICT consolidate the meeting of changing media, telephone and PC sorts out through an ordinary cabling system. The system gets to providers conventionally give web, phone and TV organizations to homes and associations through a singular optical connection. The transfer of the telephone frameworks has given enormous fiscal spurring powers to realize this blending, which executes colossal quantities of the costs related to cabling, signal course, customer foundation, changing and upkeep costs.

Chapter 11
Biometric Authentication Techniques in Online Learning Environments

Jack Curran
Open University, UK

Kevin Curran
Ulster University, UK

ABSTRACT

The deployment of online e-learning can lead to many security risks, such as confidentiality loss, exposure of critical data, availability and destruction of publicly available information services. Security and proper authentication is critical in any online learning environment because any flaws can affect perceptions of its trustworthiness. Biometric authentication is increasingly being used in the newer generation of online learning environments for authentication of remote learners. Biometrics scan unique physiological characteristics in humans to identify people. These include fingerprints, iris, retina, voice, face, gait, and odor. The authors look at the state of biometric authentication techniques applicable to online learning environments and provide a more in-depth examination of face- and iris-based authentication systems for proper identification of learners.

BACKGROUND

There is becoming close to 100% adoption of e-learning environments in education institutions. Online learning environments are by their nature subject to all the attacks that online systems are vulnerable to (Adams & Blandford, 2003). Their 'low risk' nature on roll-out however leave many of the online learning environment content management more open than usual (Chen and He, 2013). Educational institutions tend to underestimate their attack surface and implications of being penetrated (Zuev, 2012). There is also the added weakness of Denial of service attacks which render e-learning environments unavailable to students. Commons risks however are unauthorized modification or deletion of educational materials, and the equally dangerous problem unique to remote e-learning models of identity theft, impersonation,

DOI: 10.4018/978-1-5225-7724-9.ch011

and weak authentication (Ayodele et al., 2011). The roll-out of online e-learning can lead to many security risks, such as loss of confidentiality, the exposure of critical data, availability and destruction of publicly available information services (Srivastava & Sinha, 2013). Security & proper authentication is critical as a means to retain user trust in any online learning environment because any risk can affect perceptions of its trustworthiness (Weippl & Ebner, 2008). Therefore, it is crucial to try to identify any underlying factors that lead to security issues in online learning and identify the limitations of security in place (Yao et al., 2011). The next step of course is to develop mitigations for any weaknesses uncovered.

Authentication is a key aspect for online learning (Raitman et al., 2005). There are several ways to achieve this. They can use knowledge-based authentication where users enter passwords or pins. They can use token-based authentication with key card, smart-phones or some security token or they can use biometric based authentication such as fingerprints, palm print, a retinal scan, or a face scan (Alotaibi & Argles, 2011; Garfinkel & Spafford, 1996). Among these authentication methods, user logins are the simplest means for providing identity and access services while retinal or face scans are the more difficult - but seen as the stronger (Song et al., 2013). In fact, biometric authentication seems to be increasingly used in the newer generation of online learning environments for authentication (Wang et al., 2013).

Humans possess many unique physiological body and shape characteristics that distinguish one from another. Some of these biometric characteristics in humans include fingerprints, iris, retina, voice, face, Palm prints and palm veins (Li & Kot, 2013). Some are more unique and secure than others. The uniqueness of a biometric feature that nearly every human possesses is determined by how many possible combinations can exist. Such a measure maximizes between-person random variations while at the same time minimizes within-person variability (Teoh et al., 2004). A biometric with many combinatorial possibilities means the possibility of two individuals in possession of identical patterns becomes increasingly less probable, therefore making it more secure as a biometric authenticator. There are two categories of biometrics. One deals with the physiological aspect, such as patterns, prints and physical features and the other is behavioural concentrating on aspects such as typing patterns, walking gaits, mannerisms, voice and computer usage patterns (Ratha & Zhang, 2010). An example of an insecure biometric authenticator would be the colour of people's eyes. If someone had brown eyes and they were enrolled on a database, everyone else with brown eyes who were to attempt to enrol themselves would be authenticated as being the first individual which was scanned originally (Pfleeger & Pfleeger, 2007). Therefore, this is insecure as anyone with brown eyes can be an imposter to the original individual who was enrolled. If, however there was a biometric feature with the probability of having the same pattern with another human being 1/1,000,000,000, then this would be considered secure as it is almost impossible to have two individuals with the same biometric pattern.

Biometric scans can be passive or contact. Passive scanning means no physical contact is required e.g. in iris scanning, where a user is told to stand a certain distance from the camera. Contact biometrics is when contact with a physical object is needed e.g. finger print scanning, when a user is asked to place their specified finger upon a scanning surface (Farooq et al., 2007). Compared with traditional password or token-based authentication, biometrics are considered more secure when executed correctly. Password and token-based methods can only authenticate a person possesses a token or knows a password. However, biometrics can check that a person is who they claim to be. Deployment of proper biometric solutions should significantly reduce identity thefts with great benefits for the economy by eliminating passwords from the equation in place of more reliable solutions. Trust is particularly important for financial institutions and merchants as well as consumers (identify theft, account blocking inconvenience) alike so we can expect the first to deploy biometrics in large scale will be those involved in mobile payments and other financial organisations.

There are of course many biometric solutions. None are a silver bullet and one size does not fit all. The accuracy of facial recognition systems varies greatly due to factors such as lighting, angle & camera sensitivity and more. Facial techniques can also be thrown off by facial characteristics or a person simply wearing glasses will be different with sunglasses, no glasses and colour of the ambient light. Fingerprint readers likewise are affected by temperature and other factors. Fingerprint scanners in phones are deployed en mass now due to Apples Touch ID system. They have of course been on laptops for years but hardly used. The Touch ID system from Apple is quite impressive from a security perspective. Fingerprint scanners are not the solution however as we simply leave fingerprints on every surface we touch. There have been many examples of Apples' Touch ID been bypassed using scanners, latex and patience. This chapter provides an overview of popular biometric authentication approaches and provides a more in-depth examination of face and iris-based authentication systems

AUTHENTICATION

The alphanumeric password is revealing itself as unfit for purpose. As the number of devices (approaching 20 billion) and services which need access control increases, this legacy standard has reached a tipping point when measured by the vast number of identity thefts and account hijacks impacting people in recent times. There are more secure, less costly and more adoptable alternatives to the password/PINs approach. It is estimated that online "direct" fraud is costing the global economy around £60bn a year. The associated indirect costs of identity theft and recovery have not been fully quantified but is possibly many times the actual direct costs (Rane, 2014). Large e-commerce merchants believe that fraud is inevitable but understand that their prevention efforts will result in more positive customer relationships. The current most widespread authentication approach employs passwords. Passwords are weak and susceptible to many types of attacks as the security is dependent on a users' ability to keep the password secret from eavesdropping in its many forms. If we were to invent it today, we might find that it does not pass quality control for making financial transactions remotely, such as fund transfers and other payments through an Internet banking channel. There is also the question of the cost of support associated with user ID and password complexity as IT support staff often need to spend extra time dealing with authentication problems, such as helping staff reset passwords that are locked after a certain number of failed entry attempts (Dunker, 2003).

There are other clever biometric systems on the market using contextual information (location etc) in clever ways (Dunker, 2003). For instance, many are not yet familiar with keystroke dynamics. Keystroke dynamics is where keystroke logging can be analyzed. The time to get to and depress a key (seek-time), and the time the key is held-down (hold-time) can be very specific to a person, regardless of how fast they are going overall. Most people have specific letters that take them longer to find or get to than their average seek-time over all letters, but which letters those are may vary dramatically but consistently for different people. Right-handed people may be statistically faster in getting to keys they hit with their right-hand fingers than they are with their left-hand fingers. Index fingers may be characteristically faster than other fingers to a degree that is consistent for a person day-to-day regardless of their overall speed that day. Normally, all that is retained when logging a typing session is the sequence of characters corresponding to the order in which keys were pressed and timing information is discarded. Keystroke dynamic information which is normally discarded, can be used to verify or even try to determine the identity of the person who is producing those keystrokes. There are several home software and commercial software products which claim to use keystroke dynamics to authenticate a user such as BioTracker, ID Control, TypeWATCH, Authenware, Probayes and KeyTrac.

There is also voice as a biometric technique. Voice however has to be measured against both the ambient background e.g. a restaurant, a train, corner shop street or sports arena (Masala et al., 2017). Again, fingerprints taken when the finger is flat will be different when misaligned, wet, dirty or practically frozen. There really has not been much movement in trying to implement voice authentication. It does play a part in some multi factor systems. The main barrier to any widespread adoption has been the problem of aural eavesdropping. Quite simply, casual or malicious bystanders may overhear private information spoken by screen readers or users (Rane, 2014). There are some niche areas where it is adopted. Take for instance the special needs of disabled citizens. In the context of disability, the process of authentication is stressful for many users looking to access devices or services. For example, individuals who are blind have difficulties with processes of authentication such as Captcha which cut them off from bank accounts and all other online access points because they must visualise and input meaningless character sequences. The spoken option is not good. Voice authentication can play an important role here (Patel et al., 2015).

These sources of potential error create two measuring levels that biometrics algorithms build in to their calculations which are false acceptance and false rejection. If this is not managed and measured properly, it can lead to a bad user experience which has been a problem with commercialisation of such technologies in the past decade as they seek to achieve the elusive 100% accuracy rate. The objective of biometric identity authentication is to establish a bond of trust between an organization and the user who is requesting system access. More specifically, identity authentication ascertains a level of trust regarding who the user claims to be (Patel et al., 2015). It follows that the more accurate any chosen authentication method the user can present to prove his/her identity, then the stronger this bond of trust becomes. It is feasible that biometric authentication becomes the de facto form of providing credentials (although it should be combined with multi-factor methods). Hardware devices do potentially offer ideal security but often the problem is the need to carry such a device on the person (Rane, 2014). Hence the move towards making our mobile phones the de facto hardware device. One popular hardware approach for authentication is smart cards. Smart cards technology provides an excellent medium for storing biometrics. A smart card can provide a strong authentication platform in our pocket. Mobile phones and smart cards can be used for both physical and logical access authentication (Sutcu et al., 2007).

There are problems in general with hardware security tokens. Firstly, they involve additional costs, such as the cost of the token and any replacement fees. Users always need to carry the token with them. Users need multiple tokens for multiple Web sites and devices (Daugman, 2006). Finally, they do not protect fully from man-in-the-middle attacks (i.e., attacks where an intruder intercepts a user's session and steals the user's credentials by acting as a proxy between the user and the authentication device without the user's knowledge). Basically, if you lose the token, you lose control. A related area is Digital Certs on USBs or Smart Cards. Here however are problems as well as they require users to carry an additional smart card reading device or USB. They also involve additional cost, such as certifying authority's subscription cost. You need multiple certificates for different sites or devices. It can be difficult for non-authorized users to extract the private key when stored on a smart card and they require user training for certificate generation and use (Merhav, 2017).

One technique which should be combined with any biometric authentication is multi-factor authentication. Multifactor authentication reduces authentication risk by involving separate types of factors that would require an attacker to use different methods of attack, thus making a breach more difficult. Multi-factor authentication combines at least two of the following methods to strongly authenticate a user. These are something you know (typically a password/PIN), something you have (a trusted device identifier that is not easily duplicated) and something you are (your unique biometrics). In order to qualify as multi-factor authentication, two items must be combined from different categories, therefore a PIN plus a password is not actually multi-factor, since both of these are something you know. Full three factor authentication when combined with a device ID allows you to easily combine "what we have" and "what we know" with the important "who we are". This is important for future security systems.

What however is now considered best practice is to use tools like tools like Google Authenticator or an RSA token which can also prove possession. These do not involve a communication which can be as easily eavesdropped upon nor a sim-card that can be replaced. In fairness however, the 2-factor attacks just described do involve targeting an individual and some high level of skill in addition to hardware. The problem is that attacks never get worse and there is always someone out there making the hardware cheaper and the software easier to use so we can expect to see lesser skilled attackers exploiting weaknesses in multi-factor authentication - which for a brief period did what it said on the box.

Biometric solutions may provide us with the solution we need to more fully secure our online learning environments. They are playing an increasingly significant role in new online learning environments (Ayodele et al., 2011; Chen and He, 2013). Next, we examine face and iris-based authentication systems.

BIOMETRIC AUTHENTICATION SYSTEMS

The process of face scanning pertains to the ability of measuring the distance between key features and points of the face known as references. Key reference features on the face include distance between irises, length of eye brows, length of philtrum, width of nose, height and shape of eye lids and distance from eyes to mouth (Li & Kot, 2013). When scanning a detainee's face, all references on the face are combined into a biometric facial template which is enrolled in a database. This template is compared to all pre-existing facial templates and if a pre-defined threshold of measurement values is matched with a template stored on the database, then the detainee is identified as being the same person. Occasionally there may be false matches, meaning it is not the same person even though their measurements exceeded the threshold.

Several hundred million persons around the world have been enrolled in iris recognition systems for purposes such as passport-free automated border-crossings and national ID programs (BBC, 2012). A key advantage of iris recognition, in addition to its speed of matching and extreme resistance to false matches, is the stability of the iris as an internal and protected, yet externally visible organ of the eye. The process of iris scanning involves measuring the various minutiae of the iris. There are 250 points within the iris, of which can take on one shape from a plethora of varying types of shapes creating an almost unique iris pattern due to the sheer amount of combinatorial possibilities. When all minutiae of the iris are scanned and recognised, it forms a biometric template. This is stored for checking against future scans to look for a match (Daugman, n.d).

Convenience

Facial scanning may be inconvenient to detainees as they will have to remove any facial obstructions in order to scan the face such as glasses, contact lenses, hats, makeup, fake eyelashes or fake tan. New developments in facial recognition technology allow for 3D capture of a subject's face. It enables greater precision and allows for a much wider range of angles as opposed to flat RGB. Additionally, new research has seen the introduction of technology that allows for subjects to have varying facial expressions and to still be recognised as the same person previously captured in a different expression (Chen et al., 2014). Facial scanning is convenient as detainees simply have to be facing in the general direction of the scanner. This also allows increased accessibility to vision impaired individuals who may not be able to precisely orientate their face to look in a specific direction.

Iris scanning requires users to face the scanner. Their face must retain a neutral expression, otherwise the iris could be obstructed by the eyelids. Iris recognition works with clear contact lenses, eyeglasses, and non-mirrored sunglasses. Deviations in position or rotation of a subject's face are also not allowed as they can result in a failed scanned (Daugman, n.d).

Reliability

A major point to consider when comparing iris vs facial biometrics is identity theft. It is possible to have a facial biometric stolen and used against a biometric facial scanning system to trick it. An image or scan of a victim's face can be stolen from afar without the victim ever knowing. A scanner can however use thermal imaging to detect whether the facial substitution is living or not so as to render paper-based images useless (Cai et al., 2018). If a migrant were to have altered their facial appearance, a facial recognition machine may not be able to recognise them. This can be due to many reasons such as facial hair growth or removal, hairstyle change, facial injury or face altering surgery. This could mean needing to rescan detainees.

Iris scanning does not have this problem as the iris, unless permanently damaged will remain the same since the initial scan (Bleicher,2005). This can be as long as decades. Facial scanning is not as effective as other biometrics in identifying minorities when most of the subjects used in training the technology were from the majority group (Bonsor & Johnson, 2016). Iris scanning in such cases would fare better as there are a significantly lower amount of non-iris scannable minorities.

Acceptability

There may be use cases where individuals may not want to have their biometric identity stored digitally on a database. They may desire to retain their privacy. If data from a biometric database is leaked, that data may be used in identity theft (Mordini & Tzovaras, 2012). This is a key issue with regards biometrics due to the irrevocable link between biometric traits and a persistent information record concerning an individual. Unlike most other forms of recognition, biometric techniques are firmly tied to a physical body and this link between personal records and biometrics can have negative consequences for individuals (Pato & Millett, 2010). The problem of course with losing biometric data belonging to people is that they cannot get it back or change it so if you lose the data for someone's fingerprints - it is out there, and they have no way of course to change their fingerprint pattern.

An advantage of iris scanning is that it is akin to taking a photograph and can be done from 10 cm to a few meters distance. The person being identified does not have to touch any equipment that has recently been touched by a stranger. This is important for acceptability as it eliminates an objection that has been raised in some cultures against fingerprint scanners, where a finger needs to touch a surface, or retinal scanning, where the eye must be brought very close to an eyepiece (Martin, 2011).

FACE AND IRIS-BASED AUTHENTICATION CAPTURE

Iris

Iris scanning utilises infrared lights for scanning as well as the addition of digital camera technology to capture a complex and detailed iris (Barra et al., 2015). When an iris has light shone upon it, the eye will compensate for the exposure and dilate. However, inanimate objects will not be able to perform this function, or at the realism level needed to be considered an actual iris by the scanner Therefore, if someone attempts bypassing the iris scanner, they may be able to get past the first stage of recognising the iris pattern, but they will most likely fail upon the second test for life in the iris. Initially the scanner must localize the inner and outer boundaries of the iris. Subsequent subroutines detect and exclude eyelids, eyelashes, and any specular reflections that often occlude parts of the iris. The pixel set containing the iris is normalized by a rubber-sheet model to compensate for pupil dilation or constriction. This is next analyzed to extract a bit pattern encoding the information needed to compare two iris images (Daugman, 2006).

Face

Capturing an individual's face for biometric facial recognition is not as complex as iris scanning. Facial scanning can even be performed from simple 2D photographs. In most high end facial scanning systems, various hardware exists to ensure the security of the scan and to avoid circumvention using substitute fake faces (Kisku, 2016). Some items that may exist in a biometric facial scanner are a 2D camera system, infrared dot projector and thermal imaging camera. During a facial scan, various checks and processes will occur such as extracting colour information and texture detail. A dot projector will also be used to capture 3D facial features to ensure the subject is not a flat high resolution 2D image. A thermal heat map can also be used to detect if the object is living.

FACE AND IRIS-BASED AUTHENTICATION PROCESSING

Iris

Once the user's facial biometric template has been captured, an iris template is generated. This template encodes the pattern of the iris which can later be compared against all pre-existing iris template stored within the database to authenticate or identify individuals (Sutcu et al., 2007). Unlike current facial processing techniques, iris processing is remarkably fast, testing has revealed iris match comparisons on large databases are completed with millions per second on only 1 CPU core (Patel et al., 2015). Accuracy and false match rates are not sacrificed either. False match rates have been reported to be very low. Thus, with the speed of iris template processing, no other methods are required for speeding up the process (Rane, 2014).

Face

Once the user's facial biometric template has been captured, it must be compared against pre-existing templates to identify the individual. Some databases contain millions of unique facial templates (Weaver, 2006). This can take a long time therefore leading to slow authentication of users (Ratha et al., 2001). A solution involves categorising parts of each user's template face at the time of creation. This effectively divides templates into categories based on their facial features and what category each part slots into such as eye colour, ear shape, skin colour, eye shape, lip shape, gender, nose shape, face height and jaw line. Each of these individual categories when viewed individually have few user templates however, as a collective there could be millions of templates (Open University, 2018). Eventually there will be a point reached where the number of templates left to be processed based on the final category is satisfactorily low to individually processing of the remaining templates that fit within the category. This division method speeds processing time up significantly (Merhav, 2018).

FUTURE RESEARCH DIRECTIONS

Biometrics will play a role in authentication in many online learning environments in the future as biometrics-based authentication provides a real-world alternative to using passwords and pins. Biometrics can play a role in most areas where authentication of a service is necessary. It works in validating the identity of users by measuring unique physiological and behavioural characteristics of individuals. Such a measure maximizes between-person random variations while at the same time minimizes within-person variability. In contrast with passwords and pins, a biometric identifier cannot be lost, forgotten or shared. One can choose from a large list that includes finger, face, retinal scan, iris, gait, vein infrared thermogram, hand geometry and palm print or from a combination of all these identifiers termed multimodal-biometrics.

There are also moves to utilise novel biometric systems such as the Tongue. The tongue is a unique vital organ which is pretty well protected inside the mouth. Hence it is affected by external factors. The Dorsum of the tongue shows a lot of information along with its visual differences in shape, texture and pattern which is known as the tongue print. The tongue therefore exhibits rich textural patterns. (Trivandrum et al., 2018) demonstrate a Local Binary Pattern (LBP) algorithm for extracting features which are then trained by a linear Support Vector Machine (SVM) for personal identification. From a database of 136 tongue print images of 34 individuals, they achieved an accuracy of 97.05% for identification. This was one of the earliest studies where texture patterns are extracted from tongue images for biometric authentication. We can expect to see more work in this interesting area.

The biggest change in future learners using online learning environments might be the rise of the mobile device as the device of choice for biometric reading. We now see that mobile devices are increasingly the mainstay of a person's online activities. Most of the market dominating smartphones now have biometric readers or sensors already incorporated into the hardware. Biometrics will also have some impact on the home with regards the need for additional hardware costs such as scanners. There are also extra costs needed for deployment, support and maintenance. It may also not be suitable for mass-consumer deployment just yet. The main place that biometrics play a role in the home is in door entry systems. Of late there have been a few breakthroughs in biometric technology that can help make these systems even more secure. Japan's Fujitsu has miniaturised iris recognition technology which takes barely a second, meaning the home can be unlocked with just a look, and the camera can be placed at normal viewing distance, not the 10cm required by most iris scanners.

While codes and fingerprints work well, this technology means gloves can be kept on in cold weather and firing up the phone with one hand is that little bit easier. The phone itself will be able to register its owner's iris pattern for the first time, a process which takes 10 seconds, using an LED light that shines into the eye and a small infrared camera. Fujitsu hopes to persuade manufacturers, software and app developers to incorporate its invention for commercial release this year. Qualcomm have a technology called sense ID. It is a '3D' fingerprint scanner technology. It not only scans the print but also scans the depth of the ridges. You cannot therefore use a photo to bypass this authentication mechanism. This could be better than touch id. It can also scan through glass, sapphire, plastic.

Specifically, biometrics can be used by online learning environment system admins to grant users only select features in a system. Voice authentication has been trialled in some online systems case studies. However, there really has not been much movement in trying to implement voice authentication. It does play a part in some multi factor systems. The main barrier to any widespread adoption has been the problem of aural eavesdropping. Quite simply, casual or malicious bystanders may overhear private information spoken by screen readers or users. Others involved in the home automation voice command systems include GEO Semiconductor. They recently launched Granta smart home automation technology which contains voice and vision control capabilities designed for user interfaces for home security and entertainment control devices (Cericola, 2013). It was demonstrated at the 2015 Voice Biometrics Conference in San Francisco. Granta is primarily targeted at home security, smart TC and entertainment markets.

CONCLUSION

We provided here an overview of security in online learning environments which are by their nature subject to many of the attacks that standard online systems are vulnerable to also (Adams & Blandford, 2003). Their 'low risk' nature on roll-out however leave many of the online learning environment content management more open than usual (Chen and He, 2013). Educational institutions tend to underestimate their attack surface and implications of being penetrated (Zuev, 2012). There is also the added weakness of Denial of service attacks which render e-learning environments unavailable to students. Commons risks however are unauthorized modification or deletion of educational materials, and the equally dangerous problem unique to remote e-learning models of identity theft, impersonation, and weak authentication (Ayodele et al., 2011). Biometric authentication is increasingly being used in the newer generation of online learning environments for authentication.

We also looked at biometrics based authentication for online learning environments and in particular Face and iris-based authentication capture & processing procedures. We highlighted the actual scientific methods by which an Iris and a Face are scanned to capture the unique physiological details which comprise each. The discussion on the processing firstly focussed on the Iris technique where a biometric template is generated and the accuracy & speed of iris scanners. In discussing facial processing, we highlighted the need to overcome slow authentication of users in large databases through the categorisation of features such as eye colour, ear shape, skin colour, eye shape, lip shape, gender, nose shape to speeds up processing time significantly. Some religions that choose to wear a burka may not wish to unveil their face in public to a facial scanner. However, iris scanning enables these people to be biometrically identified as their eyes are still visible through a burka.

Facial injuries might affect the ability to facially recognise an individual as their face may have been altered. If the individual's iris has not been damaged, a facial injury will not be an issue. Iris recognition is also more secure. The commercially deployed iris-recognition algorithm '*IrisCode*', has an unprecedented false match rate better than 10^{-11}. The Hamming distance threshold of 0.26 used, means that up to 26% of the bits in two IrisCodes can disagree due to imaging noise (Daugman, 2006). While there are some medical and surgical procedures that can affect the colour and overall shape of the iris, the fine texture remains remarkably stable over many decades. Some iris identifications have succeeded over a period of about 30 years. Identical twins can also be uniquely identified due to different iris patterns. Identical twins are problematic for facial scanners. Iris recognition is significantly more difficult to circumvent due to the complexity within the iris. Iris recognition removes the need for physical contact with the biometric recording device, but it does require a clear line of sight of the detainee's iris (Dunker, 2003). It must be noted however that like other photographic biometric technologies, iris recognition is susceptible to poor image resolution, with related failure to enrol rates (BBC, 2012).

Biometrics based authentication do provide a robust alternative to using passwords and pins. It validates identity of users by measuring unique physiological and behavioural characteristics of individuals. Such a measure maximizes between-person random variations while at the same time minimizes within-person variability. In contrast with passwords and pins, a biometric identifier cannot be lost, forgotten or shared. One can choose from a large list that includes finger, face, retinal scan, iris, gait, vein infrared thermogram, hand geometry and palm print or from a combination of all these identifiers termed multimodal-biometrics. The biggest change in the future might be the rise of the mobile device as the device of choice for biometric reading. We now see that mobile devices are increasingly the mainstay of a person's online activities. Most of the market dominating smartphones now have biometric readers or sensors already incorporated into the hardware. Biometrics will also have some impact on the workplace with regards the need for additional hardware costs such as scanners. There are also extra costs needed for deployment, support and maintenance. It may also not be suitable for mass-consumer deployment.

While biometrics are promising, the true 'strong' solution might simply be multifactor authentication. Multifactor authentication reduces authentication risk by involving separate types of factors that would require an attacker to use different methods of attack, thus making a breach more difficult. Multi-factor authentication combines at least two of the following methods to strongly authenticate a user (1) Something you know (typically a password/PIN) (2) Something you have (a trusted device identifier that is not easily duplicated) and (3) Something you are (your unique biometrics). To qualify as multi-factor authentication, two items must be combined from different categories, therefore a PIN plus a password is not actually multi-factor, since both items are something you know. Full three factor authentication when combined with a device ID allows enterprises to easily combine "what we have" and "what we know" with the all-important "who we are", thereby integrating a core benefit to future home automation security systems. Security & proper authentication is critical as a means to retain user trust in any online learning environment because any risk can affect perceptions of its trustworthiness and biometrics properly implemented may be a good fit.

REFERENCES

Adams, A., & Blandford, A. (2003). Security and online learning: To protect or prohibit. *Usability Evaluation of Online Learning Programs*, 331-359.

Alotaibi, S. J., & Argles, D. (2011). FingerID: A new security model based on fingerprint recognition for personal learning environments (PLEs). In *Global Engineering Education Conference (EDUCON)* (pp. 142-151). IEEE 10.1109/EDUCON.2011.5773128

Ayodele, T., Shoniregun, C. A., & Akmayeva, G. (2011). Towards e-learning security: A machine learning approach. In *Information Society (i-Society), 2011 International Conference* (pp. 490-492). IEEE.

Barra, S., Casanova, A., Narducci, F., & Ricciardi, S. (2015). Ubiquitous iris recognition by means of mobile devices. *Pattern Recognition Letters, 57*, 66-73. doi:10.1016/j.patrec.2014.10.011

BBC. (2012). Black Hat: Iris scanners 'can be tricked' by hackers. *BBC Technology*. Available at: http://www.bbc.co.uk/news/technology-18997580

Bleicher, P. (2005). Biometrics comes of age: despite accuracy and security concerns, biometrics are gaining in popularity. *Applied Clinical Trials*. Available at: http://www.appliedclinicaltrialsonline.com/biometrics-comes-age

Bonsor, K., & Johnson, R. (2016). How Facial Recognition Systems Work. *HowStuffWorks*. Available at: https://electronics.howstuffworks.com/gadgets/high-tech-gadgets/facial-recognition.htm

Cai, Y., Jiang, H., Chen, D., & Huang, M. (2018). Online learning classifier based behavioral biometric authentication. *2018 IEEE 15th International Conference on Wearable and Implantable Body Sensor Networks (BSN)*, 62-65. doi: 10.1109/BSN.2018.8329659

Cericola, R. (2013). Granta Will Bring Voice Control to Home Automation. *Electronic House*. Available at: https://www.electronichouse.com/smart-home/geo-semiconductor-speechfx-and-voicevault-partner-up-for-granta

Chen, S., Pande, A., & Mohapatra, P. (2014). Sensor-assisted facial recognition: an enhanced biometric authentication system for smartphones. *Proceedings of the 12th annual international conference on Mobile systems, applications, and services (MobiSys '14)*, 109-122. 10.1145/2594368.2594373

Chen, Y., & He, W. (2013). Security Risks and Protection in Online Learning: A Survey. *The International Review of Research in Open and Distance Learning*. Available at: http://www.irrodl.org/index.php/irrodl/article/view/1632/2712

Daugman, J. (2006). Probing the uniqueness and randomness of IrisCodes: Results from 200 billion iris pair comparisons. *Proceedings of the IEEE, 94*(11), 1927–1935. doi:10.1109/JPROC.2006.884092

Daugman, J. (n.d.). Iris recognition. *International Center for Disability Resources on the Internet*. Available at http://www.icdri.org/ biometrics/ iris_biometrics.htm

Davida, G., Frankel, Y., & Matt, B. (1998). On enabling secure applications through off-line biometric identification. *IEEE Symp. Security and Privacy*, 148-157. 10.1109/SECPRI.1998.674831

Dunker, M. (2003). Don't Blink: Iris Recognition for Biometric Identification. *SANS Institute InfoSec Reading Room*. Available at http://www.sans.org/ reading_room/ whitepapers/ authentication/dont-blink-iris-recognition-biometric-identification_1341

Farooq, F., Bolle, R., Jea, T., & Ratha, N. (2007). Anonymous and revocable fingerprint recognition. *Proc. IEEE Conf. Computer Vision and Pattern Recognition*, 1-7. DOI: 10.1109/CVPR.2007.383382

Garfinkel, S., & Spafford, G. (1996). Practical Unix and Internet security. In *INET 96*. O'Reilly & Associates, Inc.

Kisku, D., Gupta, P., & Sing, J. K. (2016). *Advances in Biometrics for Secure Human Authentication and Recognition*. London: CRC Press.

Li, S., & Kot, A. (2013). Fingerprint combination for privacy protection. *IEEE Transactions on Information Forensics and Security*, *8*(2), 350–360. doi:10.1109/TIFS.2012.2234740

Martin, Z. (2011). Biometric Trends: Will emerging modalities and mobile applications bring mass adoption? *SecureIDNews*. Available at: https://www.secureidnews.com/news-item/biometric-trends-will-emerging-modalities-and-mobile-applications-bring-mass-adoption/?tag=biometrics&tag=Law_Enforcement

Masala, G., Ruiu, P., & Grosso, E. (2017). *Biometric Authentication and Data Security in Cloud Computing*. Computer and Network Security Essentials.

Merhav, N. (2017). The generalized stochastic likelihood decoder: Random coding and expurgated bounds. *IEEE Transactions on Information Theory*, *63*(8), 5039–5051. doi:10.1109/TIT.2017.2689787

Merhav, N. (2018). Ensemble Performance of Biometric Authentication Systems Based on Secret Key Generation. *2018 IEEE International Symposium on Information Theory (ISIT)*, 1455-1459. 10.1109/ISIT.2018.8437768

Mordini, E., & Tzovaras, D. (2012). *Second Generation Biometrics: the Ethical and Social Context*. Springer-Verlag.

Open University. (2018). *T215 Communication and information technologies Module Handbook*. Open University; doi:10.1007/978-94-007-3892-8

Patel, V., Ratha, N., & Chellappa, R. (2015). Cancelable Biometrics: A review. *IEEE Signal Processing Magazine*, *32*(5), 54–65. doi:10.1109/MSP.2015.2434151

Pato, J., & Millett, L. (2010). *Biometric Recognition: Challenges and opportunities*. National Academies Press.

Pfleeger, C., & Pfleeger, S. (2007). *Security in Computing* (4th ed.). Boston: Pearson Education.

Raitman, R., Ngo, L., Augar, N., & Zhou, W. (2005). Security in the online e-learning environment. In *Advanced Learning Technologies, 2005. ICALT 2005. FIFTH IEEE International Conference* (pp. 702-706). IEEE.

Rane, S. (2014). Standardization of biometric template protection. *IEEE MultiMedia, 21*(4), 94–99. doi:10.1109/MMUL.2014.65

Ratha, N., Connel, J., & Bolle, R. (2001). Enhancing security and privacy in biometrics-based authentication systems. *IBM Systems Journal, 40*(3), 614–634. doi:10.1147j.403.0614

Ratha, N., & Zhang, D. (2010). Privacy protection in high security biometrics applications. In A. Kumar & D. Zhang (Eds.), *Ethics and Policy of Biometrics* (pp. 62–69). Springer. doi:10.1007/978-3-642-12595-9_9

Song, K., Lee, S. M., & Nam, S. C. (2013). Combined biometrics for e-learning security. ISA 2-13. *ASTL, 21*, 247–251.

Srivastava, A., & Sinha, S. (2013). Information security through e-learning using VTE. *International Journal of Electronics and Computer Science Engineering, 2*(18), 528–531.

Sutcu, Y., Li, Q., & Memon, N. (2007). Protecting biometric templates with sketch: Theory and practice. *IEEE Transactions on Information Forensics and Security, 2*(3), 503–512. doi:10.1109/TIFS.2007.902022

Teoh, A., Ling, D., & Goh, A. (2004). Biohashing: Two factor authentication featuring fingerprint data and tokenised random number. *Pattern Recognition, 37*(11), 2245–2255. doi:10.1016/j.patcog.2004.04.011

Trivandrum, S., Shali, N., Zacharias, G., & Anna, P. (2018). Identification of tongue print images for forensic science and biometric authentication. *Journal of Intelligent & Fuzzy Systems, 34*(3), 1421–1426. doi:10.3233/JIFS-169437

Wang, F., Ge, B., Zhang, L., Chen, Y., Xin, Y., & Li, X. (2013). A system framework of security management in enterprise systems. *Systems Research and Behavioral Science, 30*(3), 287–299. doi:10.1002res.2184

Weaver, A. C. (2006). Biometric Authentication. *Computer, 39*(2), 96–97. doi:10.1109/MC.2006.47

Weippl, E., & Ebner, M. (2008). Security privacy challenges in e-learning 2.0. In *World Conference on E-Learning in Corporate, Government, Healthcare, and Higher Education* (Vol. 2008, No. 1, pp. 4001-4007). Academic Press.

Yao, H., & Ji, Y. (2011). Security protection for online learning of music. In *Computer Communication and Networks (ICCCN), 2011 Proceedings of 20th International Conference* (pp. 1-4). IEEE. 10.1109/ICCCN.2011.6005740

Zuev, V. (2012). E-learning security models. *Management, 7*(2), 24–28.

Compilation of References

A multi-matcher model for ear recognition. (n.d.). Available from: https://www.researchgate.net/publication/286142859_A_multi-matcher_model_for_ear_recognition

AbdelRahman, S. M., & Sabir, A. A. (2017). Strategies and Contributions of Biometrics in E-Learning. *International Journal of Computer and Information Technology, 6*(2).

Abdullah, S., Rahman, A., Abas, Z., & Saad, W. (2016). Multilayer Perceptron Neural Network in Classifying Gender using Fingerprint Global Level Features. *Indian Journal of Science and Technology*, *9*(9). doi:10.17485/ijst/2016/v9i9/84889

Abikoye & Sanni. (2017, November). Keystroke Dynamics Authentication for a Web-Based Sales and Stock Solution. *The International Journal of Computing & ICT Research.*

Abinaya, R., & Sigappi, A. (2018). Biometric Identification of a Genuine User/Imposter from Keystroke Dynamics Dataset. *International Journal of Chemtech Research*, *11*(08), 147–160. doi:10.20902/IJCTR.2018.110816

Abualgasim, S. D., & Osman, I. (2011). An Application of the Keystroke Dynamics Biometric for Securing PINs and Passwords. *World of Computer Science and Information Technology Journal*, *1*(9), 398–404.

Abuhaiba, I. S. I. (2007). Offline Signature Verification Using Graph Matching. *Turk J Elec Engin*, *15*(1), 89–104.

Adams, A., & Blandford, A. (2003). Security and online learning: To protect or prohibit. *Usability Evaluation of Online Learning Programs*, 331-359.

Adams, C., & Armstrong, A. A. (1998). A Web-based testing: A study in interesting. *World Wide Web (Bussum)*, *1*(4), 193–208. doi:10.1023/A:1019224017027

Adetoba, B. T., Awodele, O., & Kuyoro, S. O. (2016). E-learning security issues and challenges: A review. *Journal of Scientific Research and Studies*, *3*(5), 96–100.

Agulla, E. G., Rifón, L. A., Alba Castro, J. L., & Mateo, C. G. (2008). Is My Student At The Other Side? Applying Biometric Web Authentication To Elearning Environments. *Proceedings of the 8th IEEE International Conference On Advanced Learning Technologies, ICALT 2008.*

Agulla, E. G., Rifón, L. A., Castro, J. L. A., & Mateo, C. G. (2008, July). Is my student at the other side? Applying biometric web authentication to E-learning environments. In *Advanced Learning Technologies, 2008. ICALT'08. Eighth IEEE International Conference on* (pp. 551-553). IEEE.

Ahmed Awad, E. Ahmed & Issa Traore. (2007). Anomaly Intrusion Detection Based on Dynamic. *Proceedings of the 2005 IEEE Workshop on Information Assurance United States Military Academy.*

Ahmed, A. A., & Traore, I. (2014). *Biometric Recognition Based on Free-Text Keystroke Dynamics*. Academic Press.

Ahmed, A. A., & Traore, I. (2014). Biometric Recognition Based On Free-Text Keystroke Dynamics. *IEEE Transactions on Cybernetics*, *44*(4), 458–472. doi:10.1109/TCYB.2013.2257745 PMID:23757560

Ahmed, A., & Traore, I. (2005). *Detecting Computer Intrusions Using Behavioral Biometrics. Technical Report*. University of Victoria.

Ahmed, A., & Traore, I. (2007). A new biometric technology based on mouse dynamics. *IEEE Transactions on Dependable and Secure Computing*, *4*(3), 165–179. doi:10.1109/TDSC.2007.70207

Ahmed, I. (2017). *Smartphone Authentication: User experience, expectation and satisfaction*. Lappeenranta University of Technology.

Ahuja, M. S., & Chabbra, S. (2011). A Survey of Multimodal Biometrics. *International Journal of Computer Science and its Applications, 1*, 157-160.

Akila, M., Suresh Kumar, V., Anusheela, N., & Sugumar, K. (2012). A Novel Feature Subset Selection Algorithm Using Artificial Bee Colony in Keystroke Dynamics. In *Proceedings of the International Conference on Soft Computing for Problem Solving (SocProS 2011)*. Springer. 10.1007/978-81-322-0491-6_74

Akila, M., & Suresh Kumar, S. (2011). Improving Feature Extraction in Keystroke Dynamics using Optimization Techniques and Neural Network. *Proceedings of the International Conference on Sustainable Energy and Intelligent Systems (SEISCON 2011)*, 891–898. 10.1049/cp.2011.0493

Al Galib, A., & Safavi-Naini, R. (2015, January). User authentication using human cognitive abilities. In *International Conference on Financial Cryptography and Data Security* (pp. 254-271). Springer. 10.1007/978-3-662-47854-7_16

Al Solami, E., Boyd, C., Clark, A., & Ahmed, I. (2011). User-Representative Feature Selection for Keystroke Dynamics. *Proceedings on 5th International Conference on Network and System Security (NSS)*, 229-233.

Alexander, M. W., Bartlett, J. E., Truell, A. D., & Ouwenga, K. (2001). Testing in a computer technology course: An investigation of equivalency in performance between online and paper and pencil methods. *Journal of Career and Technical Education*, *18*(1), 69–80. doi:10.21061/jcte.v18i1.600

Ali, L., Tappert, C., & Qiu, M. (2015). Authentication and identification methods used in keystroke biometric systems. *International Conference on High Performance Computing and Communications (HPCC)*, 1424-1429. 10.1109/HPCC-CSS-ICESS.2015.66

Allen, J. D. (2010). *An analysis of pressure-based keystroke dynamics algorithms* (Master's thesis). Dallas, TX: Southern Methodist University.

Alotaibi, S. (2010). Using Biometrics Authentication via Fingerprint Recognition in E-exams in E-learning Environment. In *The 4th Saudi International Conference*. The University of Manchester.

Alotaibi, S. J., & Argles, D. (2011). FingerID: A new security model based on fingerprint recognition for personal learning environments (PLEs). In *Global Engineering Education Conference (EDUCON)* (pp. 142-151). IEEE 10.1109/EDUCON.2011.5773128

Alpar, O. (2014). Keystroke recognition in user authentication using ANN based RGB histogram technique. *Engineering Applications of Artificial Intelligence*, *32*, 213–217. doi:10.1016/j.engappai.2013.11.009

Alpar, O. (2017). Frequency Spectrograms for Biometric Keystroke Authentication Using Neural Network Based Classifier. *Knowledge-Based Systems*, *116*, 163–171. doi:10.1016/j.knosys.2016.11.006

Al-Saleem, S. M., & Ullah, H. (2014). Security Considerations and Recommendations in Computer-Based Testing. *The Scientific World Journal*, *2014*, 1–7. doi:10.1155/2014/562787 PMID:25254250

Alshehri, Coenen, & Bollegala. (2018). Iterative Keystroke Continuous Authentication: A Time Series Based Approach. In *Künstliche Intelligenz*. Springer. DOI: . doi:10.100713218-018-0526-z

Alves, D. D., Cruz, G., & Vinhal, C. (2014). Authentication system using behavioral biometrics through keystroke dynamics. *IEEE Symposium on Computational Intelligence in Biometrics and Identity Management (CIBIM)*, 181-184. 10.1109/CIBIM.2014.7015461

Alwi, N. H., & Fan, I. S. (2010). Threats analysis for e-learning. *International Journal of Technology Enhanced Learning*, *2*(4), 358–371. doi:10.1504/IJTEL.2010.035738

Alwi, N., & Fan, I. (2010). E-learning and information system management. *International Journal of Digital Society*, *1*(2), 148–156. doi:10.20533/ijds.2040.2570.2010.0019

Amigud, A., Arnedo-Moreno, J., Daradoumis, T., & Guerrero-Roldan, A.-E. (2016). *A Behavioral Biometrics Based And Machine Learning Aided Framework For Academic Integrity In E-Assessment.* Paper Presented At The Intelligent Networking And Collaborative Systems (Incos), 2016 International Conference On. 10.1109/INCoS.2016.16

Anderson, J., & Applebome, P. (2011). Exam Cheating On Long Island Hardly A Secret. *New York Times*.

Anderson, R. (2008). *Security Engineering: A Guide to Building Dependable Systems*. Indianapolis, IN: Wiley Publishing, Inc.

Antal Margit & Nemes Lehel. (2016). The MOBIKEY Keystroke Dynamics Password Database: Benchmark Results. In *Software Engineering Perspectives and Application in Intelligent Systems* (pp. 35–46). Boston, MA: Springer.

Antal. (2015). Keystroke Dynamics on Android Platform. Procedia Technology, 820-826.

Anuradha, S.G., Kavya, B., Akshatha, S., Jyothi, K., & Ashalatha, G. (2016). Automated Face Detection & Recognition for Detecting Impersonation of Candidate in Examination System. *International Journal of Scientific & Engineering Research, 7*(3).

Arivoli. (2013). A framework of secure biometric based online exam authentication: An alternative to traditional exam. *International Journal of Scientific & Engineering Research*, *4*(11).

Asha, S., & Chellappan, C. (2008). *Authentication Of E-Learners Using Multimodal Biometric Technology.* Paper Presented At The 2008 International Symposium On Biometrics And Security Technologies. 10.1109/ISBAST.2008.4547640

Athavale, V., Bharadwaj, S., Pamecha, M., Prabhu, A., & Shrivastava, M. (2016). *Towards deep learning in hindi ner: An approach to tackle the labelled data scarcity.* arXiv preprint arXiv:1610.09756

Atlam, H. F., Attiya, G., & El-Fishawy, N. (2017). Integration of Color and Texture Features in CBIR System. *Integration (Tokyo, Japan)*, *164*(3).

Atoum, Y., Chen, L., Liu, A. X., Hsu, S. D. H., & Liu, X. (2017). Automated Online Exam Proctoring. *IEEE Transactions on Multimedia*, *19*(7), 1609–1624. doi:10.1109/TMM.2017.2656064

Awad, Ahmed, & Traore. (2005). Anomaly Intrusion Detection based on Biometrics. *Proceedings of 6th IEEE Information Assurance Workshop*, 452- 453.

Ayodele, T., Shoniregun, C. A., & Akmayeva, G. (2011). Towards e-learning security: A machine learning approach. In *Information Society (i-Society), 2011 International Conference* (pp. 490-492). IEEE.

Azevedo, G. L. F. B. G., Cavalcanti, G. D. C., & Carvalho Filho, E. C. B. (2007). Hybrid solution for the feature selection in personal identification problems through keystroke dynamics. *IEEE International Conference on Neural Networks - Conference Proceedings*, 1947–1952. 10.1109/IJCNN.2007.4371256

Azhar, F., & Anderson, W. (2012). Predicting Single-Neuron Activity in Locally Connected Networks. *Neural Computation, 24*(10), 2655–2677. doi:10.1162/NECO_a_00343

Bailie & Jortberg. (2009). Online Learner Authentication: Verifying the Identity of Online Users. *MERLOT Journal of Online Learning and Teaching, 5*(2).

Bailie, J. L., & Jortberg, M. A. (2009). Online learner authentication: Verifying the identity of online user's. *Journal of Online Learning and Teaching / MERLOT, 5*(2), 197–207.

Bal, A., & Acharya, A. (2011, December). Biometric authentication and tracking system for online examination system. In *Recent Trends in Information Systems (ReTIS), 2011 International Conference on* (pp. 209-213). IEEE. 10.1109/ReTIS.2011.6146869

Barra, S., Casanova, A., Narducci, F., & Ricciardi, S. (2015). Ubiquitous iris recognition by means of mobile devices. *Pattern Recognition Letters, 57*, 66-73. doi:10.1016/j.patrec.2014.10.011

Bawarith, R., Basuhail, A., Fattouh, A., & Gamalel-Din, S. (2017). E-exam Cheating Detection System. *International Journal of Advanced Computer Science and Applications, 8*(4), 176–181.

BBC. (2012). Black Hat: Iris scanners 'can be tricked' by hackers. *BBC Technology*. Available at: http://www.bbc.co.uk/news/technology-18997580

Beaudin, Levy, Parrish, & Danet. (2016). An empirical study of authentication methods to secure e-learning system activities against impersonation fraud. *Online Journal of Applied Knowledge Management, 4*(1).

Benjamin, M. (1994). Vital signs of identity. *IEEE Spectrum, 31*(2), 22–30. doi:10.1109/6.259484

Bergadano, F., Gunetti, D., & Picardi, C. (2002). User Authentication through Keystroke Dynamics. *ACM Transactions on Information and System Security, 5*(4), 367–397. doi:10.1145/581271.581272

Bergadano, F., Gunetti, D., & Picardi, C. (2003). Identity Verification through Dynamic Keystroke Analysis. *Journal Intelligent Data Analysis, 7*(5), 469–496. doi:10.3233/IDA-2003-7506

Berglund, A. (2013). A framework to study learning in a complex learning environment. *Research in Learning Technology, 12*(1). doi:10.3402/rlt.v12i1.21457

Bhagat, V., & Katankar, V. (2014). Novel method for Student Authentication in Online Examination. *International Journal of Research, 1*(7), 974–979.

Bhargav-Spantzel, A., Squicciarini, A. C., Bertino, E., Modi, S., Young, M., & Elliott, S. J. (2007). Privacy preserving multi-factor authentication with biometrics. *Journal of Computer Security, 15*(5), 529–560. doi:10.3233/JCS-2007-15503

Bhatia, A., & Hanmandlu, M. (2017). Keystroke Dynamics Based Authentication Using Information Sets. *Journal of Modern Physics, 8*(9), 1557–1583. doi:10.4236/jmp.2017.89094

Biel, L., Pettersson, O., Philipson, L., & Wide, P. (2001). ECG analysis: A new approach in human identification. *IEEE Transactions on Instrumentation and Measurement, 50*(3), 808–812. doi:10.1109/19.930458

Biggio, B., fumera, Russu, P., Didaci, L., & Roli, F. (2015). Adversarial Biometric Recognition: A review on biometric system security from the adversarial machine-learning perspective. *IEEE Signal Processing Magazine, 32*(5), 31–41. doi:10.1109/MSP.2015.2426728

Biometric News Portal. (n.d.). *Fingerprint biometrics.* Retrieved from http://www.biometricnewsportal.com/fingerprint_biometrics.asp

Biometric recognition: challenges and opportunities. (2010). Retrieved from http://www.nap.edu/openbook.php?record_id=12720&page=2

Biometrics, I. I. J. S. (2015). *ISO/IEC DIS 30107–1. Information Technology - Biometric Presentation Attack Detection - Part 1: Framework.* International Organization For Standardization.

Bleha, S., Slivinski, C., & Hussien, B. (1990). Computer-Access Security Systems Using Keystroke Dynamics. *IEEE Transactions on Pattern Analysis and Machine Intelligence*, *12*(12), 1217–1222. doi:10.1109/34.62613

Bleicher, P. (2005). Biometrics comes of age: despite accuracy and security concerns, biometrics are gaining in popularity. *Applied Clinical Trials.* Available at: http://www.appliedclinicaltrialsonline.com/biometrics-comes-age

Boakye, O., & Marfo, Y. (2016). Utilizing Keystroke Dynamics as an Additional Security Measure to Password Security in Computer Web-based Applications - A Case Study of UEW. *International Journal of Computers and Applications*, *149*(5).

Bolle, R. M., Connell, J. H., Pankanti, S., Ratha, N. K., & Senior, A. W. (2004). *Guide to Biometrics.* Springer Verlag. doi:10.1007/978-1-4757-4036-3

Bolle, R. M., Connell, J. H., & Ratha, N. K. (2002). Biometric perils and patches. *Pattern Recognition*, *35*(12), 2727–2738. doi:10.1016/S0031-3203(01)00247-3

Bonneau, J. (2012). The Quest To Replace Passwords. *IEEE Symposium on Security and Privacy*, 553-567.

Bonsor, K., & Johnson, R. (2016). How Facial Recognition Systems Work. *HowStuffWorks.* Available at: https://electronics.howstuffworks.com/gadgets/high-tech-gadgets/facial-recognition.htm

Botha, R. A., Furnell, S. M., & Clarke, N. L. (2009). From desktop to mobile: Examining the security experience. *Computers & Security*, *28*(3–4), 130–137. doi:10.1016/j.cose.2008.11.001

Bours, P. (2012). Continuous keystroke dynamics: A different perspective towards biometric evaluation. *Information Security Technical Report*, *17*(1–2), 36–43. doi:10.1016/j.istr.2012.02.001

Bours, P. (2012). Continuous keystroke dynamics: A different perspective towards biometric evaluation. *Journal of Information Security of Tech. Report*, *17*(1), 36–43.

Brocardo, M. L., Traore, I., & Woungang, I. (2014). Toward a Framework for Continuous Authentication Using Stylometry. *IEEE Xplore*, 106–115.

Brown, M., & Rogers, J. (1993). User Identification via Keystroke Characteristics of Typed Names using Neural Networks. *International Journal of Man-Machine Studies*, *39*(6), 999–1014. doi:10.1006/imms.1993.1092

Cai, Y., Jiang, H., Chen, D., & Huang, M. (2018). Online learning classifier based behavioral biometric authentication. *2018 IEEE 15th International Conference on Wearable and Implantable Body Sensor Networks (BSN)*, 62-65. doi: 10.1109/BSN.2018.8329659

Caldarola, R., & MacNeil, T. (2009). Dishonesty deterrence and detection: How technology can ensure distance learning test security and validity. *Proceedings of the 8th European Conference on E-Learning*, 108-115.

Campisi, P., Maiorana, E., Lo Bosco, M., & Neri, A. (2009). User authentication using keystroke dynamics for cellular phones. *IET Signal Processing*, *3*(1), 333–341. doi:10.1049/iet-spr.2008.0171

Canales, O., Murphy, T., Zych, E., Stewart, J., Tappert, C., Castro, A., … Plains, W. (2011). A Stylometry System for Authenticating Students Taking Online Tests. P. of Student-Faculty Research Day, Ed., CSIS. Pace University.

Cano, P., Batlle, E., Kalker, T., & Haitsma, J. (2005). A Review of Audio Fingerprinting. *The Journal of VLSI Signal Processing*, *41*(3), 271–284. doi:10.100711265-005-4151-3

Carey, M., Parris, E., & Bridle, J. (1991). A speaker verification system using alphanets. *Proceedings of the International Conference on Acoustics, Speech, and Signal Processing*, 397–400.

Cartoux, J. Y., Lapreste, J. T., & Richetin, M. (1989). Face authentication or recognition by profile extraction from range images. *Proceedings of the IEEE Workshop on Interpretation of 3D Scenes*, 194-199.

Castella-Roca, J., Herrera-Joancomarti, J., & Dorca-Josa, A. (2006). *A Secure E-Exam Management System.* Paper Presented At The The First International Conference On Availability, Reliability And Security, 2006. ARES 2006. 10.1109/ARES.2006.14

Ceker, H., & Upadhyaya, S. (2015). Enhanced Recognition of Keystroke Dynamics using Gaussian Mixture Models. *Military Communications Conference, MILCOM2015-2015 IEEE*, 1305–1310. 10.1109/MILCOM.2015.7357625

Cericola, R. (2013). Granta Will Bring Voice Control to Home Automation. *Electronic House*. Available at: https://www.electronichouse.com/smart-home/geo-semiconductor-speechfx-and-voicevault-partner-up-for-granta

Chandrasekar, V., & Suresh Kumar, S. (2016). A dexterous feature selection artificial immune system algorithm for keystroke dynamics. *Stochastic Analysis and Applications*, *34*(1), 147–154. doi:10.1080/07362994.2015.1110707

Chandrasekar, V., Suresh Kumar, S., & Maheswari, T. (2016). Authentication based on keystroke dynamics using stochastic diffusion algorithm. *Stochastic Analysis and Applications*, *34*(1), 155–164. doi:10.1080/07362994.2015.1112291

Chaudhary. (2012). Performance study of Text-independent Speaker identification system using MFCC & IMFCC for Telephone and Microphone Speeches. *International Journal of Computing and Business Research, 3*(2).

Chaya Shree, G., & Jagadamba, G. (2016). Keystroke Dynamics in Authentication Scheme. *International Journal of Computer Information Systems*, *12*(3).

Checco. (2006). Keystroke Dynamics and Corporate Security. *WSTA Ticker Magazine*.

Chellappa, R., Wilson, C. L., & Sirohey, S. (1995). Human and Machine Recognition of Faces: A Survey. *Proceedings of the IEEE*, *83*(5), 705–740. doi:10.1109/5.381842

Chen, S., Pande, A., & Mohapatra, P. (2014). Sensor-assisted facial recognition: an enhanced biometric authentication system for smartphones. *Proceedings of the 12th annual international conference on Mobile systems, applications, and services (MobiSys '14)*, 109-122. 10.1145/2594368.2594373

Chen, Y., & He, W. (2013). Security Risks and Protection in Online Learning: A Survey. *The International Review of Research in Open and Distance Learning*. Available at: http://www.irrodl.org/index.php/irrodl/article/view/1632/2712

Chen, B., Bastedo, K., & Howard, W. (2018). Exploring Best Practices for Online STEM Courses: Active Learning, Interaction & Assessment Design. *Online Learning*, *22*(2). doi:10.24059/olj.v22i2.1369

Chen, C., & Lee, C. (2014). A two-factor authentication scheme with anonymity for multi-server environments. *Security and Communication Networks*, *8*(8), 1608–1625. doi:10.1002ec.1109

Chen, W., & Chang, W. (2004). Applying Hidden Markov Models to Keystroke Pattern Analysis for Password Verification. *Proceedings of IEEE International Conference on Information Reuse and Integration*, 467–474. 10.1109/IRI.2004.1431505

Chen, W.-S., Chiang, Y.-S., & Chiu, Y.-H. (2007). Biometric Verification by Fusing Hand Geometry and Palmprint. *Proc. 3rd International Conference on Intelligent Information Hiding and Multimedia Signal Processing (IIHMSP)*, 403-406. 10.1109/IIHMSP.2007.4457734

Chiasson, S., Biddle, S., & van Oorschot, P. (2012). A Second Look at the Usability of Click-Based Graphical Passwords. *Proc. ACM Symp. Usable Privacy and Security (SOUPS).*

Chiasson, S., Biddle, S., & van Oorschot, P. (2012). A Second Look at the Usability of Click-Based Graphical Passwords. *Proc. ACM Symp. Usable Privacy and Security (SOUPS).* doi:10.1109/ICEMI.2017.8265799

Choong, Y.-Y., Franklin, J.M. & Greene, K.K., 2016. Usability and Security Considerations for Public Safety Mobile Authentication. *National Institute of Standards and Technology Interagency Report 8080.* Available at: . doi:10.6028/NIST.IR.8080

Cho, S., Han, C., Hee Han, D., & Kim, H.-I. (2000). Web based Keystroke Dynamics Identity Verification using Neural Network. *Journal of Organizational Computing and Electronic Commerce, 10*(4), 295–307. doi:10.1207/S15327744JOCE1004_07

Chourasia, N. (2014). Authentication of the user by keystroke dynamics for banking transaction system. *Proceedings of International Conference on Advances in Engineering & Technology*, 41–45.

Chowdhury, G. G. (2003). Natural language processing. *Annual Review of Information Science & Technology, 37*(1), 51–89. doi:10.1002/aris.1440370103

Chugh, R. (2016). *Students Are Using'smart'spy Technology To Cheat In Exams.* Academic Press.

Chutel, P. M., & Sakhare, A. (2014). Reverse Image Search Engine using Compact Composite Descriptor. *International Journal (Toronto, Ont.), 2*(1).

Cizek, G. J. (1999). *Cheating On Tests: How To Do It, Detect It, And Prevent It.* Routledge.

Clarke, N. L., & Furnell, S. M. (2007). Authenticating mobile phone users using keystroke analysis. *International Journal of Information Security, 6*(1), 1–14. doi:10.100710207-006-0006-6

Cluskey, G. R., Ehlen, C. R., & Raiborn, M. H. (2011). Thwarting online exam cheating without proctor supervision. *Journal of Academic and Business Ethics, 4*, 1–8.

Cluskey, G. Jr, Ehlen, C. R., & Raiborn, M. H. (2011). Thwarting Online Exam Cheating Without Proctor Supervision. *Journal Of Academic And Business Ethics, 4*, 1.

Cross, T., & Palese, K. (2015). Increasing Learning: Classroom Assessment Techniques in the Online Classroom. *American Journal of Distance Education, 29*(2), 98–108. doi:10.1080/08923647.2015.1023594

Curtin, M., Tappert, C., Villani, M., Ngo, G., Simone, J., & Fort, H. S. (2006). Keystroke biometric recognition on long-text input: a feasibility study. *Proceedings of the International Workshop Scientific Computing and Computational Statistics (IWSCCS).*

Dalva, D., Revidi, I. D., & Guz, U. (2014). DQG Haka Gurkan, Extraction and Comparison of Various Prosodic Feature Sets on Sentence Segmentation Task for Turkish Broadcast News Data. *11th International Joint Conference on Computer Science and Software Engineering (JCSSE)*, 70-73.

Daniela, C., & Michal, D. (2009). Multifactor authentication based on keystroke dynamics. In *Proceedings of the International Conference on Computer Systems and Technologies and Workshop for PhD Students in Computing.* ACM.

Daniel, B. (2017). Enterprise Lecture Capture Technologies and Value to Student Learning. *International Journal of Information and Communication Technology Education*, *6*(2). doi:10.1515/ijicte-2017-0009

Darabseh, A., & Namin, S. (2015), On Accuracy of Keystroke Authentications Based on Commonly Used English Words. *2015 International Conference of the Biometrics Special Interest Group (BIOSIG)*, 1-8. 10.1109/BIOSIG.2015.7314612

Das, M. L. (2008). *A Flexible and Secure Remote Systems Authentication Scheme Using Smart Cards.* arXiv preprint arXiv:0801.0575.

Das, A. K. (2011). Analysis and improvement on an efficient biometric based remote user authentication scheme using smart cards. *IET Information Security*, *5*(3), 145–151. doi:10.1049/iet-ifs.2010.0125

Das, A. K. (2015). A Secure and Efficient User Anonymity-Preserving Three-Factor Authentication Protocol for Large-Scale Distributed Wireless Sensor Networks. *Wireless Personal Communications*, *82*(3), 1377–1404. doi:10.100711277-015-2288-3

Daugman, J. (2004). *Iris recognition and anti-spoofing countermeasures.* 7th International Biometrics Conference, London, UK.

Daugman, J. (n.d.). Iris recognition. *International Center for Disability Resources on the Internet.* Available at http://www.icdri.org/ biometrics/ iris_biometrics.htm

Daugman, J. (2004). How Iris Recognition Works. *IEEE Transactions on Circuits and Systems for Video Technology*, *14*(1), 21–30. doi:10.1109/TCSVT.2003.818350

Daugman, J. (2006). Probing the uniqueness and randomness of IrisCodes: Results from 200 billion iris pair comparisons. *Proceedings of the IEEE*, *94*(11), 1927–1935. doi:10.1109/JPROC.2006.884092

Davida, G., Frankel, Y., & Matt, B. (1998). On enabling secure applications through off-line biometric identification. *IEEE Symp. Security and Privacy*, 148-157. 10.1109/SECPRI.1998.674831

Davis, S. F., Drinan, P. F., & Gallant, T. B. (2011). Cheating. In *School: What We Know And What We Can Do*. John Wiley & Sons.

Dawson, P. (2015). Five Ways To Hack And Cheat With Bring-Your-Own-Device Electronic Examinations. *British Journal of Educational Technology*.

Dehnavi, M. K., Sharafi, S. M., & Nematbakhsh, N. (2011). Developing a E-learning Model for Tracking the Continuous Attendance of the Students. *Journal of Theoretical & Applied Information Technology, 24*(1).

Dehnavi, M. K., & Fard, N. P. (2011). Presenting a multimodal biometric model for tracking the students in virtual classes. *Procedia: Social and Behavioral Sciences*, *15*, 3456–3462. doi:10.1016/j.sbspro.2011.04.318

Dempster, A., Laird, N., & Rubin, D. (1977). Maximum likelihood from incomplete data via the EM algorithm. *Journal of the Royal Statistical Society. Series A (General)*, *39*, 1–38.

Deng, Y., & Zhong, Y. (2013). Keystroke dynamics user authentication based on Gaussian mixture model and deep belief nets. *ISRN Signal Processing*, *2013*, 1–7. doi:10.1155/2013/565183

Deng, Y., & Zhong, Y. (2015). Keystroke Dynamics User Authentication Using Advanced Machine Learning Methods. In Y. Zhong & Y. Deng (Eds.), *Recent Advances in User Authentication Using Keystroke Dynamics Biometrics* (Vol. 2, pp. 23–40). GCSR.

Desouza, E., & Fleming, M. (2003). A Comparison of In-Class and Online Quizzes on Student Exam Performance. *Journal of Computing in Higher Education*, *14*(2), 121–134. doi:10.1007/BF02940941

Dhakal, V., Feit, A. M., Kristensson, P. O., & Oulasvirta, A. (2018). Observations on Typing from 136 Million Keystrokes. *Proceedings of the SIGCHI Conference on Human Factors in Computing Systems - CHI'18*, 1–12. 10.1145/3173574.3174220

Doddington, G., Liggett, W., Martin, A., Przybocki, M. & Reynolds, D. (1998). *Sheep, goats, lambs and wolves: A statistical analysis of speaker performance in the nist 1998 speaker recognition evaluation.* ICSLP98.

Dowland, P. S., Furnell, S. M., & Papadaki, M. (2002). Keystroke Analysis as a Method of Advanced User Authentication and Response. In A. El-Hadidi (Ed.), *Ghonaimy* (pp. 215–216). Security in the Information Society. doi:10.1007/978-0-387-35586-3_17

Dunker, M. (2003). Don't Blink: Iris Recognition for Biometric Identification. *SANS Institute InfoSec Reading Room.* Available at http://www.sans.org/ reading_room/ whitepapers/ authentication/dont-blink-iris-recognition-biometric-identification_1341

Du, Y., Ives, R. W., Euer, D. M., Welch, T. B., & Chang, C.-I. (2004). One Dimensional Approach to Iris Recognition. *Proceedings of the Society for Photo-Instrumentation Engineers*, *5404*, 237–247. doi:10.1117/12.542499

EdSurge. (2015). *Udacity, Coursera and edX now claim over 24 million students.* Retrieved September 17, 2018, from https://www.edsurge.com/news/2015-09-08-udacity-coursera-and-edx-now-claim-over-24-million-students

El-Abed, M., Giot, R., Hemery, B., Shwartzmann, J.-J., & Rosenberger, C. (2011). Towards the security evaluation of biometric authentication systems. *IEEE International Conference on Security and Technology (ICSST).*

El-Gayar, M. M., Soliman, H., & meky, N. (2013). A comparative study of image low level feature extraction algorithms. *Egyptian Informatics Journal*, *14*(2), 175–181. doi:10.1016/j.eij.2013.06.003

Elsadig, Salami, Faris, & Kin. (2004). Dynamic keystroke analysis using AR model. *IEEE International Conference Industrial Technology*, 3, 1555–1560.

Eplion, D., & Keefe, T. (2005). *Best Practices For Preventing Cheating On On-Line Exams.* Paper Presented At The E-Learn: World Conference On E-Learning In Corporate, Government, Healthcare, And Higher Education.

Epp, C., Lippold, M., & Mandryk, R. L. (2011). Identifying emotional states using keystroke dynamics. *Proceedings of the 2011 Annual Conference on Human Factors in Computing Systems - CHI '11*, 715. 10.1145/1978942.1979046

Fairhurst, M., & Costa-Abreu, M. D. (2011). Using keystroke dynamics for gender identification in social network environment. *4th International Conference on Imaging for Crime Detection and Prevention*, 1-6. 10.1049/ic.2011.0124

Fan, C-I., & Lin, Y-H. (2009). Provably secure remote truly three-factor authentication scheme with privacy protection on biometrics. *Information Forensics and Security, IEEE Transactions on, 4*(4), 933-945.

Farooq, F., Bolle, R., Jea, T., & Ratha, N. (2007). Anonymous and revocable fingerprint recognition. *Proc. IEEE Conf. Computer Vision and Pattern Recognition*, 1-7. DOI: 10.1109/CVPR.2007.383382

Fayyoumi, A., & Zarrad, A. (2014). Novel Solution Based on Face Recognition to Address Identity Theft and Cheating in Online Examination Systems. *Advances in Internet of Things*, *4*, 5–12. doi:10.4236/ait.2014.42002

Feng, G., Dong, K., Hu, D., & Zhang, D. (2004). When Faces Are Combined with Palmprints: A Novel Biometric Fusion Strategy. *Proceedings of ICBA*. 10.1007/978-3-540-25948-0_95

Fenu, G., Marras, M., & Boratto, L. (2017). A Multi-Biometric System For Continuous Student Authentication In E-Learning Platforms. *Pattern Recognition Letters.*

Filho, J. M. R., & Freire, E. O. (2006). On the equalization of keystroke timing histograms. *Pattern Recognition Letters*, *27*(13), 1440–1446. doi:10.1016/j.patrec.2006.01.010

Firesmith, D. (2003). Engineering Security Requirements. *Journal of Object Technology*, *2*(1), 53–68. doi:10.5381/jot.2003.2.1.c6

Flior, E., & Kowalski, K. (2010). Continuous biometric user authentication in online examinations. *ITNG2010 - 7th International Conference on Information Technology: New Generations*, 488–492. 10.1109/ITNG.2010.250

Flom, L., & Safir, A. (1987). *Iris Recognition System*. U.S. Patent 4 641 349.

Fluck, A., & Hillier, M. (2016). Innovative Assessment With Eexams. *Paper Presented At The Australian Council For Computers In Education Conference*, Brisbane, Australia.

Franklyn-Stokes, A., & Newstead, S. E. (1995). Undergraduate Cheating: Who Does What And Why? *Studies in Higher Education*, *20*(2), 159–172. doi:10.1080/03075079512331381673

Freidman, J. (2016). *Study: Enrollment in Online Learning Up, Except at For Profits*. Retrieved September 17, 2018, from https://www.usnews.com/education/online-education/articles/2016-02-09/study-enrollment-in-online-learning-up-except-at-for-profits

Frischholz, R. W., & Dieckmann, U. (2000). Bioid: A multimodal biometric identification system. *IEEE Comput.*, *33*(2), 64–68. doi:10.1109/2.820041

Galbally, J., Marcel, S., & Fierrez, J. (2014). Biometric Antispoofing Methods: A Survey In Face Recognition. *IEEE Access: Practical Innovations, Open Solutions*, *2*, 1530–1552. doi:10.1109/ACCESS.2014.2381273

Gambhir,, A., Narke,, S., Borhade,, S., & Bokade,, G. (2014). Person Recognition Using Multimodal Biometrics. *International Journal of Emerging Technology and Advanced Engineering*.

Gao, Q. (2010). Online teaching: Do you know who is taking the final exam? In *Proceeding of Mid-Atlantic ASEE Conference*. Villanova University.

Gao, Q. (2012). Biometric Authentication to Prevent e-Cheating. *International Journal of Instructional Technology and Distance Learning*, *9*(2), 93.

Gao, Q. (2012). Using IP Addresses As Assisting Tools To Identify Collusions. *International Journal Of Business. Human Technology*, *2*(1), 70–75.

Garfinkel, S., & Spafford, G. (1996). Practical Unix and Internet security. In *INET 96*. O'Reilly & Associates, Inc.

Gauvain, J. L., & Lee, C.-H. (1994). Maximum a posteriori estimation for multivariate Gaussian mixture observations of Markov chains. *IEEE Transactions on Speech and Audio Processing*, *2*(2), 291–298. doi:10.1109/89.279278

Gawande, U., Zaveri, M., & Kapur, A. (2011). Multi-algorithmic approaches to iris recognition. *Biometric Technology Today*, *2011*(4), 8-10. doi:10.10160969-4765(11)70075-3

Gikandi, J., Morrow, D., & Davis, N. (2011). Online formative assessment in higher education: A review of the literature. *Computers & Education*, *57*(4), 2333–2351. doi:10.1016/j.compedu.2011.06.004

Giot, R., Dorizzi, B., & Rosenberger, C. (2015). A review on the public benchmark databases for static keystroke dynamics. *Journal of Computer Security*, *55*(C), 46–61. doi:10.1016/j.cose.2015.06.008

Giot, R., Dorizziy, B., & Rosenberger, C. (2011). Analysis of Template Update Strategies for Keystroke Dynamics. *Proc. IEEE Symposium Series on Computational Intelligence*, 21-28. 10.1109/CIBIM.2011.5949216

Giot, R., Hemery, B., & Rosenberger, C. (2010). Low cost and usable multimodal biometric system based on keystroke dynamics and 2D face recognition. *Proceedings of the International Conference on Pattern Recognition*, 1128–1131. 10.1109/ICPR.2010.282

Giot, R., & Rosenberger, C. (2012). A new soft biometric approach for keystroke dynamics based on gender recognition. *International Journal of Information Technology and Management*, *11*(1/2), 35. doi:10.1504/IJITM.2012.044062

Golfarelli, M., Maio, D., & Malton, D. (1997). On the error-reject trade-off in biometric verification systems. *Pattern Analysis and Machine Intelligence. IEEE Transactions on*, *19*(7), 786–796.

Gonzalez-Agulla, E., Alba-Castro, J. L., Argones-Rua, E., & Anido-Rifon, L. (2010, January). Realistic measurement of student attendance in LMS using biometrics. *Proc. of the Int. Symposium on Engineering Education and Educational Technologies (EEET'09).*

Goodkind, A., Brizan, D. G., & Rosenberg, A. (2017). Utilizing overt and latent linguistic structure to improve keystroke-based authentication. *Image and Vision Computing*, *58*, 230–238. doi:10.1016/j.imavis.2016.06.003

Gorman. (2003). Comparing Passwords, Tokens, and Biometrics for User Authentication. Proc. IEEE, 91(12), 2019-2040.

Gottschlich, C., & Huckemann, S. (2014). Separating the real from the synthetic: Minutiae histograms as fingerprints of fingerprints. *IET Biometrics*, *3*(4), 291–301. doi:10.1049/iet-bmt.2013.0065

Grabham, N. J., & White, N. M. (2008). Use of a novel keypad biometric for enhanced user identity verification. *Proceedings of the IEEE International Instrumentation and Measurement Technology Conference (IMTC '08)*, 12–16. 10.1109/IMTC.2008.4546995

Graf, F. (2002). Providing security for eLearning. *Journal of Computers & Graphics*, *26*(2), 355–365. doi:10.1016/S0097-8493(02)00062-6

Grijalva, T., Nowell, C., & Kerkvliet, J. (2006). Academic honesty and online courses. *College Student Journal*, *40*(1), 180–185.

Gu, G., & Liu, F. (2011). Contourlet domain image encryption based on chaos mapping. *Jisuanji Yingyong*, *31*(3), 771–773. doi:10.3724/SP.J.1087.2011.00771

Guillén-Gámez, F. D. (2017). Biometrics and education: A review about facial authentication software for the identification and verification of students who use virtual learning platform (LMS). *Advances in Educational Technology and Psychology*, *1*(1), 1–8.

Gunasekarn, A., McNeil, R. D., & Shaul, D. (2002). E-learning Research and applications. *Industrial and Commercial Training*, *34*(2), 44–54. doi:10.1108/00197850210417528

Gunetti, D., & Picardi, C. (2005). Keystroke analysis of free text. *ACM Transactions on Information and System Security*, *8*(3), 312–347. doi:10.1145/1085126.1085129

Gupta, A., Khanna, A., Jagetia, A., Sharma, D., Alekh, S., & Choudhary, V. (2015, October). Combining Keystroke Dynamics and Face Recognition for User Verification. In *Computational Science and Engineering (CSE), 2015 IEEE 18th International Conference on* (pp. 294-299). IEEE. 10.1109/CSE.2015.37

Guven, A., & Sogukpinar, I. (2003). Understanding users' keystroke patterns for computer access security. *Journal of Computer Security*, *22*(8), 695–706. doi:10.1016/S0167-4048(03)00010-5

Halakou, F. (2013). Feature Selection in Keystroke Dynamics. *Norwegian Information Security conference(NISK).*

Handwritten English Character Recognition Using Logistic Regression and Neural Network. (2016). *International Journal Of Science And Research*, *5*(6), 750–754. doi:10.21275/v5i6.nov164228

Heck, L. P., & Weintraub, M. (1997). Handset-dependent background models for robust text independent speaker recognition. *Proceedings of the International Conference on Acoustics, Speech, and Signal Processing*, 1071–1073. 10.1109/ICASSP.1997.596126

Hernández, J., Ortiz, A. O., Andaverde, J., & Burlak, G. (2008). *Biometrics In Online Assessments: A Study Case In High School Students.* Paper Presented At The Electronics, Communications And Computers, 2008. CONIELECOMP 2008, 18th International Conference On. 10.1109/CONIELECOMP.2008.36

Hernández, J. A., Ochoa, A., Muñoz, J., & Burlaka, G. (2006). Detecting cheats in online student assessments using Data Mining. *Conference on Data Mining*, 6, 205.

Higher Education Opportunity Act (HEOA) of 2008. (n.d.). Retrieved September 17, 2018, from https://www2.ed.gov/policy/highered/leg/hea08/index.html

Ho, B., & Kimand, H. (2015). Proposal: A Design of E-learning learning User Authentication System. *International Journal of Security and Its Applications*, *9*(1), 45–50. doi:10.14257/ijsia.2015.9.1.05

Hong, L., Jain, A. K., & Pankanti, S. (1999). Can multibiometrics improve performance? *Proc. AutoID'99.* Retrieved from http://www.griaulebiometrics.com/en-us/fingerprint_sdk http://www.computerscienceweb.com/

Hong, L., & Jain, A. K. (1998). Integrating faces and fingerprints for personal identification. *IEEE Transactions on Pattern Analysis and Machine Intelligence*, *20*(12), 1295–1307. doi:10.1109/34.735803

Höynälänmaa, T. (2015). Multiresolution analysis for compactly supported interpolating tensor product wavelets. *International Journal of Wavelets, Multresolution, and Information Processing*, *13*(02), 1550010. doi:10.1142/S0219691315500101

Huang, W., Yen, D.C., Lin, Z.X., & Huang, J.H. (2004). How to compete in a global education market effectively: A conceptual framework for designing a next generation eEducation system. *Journal of Global Information Management, 12*(20), 84-107.

Huang, G. B., Zhu, Q. Y., & Siew, C. K. (2004). Extreme Learning Machine: A New Learning Scheme of Feedforward Neural Networks. *Proceedings of 2004 International Joint Conference on Neural Networks*. 10.1109/IJCNN.2004.1380068

Huang, G. B., Zhu, Q. Y., & Siew, C. K. (2006). Extreme Learning Machine: Theory and Applications. *Neurocomputing. Elsevier*, *70*(1-3), 489–501. doi:10.1016/j.neucom.2005.12.126

Hugl, U. (2005). Tech-developments and possible influences on learning processes and functioning in the future. *Journal of American Academy of Business,* 6(20), 250-256.

Hu, J., Gingrich, D., & Sentosa, A. (2008). Authentication through Biometric Keystroke Dynamics. *IEEE International Conference on Communications*, 1556–1560. 10.1109/ICC.2008.301

Hurley, D. J., Nixon, M. S., & Carter, J. N. (2009). Force field energy functionals for ear biometrics. *Computer Vision and Image Understanding*, *98*(3), 491–512. doi:10.1016/j.cviu.2004.11.001

Hurley, D. J., Nixon, M. S., & Carter, J. N. (2011). Force field energy functionals for image feature extraction. *Image and Vision Computing*, *20*(5–6), 311–318.

Hwang, S. S., Cho, S., & Park, S. (2009a). Keystroke dynamics-based authentication for mobile devices. *Computers & Security*, *28*(1), 85–93. doi:10.1016/j.cose.2008.10.002

Idrus, S. Z. S., Cherrier, E., Rosenberger, C., & Bours, P. (2013).Soft biometrics database: a benchmark for keystroke dynamics Biometric methods. *2013 International conference of the biometrics special interest group (BIOSIG)*, 1-8.

Idrus, S. Z. S., Cherrier, E., Rosenberger, C., & Schwartzmann, J.-J. (2013). A Review On Authentication Methods. *Australian Journal of Basic and Applied Sciences*, *7*(5), 95–107.

Idrus, S., Zulkarnain, S., Cherrier, E., Rosenberger, C., & Bours, P. (2013). Soft Biometrics For Keystroke Dynamics. *International Conference on Image Analysis and Recognition (ICIAR)*, 11-18. 10.1007/978-3-642-39094-4_2

Internet of Things Service. (n.d.). *ISPRS International Journal of Geo-Information*, *5*(7), 116. doi:10.3390/ijgi5070116

Isobe, T., & Takahashi, J. (1999). Text-independent speaker verification using virtual speaker based cohort normalization. *Proceedings of the European Conference on Speech Communication and Technology*, 987–990.

Israa, A.M. (2015). Physiological Biometric Authentication Systems. *International Journal of Scientific & Technology Research, 4*(12).

Israel, S. A., Irvine, J. M., Cheng, A., Wiederhold, M. D., & Wiederhold, B. K. (2005). ECG to identify individuals. *Pattern Recognition*, *38*(1), 133–142. doi:10.1016/j.patcog.2004.05.014

Jabeen, F., & Braun, B. (2018). *Production and Perception of Prosodic Cues in Narrow & Corrective Focus in Urdu/ Hindi*. 9th International Conference on Speech Prosody 2018, Poznań, Poland.

Jagadamba & Sathish Babu. (2015). Security in Cyber-Physical Systems: Generalized Framework Based on Trust with Privacy. In *Cyber-Physical Systems: A Computational Perspective* (pp. 443-462). CRC Press, Taylor & Francis Group.

Jagadamba, G., & Sharmila, S. P. (2014). A Secured Authentication using Keystroke Dynamics. Lecture Notes in Electrical Engineering, 248, 253-260.

Jain, Hong, & Pankanti. (2000). Biometric Identification. *Communications of the ACM, 43*(2), 91-98.

Jain. (2003). Information fusion in biometrics. *Pattern Recognition Letters, 24*, 2115-2125.

Jain, A. K., Bolle, R., & Pankanti, S. (Eds.). (1999). *Biometrics: Personal Identification in Networked Society*. Dordrecht: Kluwer Academic. doi:10.1007/b117227

Jain, A. K., Ross, A., & Prabhakar, S. (2004). An Introduction to Biometric Recognition. *IEEE Transactions on Circuits and Systems for Video Technology*, *14*(1), 4–19. doi:10.1109/TCSVT.2003.818349

Jain, A., Kulkarni, G., & Shah, V. (2018). Natural Language Processing. *International Journal on Computer Science and Engineering*, *6*(1), 161–167.

Jain, A., & Ross, A. (2002). Fingerprint Mosaicking. *Proc. IEEE International Conference on Acoustics, Speech and Signal Processing (ICASSP)*, 4, IV-4064-IV-4067.

Jain, L., Monaco, J. V., Coakley, M. J., & Tappert, C. C. (2014). Passcode keystroke biometric performance on touch screen is superior to that on hardware keyboards. *International Journal of Research in Computer Applications and Information Technology*, *2*(4), 29–33.

Janakiraman, R., & Sim, T. (2007). Keystroke Dynamics in a General Setting. In S. W. Lee & S. Z. Li (Eds.), Lecture Notes in Computer Science: Vol. 4642. *Advances in Biometrics*. Berlin: Springer. doi:10.1007/978-3-540-74549-5_62

Jenkins, J. L., Grimes, M., Proudfoot, J. G., & Lowry, P. B. (2014). Improving password cybersecurity through inexpensive and minimally invasive means: Detecting and deterring password reuse through keystroke-dynamics monitoring and just-in-time fear appeals. *Journal of Information Technology for Development*, *20*(2), 196–213. doi:10.1080/02681102.2013.814040

Jin, L., Ke, X., Manuel, R., & Wilkerson, M. (2004). Keystroke Dynamics: A Software based Biometric Solution. *13th USENIX Security Symposium.*

Jitendra, J., Singh, B.K., & Ali, M.I. (2014). Voice Identification Secure System by Statistical Model of Speech Signal Using Normalization Technique. *International Journal of Engineering.*

Joosten-Ten Brinke, D., Van Bruggen, J., Hermans, H., Burgers, J., Giesbers, B., Koper, R., & Latour, I. (2007). Modeling assessment for re-use of traditional and new types of assessment. *Computers in Human Behavior*, *23*(6), 2721–2741. doi:10.1016/j.chb.2006.08.009

Jorge, M. M., Caballé, S., Xhafa, F., & Prieto-Blazquez, J. (2015). Security in online web learning assessment. *World Wide Web (Bussum)*, *18*(6), 1–24.

Josa, Pérez, & Moreno. (2017). *Using Keystroke Dynamics and context features to assess authorship in online learning environments.* Doi:10.21125/inted.2017.0819

Joyce & Gupta. (1990). Identity authentication based on keystroke latencies. *Communications of the ACM*, *33*(2), 168-176.

Joyce, R., & Gupta, G. (1990). Identity authorization based on keystroke latencies. *Communications of the ACM*, *33*(2), 168–176. doi:10.1145/75577.75582

Jung, I. Y., & Yeom, H. Y. (2009). Enhanced security for online exams using group cryptography. *IEEE Transactions on Education*, *52*(3), 340–349. doi:10.1109/TE.2008.928909

Kang, P., & Cho, S. (2015). Keystroke Dynamics-based User Authentication Using Long and Free Text Strings From Various Input Devices Keystroke Dynamics-based User Authentication Using Long and Free Text Strings From Various Input Devices. *Journal of Information Science*, *308*, 72–93. doi:10.1016/j.ins.2014.08.070

Kang, P., Hwang, S., & Cho, S. (2007). Continual Retraining of Keystroke Dynamics Based Authenticator. In S. W. Lee & S. Z. Li (Eds.), *Advances in Biometrics* (Vol. 4642). Springer. doi:10.1007/978-3-540-74549-5_125

Karim, N. A., & Shukur, Z. (2015). Review of user authentication methods in online examination. *Asian Journal of Information Technology*. doi:10.3923/ajit.2015.166-175

Karnan & Kalamani. (2009). A Feature Subset Selection in Keystroke Dynamics Using Ant Colony Optimization. *Journal of Engineering and Technology Research*, *1*(5), 72-80.

Karnan, M., & Krishnaraj, N. (2010). Bio Password-Keystroke dynamic Approach to secure Mobile Devices. *Proc. IEEE-2010, International Conference on Computational Intelligence and Computing Research*. DOI:10.1109/ICCIC

Karnan, M., & Akila, M. (2009). Identity Authentication based on Keystroke Dynamics using Genetic Algorithm and Particle Swarm Optimization. *Proceedings of 2nd IEEE International Conference on Computer Science and Information Technology*, 203–207. 10.1109/ICCSIT.2009.5234420

Karnan, M., & Akila, M. (2009). Personal Authentication based on Keystroke Dynamics using Ant Colony Optimization and Back Propagation Neural Network. *International Journal of Computer and Network Security*, *1*(2), 8–15.

Karnan, M., & Akila, M. (2010). Personal Authentication based on Keystroke Dynamics using Soft Computing Techniques. *Second International Conference on Communication Software and Networks*, 334–338. 10.1109/ICCSN.2010.50

Karnan, M., Akila, M., & Kalamani, A. (2009). Feature subset selection in keystroke dynamics using ant colony optimization. *Journal of Engineering and Technology Research, 1*(5), 72–80.

Karnan, M., Akila, M., & Krishnaraj, N. (2011). Biometric personal authentication using keystroke dynamics: A review. *Journal of Applied Soft Computing, 11*(2), 1565–1573. doi:10.1016/j.asoc.2010.08.003

Kashyap, R., & Gautam, P. (2017). Fast Medical Image Segmentation Using Energy-Based Method. *Biometrics, Concepts, Methodologies, Tools, and Applications, 3*(1), 1017–1042. doi:10.4018/978-1-5225-0983-7.ch040

Kashyap, R., & Piersson, A. (2018). *Impact of Big Data on Security. In Handbook of Research on Network Forensics and Analysis Techniques* (pp. 283–299). IGI Global. doi:10.4018/978-1-5225-4100-4.ch015

Kashyap, R., & Tiwari, V. (2017). Energy-based active contour method for image segmentation. *International Journal of Electronic Healthcare, 9*(2–3), 210–225. doi:10.1504/IJEH.2017.083165

Kaur, N., Prasad, P. W. C., Alsadoon, A., Pham, L., & Elchouemi, A. (2016, November). An enhanced model of biometric authentication in E-Learning: Using a combination of biometric features to access E-Learning environments. In *Advances in Electrical, Electronic and Systems Engineering (ICAEES), International Conference on* (pp. 138-143). IEEE.

Kennedy, J., & Eberhart, R. (1995). Particle Swarm Optimization. *Proceedings of International Conference on Neural Networks*, 1942-1948. 10.1109/ICNN.1995.488968

Ketab, S. S., Clarke, N. L., & Dowland, P. S (2017). *A Robust E-Invigilation System Employing Multimodal Biometric Authentication*. Academic Press.

KeyTrac. (2016). *How KeyTrac records your typing behavior*. Retrieved September 17, 2018, from https://www.keytrac.net/en/technology

Khairnar, C. (2015). Advance Pedagogy: Innovative Methods of Teaching and Learning. *International Journal of Information and Education Technology (IJIET), 5*(11), 869–872. doi:10.7763/IJIET.2015.V5.629

Khalaf, H. A., & Nouman, A.-H. M. (2013). Multifactor Strong Authentication Method Using Keystroke Dynamics. *International Journal of Science and Applied Information Technology*, 2(2), 31–34.

Khan, B. H. (1998). Web-based Instruction (WBI): An Introduction. *Educational Media International, 35*(2), 63–71. doi:10.1080/0952398980350202

Khan, B. H. (2000). Discussion of resources and attributes of the web for the creation of meaningful learning environments. *Cyberpsychology & Behavior, 3*(1), 17–23. doi:10.1089/109493100316193

Khlifi, Y., & El-Sabagh, H. A. (2017). A Novel Authentication Scheme for E-assessments Based on Student Behavior over E-learning Platform. *International Journal of Emerging Technologies in Learning, 12*(04), 62–89. doi:10.3991/ijet.v12i04.6478

Kiennert, C., Rocher, P., Ivanova, M., Rozeva, A., Durcheva, M., & Garcia-Alfaro, J. (2017). *Security Challenges In E-Assessment And Technical Solutions*. Paper Presented At The 2017 21st International Conference Information Visualisation (IV). 10.1109/iV.2017.70

Killourhy, K. S., & Maxion, R. A. (2012). Free vs. transcribed text for keystroke-dynamics evaluations. *Proceedings of the 2012 Workshop on Learning from Authoritative Security Experiment Results - LASER '12*, 1–8. 10.1145/2379616.2379617

Killourhy, K., & Maxion, R. (2009) Comparing anomaly-detection algorithms for keystroke dynamics. *IEEE/IFIP International Conference on Dependable Systems& Networks.*

Killourhy, K., & Maxion, R. (2009). Comparing anomaly-detection algorithms for keystroke dynamics. *International conference on Dependable Systems & Networks*, 125-134.

Kim, H. (2013). E-learning Ongoing Sesion Authentication Security. *Journal Of Security Engineering*, *10*(5), 601–608. doi:10.14257/jse.2013.10.5.08

Kim, J., Kim, H., & Kang, P. (2018). Keystroke dynamics-based user authentication using freely typed text based on user-adaptive feature extraction and novelty detection. *Applied Soft Computing*, *62*, 1077–1087. doi:10.1016/j.asoc.2017.09.045

King, C. G., Guyette, R. W., & Piotrowski, C. (2009). Online exams and cheating: An empirical analysis of business students' views. *The Journal of Educators Online*, *6*(1). doi:10.9743/JEO.2009.1.5

Kisku, D., Gupta, P., & Sing, J. K. (2016). *Advances in Biometrics for Secure Human Authentication and Recognition*. London: CRC Press.

Kołakowska, A. (2013). A review of emotion recognition methods based on keystroke dynamics and mouse movements. *6th International Conference on Human System Interactions (HSI)*, 548-555. 10.1109/HSI.2013.6577879

Kordy, B., Mauw, S., Radomirović, S., & Schweitzer, P. (2014). Attack–Defense Trees. *Journal of Logic and Computation*, *24*(1), 55–87. doi:10.1093/logcom/exs029

Korvoren, K. (1999). Creating Trust. *Proceedings of the Fourth Nordic Workshop on Secure IT System.*

Krause, M. (2014). A behavioral biometrics based authentication method for MOOC's that is robust against imitation attempts. *Proceedings of the First ACM Conference on Learning @ Scale Conference - L@S '14*, 201–202. 10.1145/2556325.2567881

Kukula, E., Sutton, M., & Elliott, S. (2010). The Human–Biometric-Sensor Interaction Evaluation Method: Biometric Performance and Usability Measurements. *IEEE Transactions on Instrumentation and Measurement*, *59*(4), 784–791. doi:10.1109/TIM.2009.2037878

Kumar, A., & Shekhar, S. (2011). Personal Identification Using Multibiometrics Rank-Level Fusion. IEEE Transactions on Systems, Man, and Cybernetics, Part C: Applications and Reviews.

Kumar, A., Wong, D. C., Shen, H. C., & Jain, A. K. (2003). *Personal verification using palmprint and hand geometry biometric*. Presented at the 4th Int. Conf. Audio- and Video-based Biometric Person Authentication, Guildford, UK. 10.1007/3-540-44887-X_78

Kumlander, D. (2008). Soft biometrical students identification method for e-learning. *Advances in Computer and Information Sciences and Engineering*, *114–118*, 114–118. doi:10.1007/978-1-4020-8741-7_21

Lee, P. M., Tsui, W. H., & Hsiao, T. C. (2015). The influence of emotion on keyboard typing: An experimental study using auditory stimuli. *PLoS One*, *10*(6), 1–16. doi:10.1371/journal.pone.0129056 PMID:26065902

Lee-Post, A., & Hapke, H. (2017). Online Learning Integrity Approaches: Current Practices and Future Solutions. *Online Learning*, *21*(1), 135–145. doi:10.24059/olj.v21i1.843

Leggetter, C., & Woodland, P. (1995). Maximum Likelihood Linear Regression for Speaker Adaptation of Hmms. *Computer Speech & Language*, *9*, 171–186. doi:10.1006/csla.1995.0010

Leggett, J., & Williams, G. (1988). Verifying identity via keystroke characteristics. *International Journal of Man-Machine Studies*, *28*(1), 67–76. doi:10.1016/S0020-7373(88)80053-1

Leggett, J., Williams, G., Usnick, M., & Longnecker, M. (1991). Dynamic identity verification via keystroke characteristics. *International Journal of Man-Machine Studies*, *35*(6), 859–870. doi:10.1016/S0020-7373(05)80165-8

Levy, Y. (2008). An empirical development of critical value factors (CVF) of online learning activities: An application of activity theory and cognitive value theory. *Computers & Education*, *51*(4), 1664–1675. doi:10.1016/j.compedu.2008.04.003

Levy, Y., & Ramim, M. (2007). *A theoretical approach for biometrics authentication of e-exams*. Nova Southeastern University.

Levy, Y., & Ramim, M. (2007). A Theoretical Approach For Biometrics Authentication Of E-Exams. *Paper Presented At The Chais Conference On Instructional Technologies Research, Raanana*, Israel.

Li, C., & Irby, B. (2008). An overview of online education: Attractiveness, benefits, challenges, concerns and recommendations. *College Student Journal*, *42*(2), 449–458.

Li, S., & Kot, A. (2013). Fingerprint combination for privacy protection. *IEEE Transactions on Information Forensics and Security*, *8*(2), 350–360. doi:10.1109/TIFS.2012.2234740

Liu, J. (2014). The BeiHang Keystroke Dynamics and baselines. *Neurocomputing*, *144*, 271–281. doi:10.1016/j.neucom.2014.04.044

Liu, X., Bowyer, K. W., & Flynn, P. J. (2005). Experimental Evaluation of Iris Recognition. *IEEE Computer Society Conference on Computer Vision and Pattern Recognition Workshops*, 158-165.

Liu, Y., Li, Y., Ma, X., & Song, R. (2017). Facial Expression Recognition with Fusion Features Extracted from Salient Facial Areas. *Sensors (Basel)*, *17*(4), 712. doi:10.339017040712

Loy, L., W. K., & Lim, C. P. (2007). Keystroke patterns classification using the ARTMAP-FD neural network. *International Conference on Intelligent Information Hiding and Multimedia Signal Processing*, 61–4. 10.1109/IIH-MSP.2007.218

Maio, D., Maltoni, D., Cappelli, R., Wayman, J. L., & Jain, A. K. (2002). Fingerprint verification competition. *Proc. Int.Conf. Pattern Recognition (ICPR)*.

Maltoni, D., Maio, D., Jain, A. K., & Prabhakar, S. (2003). *Handbook of Fingerprint Recognition*. New York: Springer-Verlag.

Manikandan & Chandra. (2016). Speaker Identification using a Novel Prosody with Fuzzy based Hierarchical Decision Tree Approach. *Indian Journal of Science and Technology*, *9*(44).

Mantyjarvi, J., Koivumaki, J., & Vuori, P. (2002). Keystroke recognition for virtual keyboard. *Proceedings of the IEEE International Conference on Multimedia and Expo (ICME '02)*, 2, 429–432. 10.1109/ICME.2002.1035630

Marsters, J. D. (2009). *Keystroke dynamics as a biometric* (Doctoral dissertation). Retrieved September 17, 2018, from https://eprints.soton.ac.uk/66795/

Martin, Z. (2011). Biometric Trends: Will emerging modalities and mobile applications bring mass adoption? *SecureIDNews*. Available at: https://www.secureidnews.com/news-item/biometric-trends-will-emerging-modalities-and-mobile-applications-bring-mass-adoption/?tag=biometrics&tag=Law_Enforcement

Masala, G., Ruiu, P., & Grosso, E. (2017). *Biometric Authentication and Data Security in Cloud Computing*. Computer and Network Security Essentials.

Matsubara, Y., Samura, T., & Nishimura, H. (2015). *Keyboard Dependency of Personal Identification Performance by Keystroke Dynamics in Free Text Typing*. Academic Press.

Matsui, T., & Furui, S. (1994). Similarity normalization methods for speaker verification based on a posteriori probability. *Proceedings of the ESCA Workshop on Automatic Speaker Recognition, Identification and Verification*, 59–62.

Matsui, T., & Furui, S. (1995). Likelihood normalization for speaker verification using a phoneme and speaker-independent model. *Speech Communication, 17*(1-2), 109–116. doi:10.1016/0167-6393(95)00011-C

Mccabe, D. L. (2005). Cheating Among College And University Students: A North American Perspective. *International Journal For Educational Integrity, 1*(1).

Mccabe, D. L., Trevino, L. K., & Butterfield, K. D. (2001). Cheating In Academic Institutions: A Decade Of Research. *Ethics & Behavior, 11*(3), 219–232. doi:10.1207/S15327019EB1103_2

Menekse, M., Stump, G., Krause, S., & Chi, M. (2013). Differentiated Overt Learning Activities for Effective Instruction in Engineering Classrooms. *Journal Of Engineering Education, 102*(3), 346–374. doi:10.1002/jee.20021

Menke, J. (2014). Implementation of Online Poster Sessions in Online and Face-to-Face Classrooms as a Unique Assessment Tool. *Journal of Chemical Education, 91*(3), 414–416. doi:10.1021/ed400665n

Merhav, N. (2017). The generalized stochastic likelihood decoder: Random coding and expurgated bounds. *IEEE Transactions on Information Theory, 63*(8), 5039–5051. doi:10.1109/TIT.2017.2689787

Merhav, N. (2018). Ensemble Performance of Biometric Authentication Systems Based on Secret Key Generation. *2018 IEEE International Symposium on Information Theory (ISIT)*, 1455-1459. 10.1109/ISIT.2018.8437768

Miao, D., Sun, Z., & Huang, Y. (2014). *Fusion of Multibiometrics Based on A New Robust Linear Programming*. Presented at the Pattern Recognition (ICPR), 2014 22nd International Conference on, Stockholm, Sweden.

Miguel, J., Caballe, S., & Prieto, J. (2013). Providing information security to MOOC: Towards effective student authentication. In *Proceedings of the IEEE International Conference on Intelligent Networking and Collaborative Systems* (pp. 289–292). IEEE. 10.1109/INCoS.2013.52

Migut, G., Koelma, D., Snoek, C. G., & Brouwer, N. (2018). Cheat Me Not: Automated Proctoring Of Digital Exams On Bring-Your-Own-Device. *Proceedings Of The 23rd Annual ACM Conference On Innovation And Technology In Computer Science Education*. 10.1145/3197091.3205813

Mikolov, T., Sutskever, I., Chen, K., Corrado, G. S., & Dean, J. (2013). Distributed representations of words and phrases and their compositionality. In Advances in neural information processing systems (pp. 3111-3119). Academic Press.

Miller, G. (1998). *WordNet: An electronic lexical database*. MIT Press.

Mitra, S., & Gofman, M. I. (2016). *Towards Greater Integrity In Online Exams*. Academic Press.

Moini, A., & Madni, A. M. (2009). Leveraging biometrics for user authentication in online learning: A systems perspective. *IEEE Systems Journal, 3*(4), 469–476. doi:10.1109/JSYST.2009.2038957

Monaco, J., Stewart, J., Cha, S., & Tappert, C. (2013). Behavioral biometric verification of student identity in online course assessment and authentication of authors in literary works. *Proceedings of IEEE Sixth International Conference on Biometrics: Theory, Applications and Systems*, 1-8. 10.1109/BTAS.2013.6712743

Monaro, M., Gamberini, L., & Sartori, G. (2017). The Detection Of Faked Identity Using Unexpected Questions And Mouse Dynamics. *PLoS One, 12*(5). doi:10.1371/journal.pone.0177851 PMID:28542248

Mondal, S., & Bours, P. (2016). Combining keystroke and mouse dynamics for continuous user authentication and identification. *IEEE International Conference on Identity, Security and Behavior Analysis, ISBA*, 1–8. 10.1109/ISBA.2016.7477228

Mondal, S., & Bours, P. (2017a). A study on continuous authentication using a combination of keystroke and mouse biometrics. *Neurocomputing, 230*, 1–22. doi:10.1016/j.neucom.2016.11.031

Mondal, S., & Bours, P. (2017b). Person Identification by Keystroke Dynamics Using Pairwise User Coupling. *IEEE Transactions on Information Forensics and Security*, *12*(6), 1319–1329. doi:10.1109/TIFS.2017.2658539

Monrose & Aviel. (2000). Keystroke dynamics as a biometric for authentication. *Journal Future Generation Computer Systems, 15*(4), 351-359.

Monrose, F., Reiter, M. K., & Wetzel, S. (2002). Password hardening based on keystroke dynamics. *International Journal of Information Security*, *1*(2), 69–83. doi:10.1007102070100006

Monrose, F., & Rubin, A. (1997). Authentication via keystroke dynamics. *Proceedings of the 4th ACM Conference on Computer and Communications Security*, 48-56.

Monrose, F., & Rubin, A. D. (2000). Keystroke dynamics as a biometric for authentication. *Future Generation Computer Systems*, *16*(4), 351–359. doi:10.1016/S0167-739X(99)00059-X

Montalvao Filho, J., & Freire, E. (2000). Multimodal biometric fusion–joint typist (keystroke) and speaker verification. *Telecommunications Symposium, 2006 International*, 609–614.

Montalvao, Almeida, & Freire. (2006). Equalization of keystroke timing histograms for improved identification performance. *Telecommunications Symposium, 2006 International*, 560–565.

Mordini, E., & Tzovaras, D. (2012). *Second Generation Biometrics: the Ethical and Social Context*. Springer-Verlag.

Moskovitch, R., Feher, C., Messerman, A., Kirschnick, N., Mustafic, T., Camtepe, A., . . . Elovici, Y. (2009). Identity theft, computers and behavioral biometrics. *IEEE International conference on intelligence and security informatics*, 155–160. 10.1109/ISI.2009.5137288

Moten, J., Fitterer, A., Brazier, E., Leonard, J., & Brown, A. (2013). Examining online college cyber cheating methods and prevention measures. *Electronic Journal of E-Learning*, *11*(2), 139–146.

Mulionoa, Y., Hamb, H., & Darmawanb, D. (2018). Keystroke Dynamic Classification using Machine Learning for Password Authorization. *Procedia Computer Science*, *135*, 564–569. doi:10.1016/j.procs.2018.08.209

Mungai, P. K., & Huang, R. (2017). Using keystroke dynamics in a multi-level architecture to protect online examinations from impersonation. *2017 IEEE 2nd International Conference on Big Data Analysis (ICBDA),* 622-627.

Muthuramalingam, A., Gnanamanickam, J., & Muhammad, R. (2018). Optimum Feature Selection Using Firefly Algorithm for Keystroke Dynamics. In A. Abraham, P. Muhuri, A. Muda, & N. Gandhi (Eds.), *Intelligent Systems Design and Applications. ISDA 2017. Advances in Intelligent Systems and Computing* (Vol. 736). Cham: Springer. doi:10.1007/978-3-319-76348-4_39

Nagaratna, B. (2018). Dialect Identification U sing Spectral a nd Prosodic Features on Single and Ensemble Classifiers, Research Article - Special Issue - Computer Engineering And Computer Science. *Arabian Journal for Science and Engineering*, *43*, 4289–4302.

Nahin, A. F. M. N. H., Alam, J. M., Mahmud, H., & Hasan, K. (2014). Identifying emotion by keystroke dynamics and text pattern analysis. *Behaviour & Information Technology*, *33*(9), 987–996. doi:10.1080/0144929X.2014.907343

Ness, J. (2017). *Presentation Attack and Detection in Keystroke Dynamics* (Master's thesis). NTNU.

Ness, J. (2017). Presentation Attack And Detection. In *Keystroke Dynamics*. NTNU.

Neves, J., Moreno, J., & Proença, H. (2017). *QUIS-CAMPI: an annotated multi-biometrics data feed from surveillance scenarios*. IET Biometrics. doi:10.1049/iet-bmt.2016.0178

Newman, M., Gillick, L., Ito, Y., McAllaster, D., & Peskin, B. (1996). Speaker verification through large vocabulary continuous speech recognition. *Proceedings of the International Conference on Spoken Language Processing*, 2419–2422. 10.1109/ICSLP.1996.607297

Newton, D. (2015). *Cheating in online classes is now big business.* Retrieved September 17, 2018, from https://www.theatlantic.com/education/archive/2015/11/cheating-through-onlinecourses/413770/

Nick, B. (2005). *Username and password verification through keystroke dynamics* (PhD Thesis). West Virginia University, ProQuest Dissertations Publishing.

Nita, S. L., & Mihailescu, M. I. (2018). Proposing a Secure E-learning System Based on Biometric Authentication and Homomorphic Encryption. *eLearning & Software for Education, 4.*

Nixon, M. A. (2004). Cheating In Cyberspace: Maintaining Quality In Online Education. *AACE Journal*, *12*(1), 85–99.

Noguera, Guerrero-Roldán, & Rodríguez. (2016). Assuring authorship and authentication across the e-assessment process. *International Computer Assisted Assessment Conference*, 86–92.

Northcutt, C. G., Ho, A. D., & Chuang, I. L. (2015). *Detecting And Preventing "Multiple-Account"*. Cheating In Massive Open Online Courses.

O'Gorman, L. (2003). Comparing passwords, tokens, and biometrics for user authentication. *Proceedings of the IEEE, 91*(12), 2019–2040. doi:10.1109/JPROC.2003.819605

Obaidat, M. S., & Sadoun, B. (1998). *Keystroke Dynamics*. Keystroke Dynamics Based Authentication.

Obaidat, M., & Macchairolo, D. (1993). An online neural network system for computer access security. *IEEE Transactions on Industrial Electronics*, 40235–40242.

Obaidat, M., & Sadoun, B. (1997). Verification of Computer User Using Keystroke Dynamics. *IEEE Transactions on Systems, Man, and Cybernetics*, *27*(2), 261–269. doi:10.1109/3477.558812 PMID:18255865

Obeidallah, R., Ahmad, A. A., Farouq, F., & Awad, S. (2015). Students authentication in e-assessment sessions: A theoretical biometric model for smartphone devices. *International Journal of Business Information Systems*, *19*(4), 450–464. doi:10.1504/IJBIS.2015.070204

Okada, A., Scott, P., & Mendonça, M. (2015). Effective web videoconferencing for proctoring online oral exams: A case study at scale in Brazil. *Open Praxis*, *7*(3). doi:10.5944/openpraxis.7.3.215

Oloyede, M. O., & Hancke, G. P. (n.d.). *Unimodal and Multimodal Biometric Sensing Systems-A Review*. doi:10.1109/ACCESS.2016.2614720

Oloyede, M. O., & Hancke, G. P. (2016). *Unimodal and Multimodal Biometric Sensing Systems: A Review* (Vol. 4). IEEE Access.

Omar Zughoul, O. (2013). Privacy and Security in Online Examination Systems. *IOSR Journal Of Computer Engineering*, *10*(4), 63–70. doi:10.9790/0661-01046370

Open University. (2018). *T215 Communication and information technologies Module Handbook*. Open University; doi:10.1007/978-94-007-3892-8

Pal, M., & Saha, G. (2015). On Robustness Of Speech Based Biometric Systems Against Voice Conversion Attack. *Applied Soft Computing*, *30*, 214–228. doi:10.1016/j.asoc.2015.01.036

Panchumarthy, R., Subramanian, R., & Sarkar, S. (2012). Biometric evaluation on the cloud: A case study with human-id gait challenge. *Proceedings of the International Conference on Utility and Cloud Computing*, 219-222. 10.1109/UCC.2012.49

Panda & Padhy. (2007). Comparison of Particle Swarm Optimization and Genetic Algorithm for TCSC-based Controller Design. *World Academy of Science, Engineering and Technology*, *27*, 497–505.

Patel, V., Ratha, N., & Chellappa, R. (2015). Cancelable Biometrics: A review. *IEEE Signal Processing Magazine*, *32*(5), 54–65. doi:10.1109/MSP.2015.2434151

Patil, R. A., & Renke, A. L. (2016). Keystroke Dynamics for User Authentication Identification by using Typing Rhythm. *International Journal of Computers and Applications*.

Pato, J., & Millett, L. (2010). *Biometric Recognition: Challenges and opportunities*. National Academies Press.

Patricia, A., & Robert, I. (2009). Student Identity and Authentication in Distance Education: A Primer for Distance Learning Administrators. *Continuing Higher Education Review, 73*.

Pattnaik & Dash. (2012). A Study on Prosody Analysis. *International Journal of Computational Engineering Research, 2*(5), 1594-1599.

Paula, M. V. S., Kinto, E. A., Hernandez, E. D. M., & Carvalho, T. (2005). User Authentication based on Human Typing Pattern with Artificial Neural Networks and Support Vector Machine. Academic Press.

Paulmann, S., Vrijders, B., Weinstein, N., & Vansteenkiste, M. (2018). How parents motivate their children through prosody. *9th International Conference on Speech Prosody*, 13-16. 10.21437/SpeechProsody.2018-51

Penteado, B. E., & Marana, A. N. (2009, May). A video-based biometric authentication for E-learning web applications. In *International Conference on Enterprise Information Systems* (pp. 770-779). Springer. 10.1007/978-3-642-01347-8_64

Peres, P., Lima, L., & Lima, V. (2014). B-learning quality: dimensions, criteria and pedagogical approach. *European Journal of Open, Distance and e-Learning, 17*(1), 56-75.

Perez, G., Tappert, C. C., Bours, P., Mondal, S., Morales, A., Fierrez, J., & Ortega-garcia, J. (2015). *One-handed Keystroke Biometric Identification Competition*. Academic Press.

Pfleeger, C., & Pfleeger, S. (2007). *Security in Computing* (4th ed.). Boston: Pearson Education.

Phillips, P. J., Flynn, P. J., Scruggs, T., Bowyer, K. W., Chang, J., Hoffman, K., . . . Worek, W. (2005). Overview of the face recognition grand challenge. *IEEE Computer Society Conference on Computer Vision and Pattern Recognition, CVPR 2005*, 947–954.

Pleva, M., Bours, P., Hladek, D., & Juhar, J. (2016). Using current biometrics technologies for authentication in e-learning assessment. In *Emerging eLearning Technologies and Applications (ICETA), 2016 International Conference on* (pp. 269-274). IEEE. 10.1109/ICETA.2016.7802043

Pointner, P. (2017). Hybrid tech: the future of biometric ID verification? *Biometric Technology Today, 2017*(8), 8-10. doi:10.10160969-4765(17)30183-2

Prabhakar, S., Pankanti, S., & Jain, A. K. (2003). Biometric recognition: Security and privacy concerns. *IEEE Journals on Security & Privacy*, *1*(2), 33–42. doi:10.1109/MSECP.2003.1193209

Pun, K., & Moon, Y. (2004). Recent advances in ear biometrics. *Proc. 6th Internat. Conf. on Automatic Face and Gesture Recognition*, 164–169. 10.1109/AFGR.2004.1301525

Qwaider, W. Q. (2017). Information Security and Learning Content Management System (LCMS). *International Journal of Advanced Computer Science and Applications*, *8*(11), 588–593.

Raitman, R., Ngo, L., Augar, N., & Zhou, W. (2005). Security in the online e-learning environment. In *Advanced Learning Technologies, 2005. ICALT 2005. FIFTH IEEE International Conference* (pp. 702-706). IEEE.

Ramachandra, R., & Busch, C. (2017). Presentation Attack Detection Methods For Face Recognition Systems: A Comprehensive Survey. *ACM Computing Surveys*, *50*(1), 1–37. doi:10.1145/3038924

Raman, M., & Levy, Y. (2006). Securing e-learning system: A case of insider cyber attacks and novice IT management in a small university. *Journal of Cases on Information Technology*, *8*(4), 24–34. doi:10.4018/jcit.2006100103

Ramim, M. M., & Levy, Y. (2007). *Towards A Framework Of Biometrics Exam Authentication In E-Learning Environments*. Academic Press.

Ramirez, J., Segura, J. C., Benítez, C., de la Torre, Á., & Rubio, A. (2004). Efficient voice activity detection algorithms using long-term speech information. *Speech Communication*, *2004*(3-4), 271–287. doi:10.1016/j.specom.2003.10.002

Ramnial, H., Panchoo, S., & Pudaruth, S. (2016). Authorship Attribution Using Stylometry And Machine Learning Techniques. *International Journal of Intelligent Systems Technologies and Applications*, 113–125.

Ramu, T., & Arivoli, T. (2013). A framework of secure biometric based online exam authentication: An alternative to traditional exam. *Int J Sci Eng Res*, *4*(11), 52–60.

Ramu, T., & Arivoli, T. (2013). A Framework of Secure Biometric Based Online Exam Authentication: An Alternative to Traditional Exam. *International Journal of Scientific & Engineering Research*, *4*(11), 52–60.

Ramu, T., & Arivoli, T. (2013). a Framework of Secure Biometric Based Online Exam Authentication: An Alternative To Traditional Exam. *International Journal of Scientific and Engineering Research*, *4*(11), 52–60.

Rane, S. (2014). Standardization of biometric template protection. *IEEE MultiMedia*, *21*(4), 94–99. doi:10.1109/MMUL.2014.65

Ratha, N. K., Connell, J. H., & Bolle, R. M. (2001). *An analysis of minutiae matching strength*. Audio- and Video-Based Biometric Person Authentication. doi:10.1007/3-540-45344-X_32

Ratha, N., Connell, J., & Bolle, R. (2001). Enhancing security and privacy in biometricsbased authentication systems. *IBM Systems Journal*, *40*(3), 614634. doi:10.1147j.403.0614

Ratha, N., & Zhang, D. (2010). Privacy protection in high security biometrics applications. In A. Kumar & D. Zhang (Eds.), *Ethics and Policy of Biometrics* (pp. 62–69). Springer. doi:10.1007/978-3-642-12595-9_9

Rejito, J., Abdullahi, A. S., Setiana, D., & Ruchjana, B. N. (2017). Image indexing using color histogram and k-means clustering for optimization CBIR in image database. *Journal of Physics: Conference Series*, *893*(1), 012055. doi:10.1088/1742-6596/893/1/012055

Revett, K., Gorunescu, F., Gorunescu, M., Magalhaes, S., & Santos, H. (2007). A machine learning approach to keystroke dynamics based user authentication. *International Journal of Electronic Security Digital Forensics*, *1*(1), 55–70. doi:10.1504/IJESDF.2007.013592

Reynolds, D. A. (1995). Speaker Identification and Verification using Gaussian Mixture Speaker Models. *Speech Communication*, *17*(1-2), 91–108. doi:10.1016/0167-6393(95)00009-D

Reynolds, D. A., Rose, R. C., & Smith, M. J. T. (1992). PC-based TMS320C30 implementation of the Gaussian mixture model text-independent speaker recognition system. *Proceedings of the International Conference on Signal Processing Applications and Technology*, 967–973.

Reynolds, D., Quatieri, T. F., & Dunn, R. B. (2000). Speaker verification using adapted Gaussian Mixture models. *Digital Signal Processing*, *10*(1-3), 19–41. doi:10.1006/dspr.1999.0361

Rodchua, S. (2017). Effective Tools and Strategies to Promote Academic Integrity in e-Learning. *International Journal of E-Education, E-Business, E-Management and E-Learning*, *7*(3), 168–179. doi:10.17706/ijeeee.2017.7.3.168-179

Rodchua, S., Yiadom-Boakye, G., & Woolsey, R. (2011). Student Verification System for Online Assessments: Bolstering Quality and Integrity of Distance Learning. *Journal of Information Technology*, *27*(3), 1–8.

Rogers, C. (2006). Faculty perceptions about e-cheating during online testing. *Journal of Computing Sciences in Colleges*, *22*(2), 206–212.

Romain, Dorizzi, & Rosenberger. (2015). A review on the public benchmark databases for static keystroke dynamics. *Computers & Security*, 46–61.

Romain, G., Mohamad, E.-A., Baptiste, H., & Christophe, R. (2011). Unconstrained keystroke dynamics authentication with shared secret. *Computers & Security, Elsevier*, *30*(6-7), 427–445. doi:10.1016/j.cose.2011.03.004

Rosenberg, A. E., DeLong, J., Lee, C. H., Juang, B. H., & Soong, F. K. (1992). The use of cohort normalized scores for speaker verification. *International Conference on Speech and Language Processing*, 599–602.

Rosenberg, A. E., & Parthasarathy, S. (1996). Speaker background models for connected digit password speaker verification. *Proceedings of the International Conference on Acoustics, Speech, and Signal Processing*, 81–84. 10.1109/ICASSP.1996.540295

Roth, J., Liu, X., Ross, A., & Metaxas, D. (2013). Biometric authentication via keystroke sound. *Proceedings - 2013 International Conference on Biometrics, ICB 2013*. 10.1109/ICB.2013.6613015

Rowe, N. C. (2004). *Cheating In Online Student Assessment: Beyond Plagiarism*. Academic Press.

Rudrapal, D., Das, S., Debbarma, S., Kar, N., & Debbarma, N. (2012). Voice Recognition and Authentication as a Proficient Biometric Tool and its Application in Online Exam for P. H People. *International Journal of Computers and Applications*, *39*(12), 975–8887. doi:10.5120/4870-7297

Ruiz, J. G., Mintzer, M. J., & Leipzig, R. M. (2006). The impact of e-learning in medical education. *Academic Medicine*, *81*(3), 207–212. doi:10.1097/00001888-200603000-00002 PMID:16501260

Rybnik, M., Tabedzki, M., & Saeed, K. (2008). A keystroke dynamics based system for user identification. *2008 7th Computer Information Systems and Industrial Management Applications,* 225-230. doi. doi:10.1109/CSIM.2008.8

Rybnik, M., Tabedzki, M., & Saeed, K. (2008). A keystroke dynamics-based system for user identification. Proceedings of 7th Computer Information Systems and Industrial Management Applications, 225–230. doi:10.1109/CISIM.2008.8

Saaty, T. L. (1996). *Decision Making with Dependence and Feedback- The Analytic Network Process*. Pittsburgh, PA: RWS Publications.

SACS, Standards for Academic and Student Affairs. (2012). Retrieved September 17, 2018, from http://www.fau.edu/iea/pdf/accreditation/as_i2-1.pdf

Saevanee, H., & Bhatarakosol, P. (2008). User authentication using combination of behavioural biometrics over the touchpad acting like touch screen of mobile device. *IEEE International Conference on Computer and Electrical Engineering*, 82–86.

Saggio, G., Costantini, G., & Todisco, M. (2011). Cumulative and Ratio Time Evaluations in Keystroke Dynamics to Improve the Password Security Mechanism. *Journal of Computer and Information Technology*, *1*(2), 2–11.

Sametinger, J., Schlöglhofer, R. & Sametinger, J. (2012). *Secure and usable authentication on mobile devices Secure and Usable Authentication on Mobile Devices*. Academic Press.

Sanchez-Reillo, R., Sanchez-Avila, C., Gonzalez-Marcos, A., Sanchez-Reillo, R., & Gonzalez-Marcos, A. (2000). Biometric identification through hand geometry measurements. *IEEE Transactions on Pattern Analysis and Machine Intelligence*, *22*(10), 1168–1171. doi:10.1109/34.879796

Sarika, S. (2015). Survey Paper on Automatic Speaker Recognition Systems, International Journal Of Engineering And. *Computer Science*, *4*(3), 10895–10898.

Sarkar, A. K., & Umesh, S. (2010). Investigation of Speaker- Clustered UBMs based on Vocal Tract Lengths and MLLR matrices for Speaker Verification. *The Speaker and Language Recognition Workshop*.

Sarrayrih, M. A., & Ilyas, M. (2013). Challenges of online exam, performances and problems for online university exam. *International Journal of Computer Science Issues*, *10*(1), 439.

Savchenko, A. (2013). Probabilistic neural network with homogeneity testing in recognition of discrete patterns set. *Neural Networks*, *46*, 227–241. doi:10.1016/j.neunet.2013.06.003

Scarfone, K., & Mell, P. (2007). *Guide to intrusion detection and prevention systems (IDPS)*. NIST special publication, 800(2007): 94.

Schlenker, A., & Sarek, M. (2013). Neural Networks in Keystroke Dynamics for Multi-Factor Authentication in. *Measurement*, 109–112.

Sewell, J., Frith, K., & Colvin, M. (2010). Online assessment strategies: A primer. *Journal of Online Learning and Teaching / MERLOT*, *6*(1), 297–305.

Shaffer, L. H. (1978). Timing in the motor programming of typing. *The Quarterly Journal of Experimental Psychology*, *30*(2), 333–345.

Shanmugapriya, D., & Padmavathi, G. (2011). An Efficient Feature Selection Technique for User Authentication using Keystroke Dynamics. *IJCSNS*, *11*(10), 191.

Shanmugapriya, D., & Padmavathi, G. (2011). An Efficient Feature Selection Technique for User Authentication using Keystroke Dynamics. *International Journal of Computer Science and Network Security*, *11*(10), 191–195.

Sharma. (2015). Feature Selection Technique Using Ant Colony Optimization on Keystroke Dynamics. *International Journal of Scientific & Engineering Research*, 668-678.

Sheard, J., Dick, M., Markham, S., Macdonald, I., & Walsh, M. (2002). *Cheating And Plagiarism: Perceptions And Practices Of First Year IT Students*. Paper Presented At The ACM SIGCSE Bulletin. 10.1145/544414.544468

Sheng, Phoha, & Rovnyak. (2005). A Parallel Decision Tree-Based Method for User Authentication Based on Keystroke Patterns. *IEEE Transactions on Systems, Man, and Cybernetics Part B: Cybernetics, 35*(4). Available: http://www.ucertify.com/certifications//ic3.html http://polinance.blogspot.in/2010/06/keystroke-dynamics-how-unique-keyboard.html

Shree, C. G., & Jagadamba, G. (2016). A Three Factor Authentication System Using Keystroke Dynamic. In *Proc. of National Conference on recent trends in communication and computing*. Davengere.

Shukla, P., & Solanki, R. (2013). Web Based Keystroke Dynamics Application for Identifying Emotional State. *International Journal of Advanced Research in Computer and Communication Engineering*, *2*(11), 4489–4493.

Si, Li, & Tan. (2017). Transformation of Prosody in Voice Conversion. *Proceedings of APSIPA Annual Summit and Conference*.

Sim, H. M., Asmuni, H., Hassan, R., & Othman, R. M. (2014). Multimodal Biometrics: Weighted Score Level Fusion Based on Non-Ideal Iris and Face Images. Expert Systems with Applications.

Sindre, G., & Vegendla, A. (2015). E-Exam Versus Paper Exams: A Comparative Analysis Of Cheating Related Security Threats And Countermeasures. *Norwegian Information Security Conference (NISK)*, Ålesund, Norway.

Singh, D., Jaggi, B., Nayyar, H., & Kumar, A. (2017). *Presskey- A Keystrokes Dynamics Based Authentication System*. Academic Press.

Singh, Jaggi, Nayyar, & Kumar. (2017). Presskey-A Keystrokes Dynamics Based Authentication System. *International Journal of Advanced Research in Computer Science*, *8*(5), 2665–2670.

Singh, R., Vatsa, M., & Noore, A. (2008). Face recognition with disguise and single gallery images. *Image and Vision Computing*. doi:10.1016/j.imavis.2007.06.010

Song, K., Lee, S. M., & Nam, S. (2013). Combined biometrics for E-learning security. ISA 2-13. *ASTL*, *21*, 247–251.

Song, K., Lee, S. M., & Nam, S. C. (2013). Combined biometrics for e-learning security. ISA 2-13. *ASTL*, *21*, 247–251.

Srikanth, M., & Asmatulu, R. (2014). Modern Cheating Techniques, Their Adverse Effects On Engineering Education And Preventions. *International Journal of Mechanical Engineering Education*, *42*(2), 129–140. doi:10.7227/IJMEE.0005

Srinivasan, H., Beal, M. J., & Srihari, S. N. (2005, May). Machine learning approaches for person identification and verification. In Sensors, and Command, Control, Communications, and Intelligence (C3I) Technologies for Homeland Security and Homeland Defense IV (Vol. 5778, pp. 574-587). International Society for Optics and Photonics. doi:10.1117/12.601987

Srivastava, H.A. (2013). Comparison Based Study on Biometrics for Human Recognition. *IOSR Journal of Computer Engineering, 15*, 22-29.

Srivastava, A., & Sinha, S. (2013). Information security through e-learning using VTE. *International Journal of Electronics and Computer Science Engineering*, *2*(18), 528–531.

Stefan, D., & Yao, D. (2010). Keystroke-dynamics authentication against synthetic forgeries. In *2010 6th International Conference on In Collaborative Computing: Networking, Applications and Worksharing (CollaborateCom)* (pp. 1-8). IEEE.

Stefan, D., Shu, X., & Yao, D. (2012). Robustness of keystroke-dynamics based biometrics against synthetic forgeries. *Computers & Security*, *31*(1), 109–121. doi:10.1016/j.cose.2011.10.001

Stefan, D., & Yao, D. (2010). Keystroke-dynamics authentication against synthetic forgeries. *6th International Conference on Collaborative Computing: Networking, Applications and Work sharing (CollaborateCom 2010)*, 1-8. doi: 10.4108/icst.collaboratecom.2010.16

Stewart, J. C., Monaco, J. V., Cha, S. H., & Tappert, C. C. (2011). An investigation of keystroke and stylometry traits for authenticating online test takers. *2011 International Joint Conference on Biometrics, IJCB 2011*. 10.1109/IJCB.2011.6117480

Stockton, Lisowski, Press, & Shapiro. (1980). *Authentication by Keystroke Timing: Some Preliminary Results*. Santa Monica, CA: RAND Corporation. Retrieved from https://www.rand.org/pubs/reports/R2526.html

Štruc, V., & Pavešić, N. (2009). Gabor-based kernel partial-least-squares discrimination features for face recognition. *Informatica, 20*(1), 115–138.

Sutcu, Y., Li, Q., & Memon, N. (2007). Protecting biometric templates with sketch: Theory and practice. *IEEE Transactions on Information Forensics and Security, 2*(3), 503–512. doi:10.1109/TIFS.2007.902022

Syed Idrus, S. Z., Cherrier, E., Rosenberger, C., & Bours, P. (2014). Soft biometrics for keystroke dynamics: Profiling individuals while typing passwords. *Computers & Security, 45*, 147–155. doi:10.1016/j.cose.2014.05.008

Takuya, O., Nobuaki, M., & Daisuke, S. (2018). Prosodic Comparison of Utterances without Extracting Fundamental Frequencies based on Vocalized Subharmonic Summation. *9th International Conference on Speech Prosody 2018*, 172-176.

Talavera Franco, R. (2014). *Facial biometrics: A solution or a problem for MOOCs assessments?* MOOCstream.

Tang, L., Yang, C., & Wu, Z. (2010). User Authentication based on Force-Sensing Keypad by Fuzzy C-Means Clustering Method. *Journal of Computer Information Systems, 6*(11), 3659–3667.

Tan, Z. (2014). A user anonymity preserving three-factor authentication scheme for telecare medicine information systems. *Journal of Medical Systems, 38*(3), 1–9. doi:10.100710916-014-0016-2 PMID:24643750

Tan, Z. W. (2013). An efficient biometric-based authentication scheme for telecare medicine information systems. *Przegl. Elektrotech., 89*(5), 200–204.

Teh, Teoh, & Yue. (2013). A Survey of Keystroke Dynamics Biometrics. *The Scientific World Journal*. doi:10.1155/2013/408280

Teoh, A., Ling, D., & Goh, A. (2004). Biohashing: Two factor authentication featuring fingerprint data and tokenised random number. *Pattern Recognition, 37*(11), 2245–2255. doi:10.1016/j.patcog.2004.04.011

Thanganayagam, R., & Thangadurai, A. (2015). Fusion approach on keystroke dynamics to enhance the performance of password authentication. *International Conference on Electrical, Computer and Communication Technologies*, 1–6. 10.1109/ICECCT.2015.7226123

Thiemann, T. (2016). Picking the right path to mobile biometric authentication. *Biometric Technology Today, 2016*(2), 5-8. doi:10.10160969-4765(16)30034-0

Traoré, I., Awad, A., & Woungang, I. (2017). *Ensuring Online Exam Integrity Through Continuous Biometric Authentication. In Information Security Practices: Emerging Threats and Perspectives* (pp. 73–81). Cham: Springer; doi:10.1007/978-3-319-48947-6

Traoré, I., Nakkabi, Y., Saad, S., Sayed, B., Ardigo, J. D., & De Faria Quinan, P. M. (2017). *Ensuring Online Exam Integrity Through Continuous Biometric Authentication. In Information Security Practices: Emerging Threats And Perspectives* (pp. 73–81). Springer International Publishing.

Traore, I., Traore, I., & Ahmed, A. A. E. (2011). *Continuous Authentication Using Biometrics: Data, Models, And Metrics*. IGI Publishing.

Traore, I., Woungang, I., Obaidat, M. S., Nakkabi, Y., & Lai, I. (2012). Combining Mouse and Keystroke Dynamics Biometrics for Risk-Based Authentication in Web Environments. *Fourth International Conference on Digital Home*, 138-145. 10.1109/ICDH.2012.59

Trivandrum, S., Shali, N., Zacharias, G., & Anna, P. (2018). Identification of tongue print images for forensic science and biometric authentication. *Journal of Intelligent & Fuzzy Systems*, *34*(3), 1421–1426. doi:10.3233/JIFS-169437

Tsai, C. J., Peng, C. C., Chiang, M. L., Chang, T. Y., Tsai, W. J., & Wu, H. S. (2015). Work in progress: A new approach of changeable password for keystroke dynamics authentication system on smart phones. *Proceedings of the 2014 9th International Conference on Communications and Networking in China, CHINACOM 2014*, 353–356. 10.1109/CHINACOM.2014.7054316

Tsimperidis, I., Arampatzis, A., & Karakos, A. (2018). Keystroke dynamics features for gender recognition. *Digital Investigation*, *24*, 4–10. doi:10.1016/j.diin.2018.01.018

U.S. Higher Education System. (2016). *In the National Science Foundation*. Retrieved September 17, 2018, from https://www.nsf.gov/statistics/2016/nsb20161/#/report/chapter-2/

Ullah, A., Xiao, H., & Barker, T. (2016). *A Classification Of Threats To Remote Online Examinations*. Paper Presented At The 2016 IEEE 7th Annual Information Technology, Electronics And Mobile Communication Conference (IEMCON). 10.1109/IEMCON.2016.7746085

Ullah, A., Xiao, H., Lilley, M., & Barker, T. (2013). *Design, Privacy And Authentication Of Challenge Questions In Online Examinations*. Paper Presented At The 2013 IEEE Conference On E-Learning, E-Management And E-Services. 10.1109/IC3e.2013.6735964

Umphress, D., & Williams, G. (1985). Identity verification through keyboard characteristics. *International Journal of Man-Machine Studies*, *23*(3), 263–273. doi:10.1016/S0020-7373(85)80036-5

Vegendla, A., Søgaard, T. M., & Sindre, G. (2016). Extending HARM To Make Test Cases For Penetration Testing. In J. Krogstie, H. Mouratidis, & J. Su (Eds.), *Advanced Information Systems Engineering Workshops: Caise 2016 International Workshops* (pp. 254-265). Cham: Springer International Publishing. 10.1007/978-3-319-39564-7_24

Vera-Rodriguez, R., Mason, J., Fierrez, J., & Ortega-Garcia, J. (2012). Comparative analysis and fusion of spatio-temporal information for footstep recognition. *IEEE Transactions on Pattern Analysis and Machine Intelligence*, 99.

Vidyadhara Raju, V. V., Gangamohan, P., Suryakanth, V. G., & Vuppala, A. K. (2016). Application of prosody modification for Speech Recognition in different Emotion conditions. *IEEE Region 10 Conference (TENCON) — Proceedings of the International Conference*, 951-954.

Vrankuli, A. (2013). *Coursera looks to verify online student identity with photo, keystroke dynamics*. Retrieved September 17, 2018 from https://www.biometricupdate.com/201301/coursera-looks-to-verify-online-student-identity-with-photo-keystroke-dynamics

Vuuren, S. (1999). *Speaker Verification in a Time-Feature Space* (Ph.D. thesis). Oregon Graduate Institute.

Wang, Y., & Payne, E. (2018). Investigating prosody in music and speech. *9th International Conference on Speech Prosody 2018*, 47-551.

Wang, F., Ge, B., Zhang, L., Chen, Y., Xin, Y., & Li, X. (2013). A system framework of security management in enterprise systems. *Systems Research and Behavioral Science*, *30*(3), 287–299. doi:10.1002res.2184

Wang, L., & Geng, X. (2009). *Behavioral biometrics for human identification, medical information science reference*. Hershey, PA: IGI Global.

Wankhede & Verma. (2014). Keystroke Dynamics Authentication System Using Neural Network. *International Journal of Innovative Research & Development*.

Wayman, J. L. (2001). Fundamentals of biometric authentication technologies. *International Journal of Image and Graphics*, *1*(1), 93–113. doi:10.1142/S0219467801000086

Weaver, A. C. (2006). Biometric Authentication. *Computer*, *39*(2), 96–97. doi:10.1109/MC.2006.47

Weippl, E., & Ebner, M. (2008). Security privacy challenges in e-learning 2.0. In *World Conference on E-Learning in Corporate, Government, Healthcare, and Higher Education* (Vol. 2008, No. 1, pp. 4001-4007). Academic Press.

Wiklund, M., Mozelius, P., Westin, T., & Norberg, L. (2016). Biometric Belt And Braces For Authentication In Distance Education. *Proceedings Of The European Conference On E-Learning, ECEL.*

Wong, F., Wong, M.H., Supian, A.S.M., Ismail, A.F., Kin, L.W., & Soon, O.C. (2011). Enhanced User Authentication through Typing Biometrics with Artificial Neural Networks and k-Nearest Neighbor Algorithm. *Conference Record of the Thirty-Fifth Asilomar Conference on Signals, Systems and Computers, 2,* 911–915.

Xiao, H., Ji, W., & Ullah, A. (2011). Authentication of Students and Students' Work in E-learning. *Report for the Development Bid of Academic Year, 2010*, 11.

Yadav, J., Pandey, K., Gupta, S., & Sharma, R. (2017). Keystroke dynamics based authentication using fuzzy logic. *2017 Tenth International Conference on Contemporary Computing (IC3)*, 1-6. 10.1109/IC3.2017.8284338

Yan, P., & Bowyer, K. W. (2005a). Empirical evaluation of advanced ear biometrics. Empirical Evaluation Methods in Computer Vision (EEMCV 2005), San Diego, CA.

Yan, P., & Bowyer, K. W. (2005b). Ear Biometrics Using 2D and 3D Images. *Proc. IEEE Conf. on Computer Vision and Pattern Recognition Workshop Advanced 3D Imaging for Safety and Security.*

Yan, P., & Bowyer, K.W. (2005c). Multi-Biometrics 2D and 3D Ear Recognition. *Proc. Audio and Video-Based Biometric Person Authentication*, 503–512.

Yang, W., Li, Y., & Liao, Q. (2014). Fast and Robust Personal Identification by Fusion of Finger Vein and Finger-Knuckle-Print Images. *Applied Mechanics and Materials, 556-562*, 5085-5088. doi: 10.4028/www.scientific.net/amm.556-562.5085

Yang, X., Zhou, Y., Zhang, T., Shu, G., & Yang, J. (2008). Gait recognition based on dynamic region analysis. *Signal Processing*, *88*(9), 2350–2356. doi:10.1016/j.sigpro.2008.03.006

Yao, H., & Ji, Y. (2011). Security protection for online learning of music. In *Computer Communication and Networks (ICCCN), 2011 Proceedings of 20th International Conference* (pp. 1-4). IEEE. 10.1109/ICCCN.2011.6005740

Yoon, E. J., & Yoo, K. Y. (2005). A new efficient fingerprint-based remote user authentication scheme for multimedia systems. *9th Int. Conf. Knowledge-Based & Intelligent Information & Engineering Systems (KES 2005)*, 332–338. 10.1007/11553939_48

Yu, C., & Tsao, C. C. (2003). Web Teaching: Design, security and legal issues. *Delta Pi Epsilon Journal*, *45*(3), 191–203.

Yu, E., & Cho, S. (2003). Novelty Detection Approach for Keystroke Dynamics Identity Verification. *Intelligent Data Engineering and Automated Learning*, *2690*, 1016–1023. doi:10.1007/978-3-540-45080-1_143

Yu, E., & Cho, S. (2004). Keystroke dynamics identity verification—its problems and practical Solutions. *Computers & Security*, *23*(5), 428–440. doi:10.1016/j.cose.2004.02.004

Yu, J., Wang, G., Mu, Y., & Gao, W. (2014). W. An efficient generic framework for three-factor authentication with provably secure instantiation. *IEEE Transactions on Information Forensics and Security*, *9*(12), 2302–2313. doi:10.1109/TIFS.2014.2362979

Zahid, S., Shahzad, M., Khayam, S. A., & Farooq, M. (2009). Keystroke-Based User Identification on Smart Phones. In E. Kirda, S. Jha, & D. Balzarotti (Eds.), Lecture Notes in Computer Science: Vol. 5758. *Recent Advances in Intrusion Detection. RAID 2009* (pp. 224–243). Berlin: Springer. doi:10.1007/978-3-642-04342-0_12

Zamonsky, G., & Pedernera et al. (2010). Revisiting Clustering Methods to their Application on Keystroke Dynamics for Intruder Classification. *Proceedings of 2010 IEEE Workshop on Biometric Measurements and Systems for Security and Medical Applications (BIOMS)*, 36-40. 10.1109/BIOMS.2010.5610443

Zhang, D., & Shu, W. (1999). Two novel characteristic in palmprint verification:Datum point invariance and line feature matching. *Pattern Recognition*, *32*(4), 691–702. doi:10.1016/S0031-3203(98)00117-4

Zhang, H., & Shu, H. (1989). Effects of phonetic components on naming compound characters [in Chinese]. *Acta Psychologica Sinica*, *21*, 284–289.

Zhang, Z.-Y., Chuang-bai, X., Zhao, D.-Q., Hong, S., Kai, X., & Tian, G.-Z. (2011). Identity Authentication System based on Improved PR-RP Model. *Proceedings of 2nd International Conference on Advanced Computer Control (ICACC)*, 65-69.

Zhao, W., Chellappa, R., Phillips, P. J., & Rosenfeld, A. (2003). Face recognition: A literature survey. *ACM Computing Surveys*, *35*(4), 399–458. doi:10.1145/954339.954342

Zhe, J., Andrew, T. S., & Connie, T. (2008). Typing dynamics biometric authentication through fuzzy logic. *International Symposium on Information Technology*, 1–6.

Zhonghu, L., Lin, Z., Luling, W., Junhong, Y., & Jinming, W. (2017, October). Research on adaptive SIFT algorithm in image matching of pipeline inner surface visual measurement. In *Electronic Measurement & Instruments (ICEMI), 2017 13th IEEE International Conference on* (pp. 305-309). IEEE.

Zhong, Y., Deng, Y., & Jain, A. K. (2012). Keystroke dynamics for user authentication. In *IEEE Computer Society Conference on Computer Vision and Pattern Recognition Workshops* (pp. 117-123). IEEE. DOI: 10.1109/CVPRW.2012.6239225

Zhong, Y., Deng, Y., & Jain, A. K. (2012). Keystroke dynamics for user authentication. *Proc. IEEE Comput. Soc. Conf. Comput. Vis.PatternRecognit. Workshops (CVPRW)*, 117–123.

Zhong, Y., & Peng, X. (2016). Multi-level fingerprint continuous classification for large-scale fingerprint database using fractal analysis. *International Journal of Biometrics*, *8*(2), 115. doi:10.1504/IJBM.2016.077829

Zuev, V. (2012). E-learning security models. *Management*, *7*(2), 24–28.

About the Contributors

A. V. Senthil Kumar has to his credit 7 Book Chapters, 95 papers in International Journals, 2 papers in National Journals, 25 papers in International Conferences, 5 papers in National Conferences, and edited four books in Data Mining, Mobile Computing, Fuzzy Expert Systems and Web Mining (IGI Global, USA). He is an Editor-in-Chief for International Journal titled "International Journal of Data Mining and Emerging Technologies", "International Journal of Image Processing and Applications", "International Journal of Advances in Knowledge Engineering & Computer Science", "International Journal of Advances in Computers and Information Engineering" and "International Journal of Research and Reviews in Computer Science".. Key Member for India, Machine Intelligence Research Lab (MIR Labs).

* * *

Kamaraj A. received his B.E. degree in Electronics and Communication Engineering from Bharathiar University, Coimbatore, Tamil Nadu, India in 2003. He completed his Post Graduation from Anna Universiy, Chennai in the field of VLSI Design in 2006. Currently he is an Assistant Professor (Sr. Grade) in the Department of Electronics and Communication Engineering, Mepco Schlenk Engineering College, Sivakasi, India. He is pursuing his Ph.D in Anna University, Chennai. His research interests include Digital circuits and logic design, Reversible logic & synthesis and advanced computing techniques. During his 11 years of teaching career, He has published 10 papers in International Journals and 22 papers in National & International Conferences. He has been Life member of IETE and ISTE.

B. Sathish Babu, Ph.D. (Electrical Communication Engineering) is a Professor in the Department of Computer Science and Engineering, RV college of Engineering, Bengaluru, Karnataka. With the teaching and research experience of over three decades, he has published many research papers in national/international journals/conferences and many book chapters. He is also the co-author of 'Mobile and Wireless Application Security' and 'Communication Protocol Engineering' book. His research interests include cognitive agents based control solutions for networks, grid computing, cloud computing scheduling and security issues, context-aware trust issues in ubiquitous computing, privacy issues in WSN and opportunistic computing.

Kalyana Sundaram C. received his B.E. degree in Electronics and Communication Engineering from Madurai Kamarajar University, Madurai, Tamil Nadu, India, in 2003. He completed his postgraduation from Anna University, Chennai, in the field of VLSI design in 2006. Currently he is an Assistant Professor in the Department of Electronics and Communication Engineering, Mepco Schlenk Engineering College, Sivakasi, India. He is pursuing his Ph.D. in Anna University, Chennai. His research interests include digital circuits and logic design, VLSI Biomedical Signal Processing and advanced computing techniques. During his 12 years of teaching experience, he has published papers in four Scopus Indexed international journals, 14 international and 8 national conferences. He has been member of IEEE, IETE and ISTE.

Kevin Curran is a Professor of Cyber Security, Executive Co-Director of the Legal innovation Centre and group leader of the Ambient Intelligence & Virtual Worlds Research Group at Ulster University. He is also a senior member of the IEEE. Prof Curran is perhaps most well-known for his work on location positioning within indoor environments and Internet security. His expertise has been acknowledged by invitations to present his work at international conferences, overseas universities and research laboratories. He is a regular contributor on TV & radio and in trade and consumer IT magazines.

Shanmugapriya D. is an Assistant Professor and Head, Department of Information Technology.

Padmavathi Ganapathi is a Professor in the Department of Computer Science.

Jagadamba G., Ph.D., is an Assistant Professor in Department of Information Science and Engineering, Siddaganga Institute of Technology, Tumakuru, Karnataka. With the teaching and research experience of over 15 years she has published book chapters and research papers in national/international journals/conferences. Her research interests include context-aware adaptive trust evaluation and modelling, security issues in ubiquitous computing environment and authentication schemes based on keystroke dynamics, trust and context-based security to provide adaptive security in many applications such as e-commerce, m-banking, healthcare and so on.

Mukta Goyal is Ph.D. (CSE) from Jaypee Institute of Information Technology, Noida. She received her under Graduation (CSE) from Govt. college of engineering, Trivandrum, Kerala University and Masters (CSE) from Banasthali University. She has more than 17 years of teaching experience.

Ramgopal Kashyap's areas of interest are image processing, pattern recognition, and machine learning. He has published many research papers, and book chapters in international journals and conferences like Springer, Inderscience, Elsevier, ACM, and IGI-Global indexed by Science Citation Index (SCI) and Scopus (Elsevier). He has Reviewed Research Papers in the Science Citation Index Expanded, Springer Journals and Editorial Board Member and conferences programme committee member of the IEEE, Springer international conferences and journals held in countries: Czech Republic, Switzerland, UAE, Australia, Hungary, Poland, Taiwan, Denmark, India, USA, UK, Austria, and Turkey. He has written many book chapters published by IGI Global, USA.

Rajalakshmi Krishnamurthi obtained her Ph. D in Computer Science and Engineering from Jaypee Institute of Information Technology Noida, India. She did her M. E (CSE), and B. E. (EEE) from Bharathiar University, Coimbatore, India. She is currently an Assistant Professor (Senior Grade) with the Department of Computer Science and Engineering, Jaypee Institute of Information Technology, Noida. Her current research interests include Mobile computing, Wireless networks, Cloud Computing, Pervasive Computing.

Rathi M. was born in Tamilnadu, India in 1980. She received her Master of Computer Applications (MCA) degree from Bharathiar University in 2003 and M.Phil Computer Science from Bharathiar University in 2005. She has 11+ years of teaching experience and currently working as Assistant Professor (SG) in the Department of Computer Technology. Her research interests include keystroke dynamic based biometric authentication, data mining etc.

Anandhavalli Muniasamy completed Ph.D. in the year 2010 from Sikkim Manipal University of Computer Science Engineering, India. She started her career as a lecturer at Kamaraj University (1998), India, and then worked at SRM University, India and later joined as a Reader in Sikkim Manipal Institute of Technology (SMIT), India. Her areas of interest are data mining and soft computing. She has published more than 20 research publications in highly reputed international journals and with the h index of 5, and with the total citations around 135. She had attended and presented in more than 15 national & international conferences. She served as a principal investigator for the funded project from All India Council for Technical Education (AICTE) and served as a manuscript reviewer for renowned international journals for past 7 years, and Ph.D. thesis reviewer for several universities. She has been invited to give guest lectures and serving as editorial board member for a few journals and conferences. Currently, she is an Associate Professor, at the College of Computer Science, King Khalid University, Abha, Kingdom of Saudi Arabia since 2011.

S. Selva Nidhyananthan received his B.E degree in Electronics and Communication Engineering from Manonmanium Sundaranar University, Tirunelveli, in 1999, M.E. degree in Communication Systems from Anna University, Chennai, in 2005 and Ph.D degree in Speech Processing from Anna University, Chennai, in 2014. He has 15 years of teaching experience and he is currently working as Associate Professor in the department of Electronics and Communication Engineering at Mepco Schlenk Engineering College, Sivakasi, Tamil Nadu. He has published more than 82 papers in refereed International Journals and conferences. His current research interest includes Signal Processing and Image Processing. He is a member of IETE, IAENG and ISTE.

Guttorm Sindre (b. 1964) has been a full Professor at the NTNU since 2003. He got his master degree from NTH, Trondheim, Norway in 1988, and his PhD from the same university in 1990. His main research interests are conceptual modeling, software requirements engineering, security requirements, and IT didactics. He has been involved in several national and international research projects and has published more than 100 papers in peer reviewed international journals, anthologies, and conferences. From Dec. 2016 he has also been centre leader for Excited (Excellent IT Education), a centre for excellence in education funded by NOKUT for the period 2016-2021.

Aparna Vegendla is a PhD candidate in information systems group at Norwegian University of Science and Technology. She received her MS degree in Computer Engineering at Pusan National University in 2014, and her B-tech degree in Computer Science from Jawaharlal Nehru Technological University Kakinada in 2011. Her research interests include information systems security, e-exams security, security modeling, digital ecosystems, and requirements engineering.

Index

A

B

C

D

E

F

H

I

K

L

M

N

O

P

S

T

U

V

W

www.ingramcontent.com/pod-product-compliance
Lightning Source LLC
LaVergne TN
LVHW062145200926
848484LV00007B/1860